*Biogeochemistry of
Gulf of Mexico
Estuaries*

Biogeochemistry of Gulf of Mexico Estuaries

Edited by

Thomas S. Bianchi
Tulane University

Jonathan R. Pennock
University of Alabama
Dauphin Island Sea Lab

Robert R. Twilley
University of Southwestern Louisiana

John Wiley & Sons, Inc.

New York / Chichester / Weinheim / Brisbane / Singapore / Toronto

This book is printed on acid-free paper. ∞

Copyright © 1999 by John Wiley & Sons, Inc. All rights reserved.

Published simultaneously in Canada.

Library of Congress Cataloging-in-Publication Data:

Biogeochemistry of Gulf of Mexico estuaries / edited by Thomas S.
  Bianchi, Jonathan R. Pennock, Robert R. Twilley.
        p.   cm.
    Includes index.
    ISBN 0-471-16174-8 (cloth : alk. paper)
      1. Biogeochemical cycles—Mexico, Gulf of.   2. Estuarine ecology—
  Mexico, Gulf of.   I. Bianchi, Thomas S.   II. Pennock, Jonathan R.,
  1956–■■.   III. Twilley, Robert R., 1952–■■.
  QH344.B5725   1998
  577.7′86364—dc21   98-15977

Printed in the United States of America.

10 9 8 7 6 5 4 3 2 1

Contents

Preface

Estuaries bordering the Gulf of Mexico are as diverse geologically, hydrodynamically, and biogeochemically as any in the world, yet, until recently, they have received little scientific attention. Recent studies, however, have provided significant insight into how these warm-temperate and sub-tropical ecosystems function in comparison with better-studied and primarily cool-temperate systems. This book was written because there is no current literature that provides a comprehensive assessment of biogeochemical processes in these interesting and important ecosystems. Moreover, there are few books in the ecological literature that address the potential importance of warm-temperate and sub-tropical estuarine ecosystems in global biogeochemical cycling.

Our main objectives in this book are to provide the reader with a comprehensive overview of what is known about biogeochemical processes—and the factors that regulate them—in warm-temperate and sub-tropical estuarine systems such as those found in the Gulf of Mexico. To provide this overview, we solicited submissions from a diverse group of multi-disciplinary scientists from all regions of the Gulf coast. From their input, we compiled a comprehensive view of the biogeochemical dynamics of estuaries in this region of the world and provided a framework to examine how these ecosystems compare with the better-studied temperate estuaries of the world. As such, this book is designed to be used both as a reference source and as a supplementary textbook for graduate and senior undergraduate courses.

The book consists of the following five sections: I. Physical Characteristics; II. Nutrient Dynamics; III. Organic Matter Cycling; IV. Trace Element/Organic Cycling, and V. Summary. In Section I we provide the reader with a foundational background on the geomorphology of Gulf of Mexico estuaries.

This section also contains information on how hydrography, residence times, and sediment inputs vary in estuaries across the Gulf region and how these processes may be important controlling variables for these systems. In Section II we discuss how the nutrient dynamics (i.e., loadings and behavior) and plankton processes in these estuaries differ between systems that receive significant riverine inputs (e.g., Mobile Bay and the Mississippi River) or groundwater discharge (e.g., Celestun and Dzilam Lagoons) and those that receive minimal fresh-water inputs (e.g., Florida Bay). In addition, as a result of the generally shallow nature of these systems and the extensive wetlands surrounding many of these estuaries, we include chapters that examine the roles of benthic-pelagic and wetland/pelagic coupling on biogeochemical cycling in these systems. Section III of the book deals primarily with organic matter cycling in these estuaries. This section is divided into two parts that address the nature of particulate and dissolved organic matter (POM and DOM) in Gulf estuaries and discuss how differences among systems affect rates of cycling. In Section IV we address trace element cycling, which includes trace metal behavior, the importance of DOM complexation, particulate scavenging, and accumulation in sediments. Finally, in Section V we have provided an overview of each section and address how biogeochemical processes discussed throughout the book will affect the management of these estuaries in the future.

Acknowledgments

Our special thanks go to the authors of each chapter, who strongly believed in our overall goal of providing a comprehensive and comparative overview of recent research on Gulf of Mexico estuaries. We would also like to thank each of the external reviewers of the individual chapters, listed here in alphabetical order: Jeffrey Chanton, William M. Landing, Lawrence R. Pomeroy, Lawrence P. Sanford, and Robert G. Wetzel. We are also grateful to Michael Dagg for sponsoring a meeting for all authors to meet and discuss plans for the book at the Louisiana Universities Marine Consortium (LUMCON), and to Amy Bennett, Erika Engelhaupt, Corey Lambert, and Siddhartha Mitra for their invaluable assistance in proofreading the final drafts of each chapter. Rick Miller of the NASA Stennis Space Center kindly provided the SeaWiFS satellite image of the Gulf of Mexico, which was used on the front cover of the book. We would also like to thank Claudia Muñoz for her assistance with illustrations.

Finally, we would also like to thank John Wiley & Sons, especially Phil Manor and Rose Leo Kish, who helped us "jump" some hurdles and expedite the editorial process along the way.

Contributors

DR. M. BASKARAN—Department of Oceanography, Texas A&M University, Galveston, TX 77553, TEL: (409) 740-4706, FAX: (409) 740-4786, e-mail: Baskaran@tamug.tamu.edu

DR. THOMAS S. BIANCHI—Department of E.E.O. Biology, Tulane University, New Orleans, LA 70118, TEL: (504) 862-8000, ext. 1557, FAX: (504) 862-8706, e-mail: TBianch@mailhost.tcs.tulane.edu

DR. JOSEPH N. BOYER—Southeastern Environmental Research Program, Florida International University, Miami, FL 33199, TEL: (305) 348-4076, FAX: (305) 348-4096, e-mail: boyerj@servax.fiu.edu

DR. DANIEL L. CHILDERS—Southeastern Environmental Research Program and Department of Biological Sciences, Florida International University, University Park, Miami, FL 33199, TEL: (305) 348-3101, FAX: (305) 348-4096, e-mail: childers@fiu.edu

DR. LUIS A. CIFUENTES—Department of Oceanography, Texas A&M University, College Station, TX 77843-3146, TEL: (409) 845-3380, FAX: (409) 862-3172, e-mail: cifuentes@ocean.tamu.edu

DR. RICHARD B. COFFIN—Naval Research Laboratories, Environmental Quality Sciences Division, 4555 Overlook Dr., SW, Washington, DC 20375, TEL: (202) 767-0065, e-mail: rcoffin@ccsalpha2.nrl.navy.mil

DR. FRANCISCO A. COMIN—Departamento de Ecologia, Universidad de Barcelona, Avda Diagonal 645, Barcelona 08028, Spain, TEL: (34) 402 15 09, FAX: (34) 411 15 09, e-mail: comin@porthos.bio.ub.es

MS. JEAN COWAN—Dauphin Island Sea Lab, P.O. Box 369-370, Dauphin Island, AL 36528, TEL: (334) 861-7586, FAX: (334) 861-7540, e-mail: jwcowan@jaguar1.usouthal.edu

MR. STEPHEN E. DAVIS III—Southeast Environmental Research Program and Department of Biological Sciences, Florida International University, University Park, Miami, FL 33199, TEL: (305) 348-1576, FAX: (305) 348-4096, e-mail: sdavis04@fiu.edu

DR. PETER M. ELDRIDGE—Aquatic Studies, Texas Parks and Wildlife Department, 3000 South IH 35, Austin, TX 78704, TEL: (512) 912-7027, FAX: (512) 707-1358, e-mail: peter.eldridge@tpwd.state.tx.us

DR. GARY GILL—Department of Oceanography, Texas A&M University, Galveston, TX 77553, TEL: (409) 740-4710, FAX: (409) 740-4853, e-mail: gill@tamug.tamu.edu

DR. LAODONG GUO—Department of Oceanography, Texas A&M University, Galveston, TX 77551, TEL: (409) 740-4772, FAX: (409) 740-4786, e-mail: guol@tamug.tamu.edu

DR. JORGE A. HERRERA-SILVEIRA—CINVESTAV-IPN, Merida, Apdo. Postal 73 CORDEMEX, Merida, Yucatan, Mexico, TEL: 52 99 81 29 05, FAX: 52 99 81 29 17: e-mail: jherrera@kin.cieamer.conacwt.mx

DR. RICHARD L. IVERSON—Department of Oceanography, Florida State University, Tallahassee, FL 32306-3048, TEL: (850) 644-6700, FAX: (850) 644-2581, e-mail: iverson@ocean.fsu.edu

DR. BRENT A. McKEE—Department of Geology, Tulane University, New Orleans, LA 70118, TEL: (504) 862-3167, FAX: (504) 865-5199, e-mail: bmckee@mailhost.tcs.edu

DR. JAY C. MEANS—Department of Chemistry, Western Michigan University, Kalamazoo, MI 49008, TEL: (616) 387-2923, FAX: (616) 387-2909, e-mail: means@wmich.edu

DR. TINA MILLER-WAY—Department of Natural Sciences, University of Mobile, P.O. Box 13220, Mobile, AL 36663, TEL: (334) 675-5990, e-mail: tmiller@jaguarl.usouthal.edu

DR. PAUL A. MONTAGNA—University of Texas Marine Science Institute, 750 Channelview Dr, Port Aransas, TX 78373-1267, TEL: (512) 749-6779, FAX: (512) 749-6777, e-mail: paul@utmsi.zo.utexas.edu

DR. JEFF MORIN—Department of Oceanography, Texas A&M University, College Station, TX 77843-3146, TEL: (409) 845-6939, FAX: (409) 862-3172, e-mail: morin@nitro. Tamu.edu

DR. BEHZAD MORTAZAVI—Department of Oceanography, Florida State University, Tallahassee, FL 32306-3048, TEL: (850) 644-6700, FAX: (850) 644-2581, e-mail: mortazavi@ocean.fsu.edu

DR. JONATHAN R. PENNOCK—University of Alabama, Dauphin Island Sea Lab, P.O. Box 369-370, Dauphin Island, AL 36528, TEL: (334) 861-7531, FAX: (334) 861-7540, e-mail: jpennock@jaguar1.usouthal.edu

MR. GARY POWELL—Texas Water Development Board, 1700 N. Congress Ave., Austin TX 78711-3231, TEL: (512) 936-0815, FAX: (512) 936-0816, e-mail: glpowell@twdb.state.tx.us

DR. NANCY N. RABALAIS—Louisiana Universities Marine Consortium, 8124 Hwy 56, Chauvin, LA 70344, TEL: (504) 851-2800, FAX: (504) 851-2874, e-mail: nrabalais@lumcon.edu

MR. M. RAVICHANDRAN—Department of Civil, Environmental and Architectural Engineering, University of Colorado at Boulder, Campus Box 428, Boulder, CO 80309-0428, TEL: (303) 492-2910, FAX: (303) 447-2505, e-mail: ravicham@ucsub.colorado.edu

DR. VICTOR RIVERA-MONROY—Department of Biology, University of Southwestern Louisiana, Box 42451, Lafayette, LA 70504, TEL: (318) 482-5253, FAX: (318) 482-5834, e-mail: riverav@usl.edu

DR. PETER H. SANTSCHI—Department of Oceanography, Texas A&M University, Galveston, TX 77553, TEL: (409) 740-4476, FAX: (409) 740-4786, e-mail: santschi@tamug.tamu.edu

DR. WILLIAM W. SCHROEDER—Marine Science Program, The University of Alabama and Dauphin Island Sea Lab, P.O. Box 369-370, Dauphin Island, AL 36528, TEL: (334) 861-7528, FAX: (334) 861-7540, e-mail: wschroed@jaguarl.usouthal.edu

DR. ALAN SHILLER—Center for Marine Sciences, The University of Southern Mississippi, Stennis Space Center, MS 39529, TEL: (601) 688-1178, FAX: (601) 688-1121, e-mail: ashiller@whale.st.usm.edu

DR. RUBEN S. SOLIS—Texas Water Development Board, 1700 N. Congress Ave., Austin, TX 78711-3231, TEL: (512) 936-0823, FAX: (512) 936-0889, e-mail: rsolis@twdb.state.tx.us

DR. R. EUGENE TURNER—Department of Oceanography and Coastal Sciences, Louisiana State University, Baton Rouge, LA 70803, TEL: (504) 388-6454, FAX: (504) 388-6326, e-mail: turner@wr3600.cwr.lsu.edu

DR. ROBERT R. TWILLEY—Department of Biology, University of Southwestern Louisiana, P.O. Box 42451, Lafayette, LA 70504, TEL: (318) 482-6146, FAX: (318) 482-5834, e-mail: rtwilley@usl.edu

DR. LIANG-SAW WEN—Department of Oceanography, Texas A&M University, Galveston, TX 77553, TEL: (409) 740-4510, FAX: (409) 740-4786, e-mail: wensl@tamug.tamu.edu

DR. TERRY E. WHITLEDGE—Marine Science Institute, University of Texas at Austin, Port Aransas, TX 78373-5015, TEL: (512) 749-6769, FAX: (512) 749-6777, e-mail: terry@utmsi.zo.utexas.edu

DR. WILLIAM J. WISEMAN, JR.—Coastal Studies Institute, Louisiana State University, Baton Rouge, LA 70803, TEL: (504) 388-2955, FAX: (504) 388-2520, e-mail: bill@emrys.csi.lsu.edu

Section I

Physical Characteristics

Chapter 1

Geology and Hydrodynamics of Gulf of Mexico Estuaries

William W. Schroeder and William J. Wiseman, Jr.

INTRODUCTION

Estuaries have been a part of the geologic record for at least the past 200×10^6 yr (Williams 1960; Hudson 1963a, 1963b). However, modern estuaries are recent features, having formed over the past 5000 to 6000 years during the stable interglacial period of the middle to late Holocene, which followed the extensive rise in sea level at the end of the Pleistocene (Nichols and Biggs 1985). Geologically, individual estuaries are considered ephemeral features. Upon formation, most begin to fill with sediments and, in the absence of sea level changes, would have life spans of only a few thousand to tens of thousands of years (Emery and Uchupi 1972; Schubel and Hirschberg 1978).

The physiography of estuaries varies widely. Fairbridge (1980) presents a classification system composed of seven basic physiographic types: coastal plain (drowned river valley), bar-built, delta, blind, ria, tectonic, and fjord. The majority of estuaries in the Gulf of Mexico fall into a bar-built and coastal plain categories or are a combination of these two types. Deltas and periodically blind estuaries are also present. In addition, in numerous areas,

Biogeochemistry of Gulf of Mexico Estuaries, Edited by Thomas S. Bianchi, Jonathan R. Pennock, and Robert R. Twilley.
ISBN 0-471-16174-8 © 1999 John Wiley & Sons, Inc.

marine processes have built extensive barrier island systems parallel to the coastline and across the mouths of small to medium-sized coastal plain estuaries, forming extensive lagoonal complexes. It is essential to keep in mind that although the geomorphic nature of individual estuaries/lagoons tends to remain somewhat steady over time, it is not a static characteristic, but rather a dynamic one that is constantly adjusting to remain in equilibrium with geophysical processes. Changes can occur as a function of both short-term (e.g., seasonal fluctuations in sediment supply, fresh-water input, and/or wind velocity and interannual recurrence of major storm events) and long-term (e.g., decadal or longer time changes in climate and secular sea level) processes.

In oceanographic terms, estuaries have been defined as semi-enclosed coastal bodies of water that have a free connection with the open sea and within which sea water is measurably diluted with fresh water derived from land drainage (Pritchard 1967). Estuaries that meet this definition have fresh-water inflow that is greater than fresh-water loss due to evaporation and are considered normal (or positive) estuaries. In estuaries where evaporation exceeds fresh-water inflow, hypersaline conditions are produced; these estuaries are classified as inverse (or negative) estuaries. The critical process associated with normal estuaries is the mixing of sea water with fresh water. Depending on the strength of the mixing processes (e.g., river flow, tidal and wind currents, and wind/wave turbulence), this interaction can result in horizontal and vertical gradients ranging from well-mixed to highly stratified estuaries (Chapter 2, this volume). Pritchard (1955) developed an estuarine classification scheme based on the relative importance of advective (the flux of salt and water) and diffusive (the flux of salt) processes. This scheme identifies four types of estuaries: salt wedge, partially mixed, vertically homogeneous and sectionally homogeneous. It is extremely important to recognize that Pritchard's classification is a short-term, dynamic scheme because estuaries can change from one type to another on seasonal (e.g., fresh-water inflow), multi-day (e.g., winter cold fronts or tropical disturbances), and even tidal (e.g., Mississippi River delta; see the discussion below) cycles. Estuaries in the Gulf of Mexico fit into combinations of all four types.

GEOLOGY

Geologic Framework of the Gulf of Mexico

The Gulf of Mexico is located between approximately 18.2° to 30.4° north latitude and 81.0° to 97.9° west longitude at the southeastern boundary of North America (Fig. 1-1). It covers an area of more than 1.5×10^6 km^2 and attains water depth in excess of 3800 m over its abyssal plain. Together with the Caribbean Sea it forms the "American Mediterranean," a relatively shallow, marginal sea in the western north Atlantic. The formation of the Gulf of Mexico basin was initiated in Late Triassic time (over 200×10^6 yr BP) by

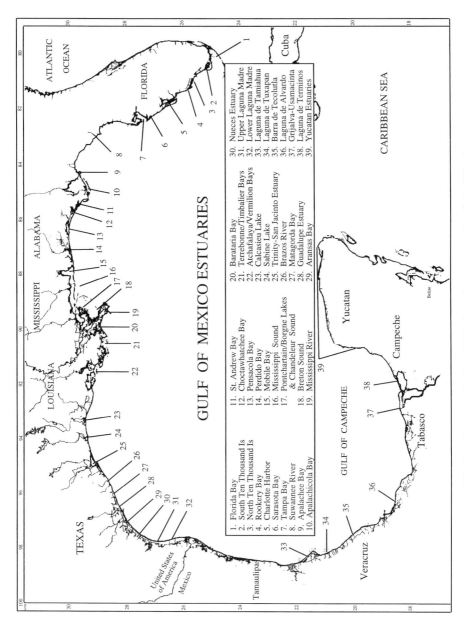

GULF OF MEXICO ESTUARIES

1. Florida Bay	11. St. Andrew Bay	20. Barataria Bay
2. South Ten Thousand Is	12. Choctawhatchee Bay	21. Terrebonne/Timbalier Bays
3. North Ten Thousand Is	13. Pensacola Bay	22. Atchafalaya/Vermilion Bays
4. Rookery Bay	14. Perdido Bay	23. Calcasieu Lake
5. Charlotte Harbor	15. Mobile Bay	24. Sabine Lake
6. Sarasota Bay	16. Mississippi Sound	25. Trinity-San Jacinto Estuary
7. Tampa Bay	17. Pontchartain/Borgne Lakes	26. Brazos River
8. Suwannee River	& Chandeleur Sound	27. Matagorda Bay
9. Apalachee Bay	18. Breton Sound	28. Guadalupe Estuary
10. Apalachicola Bay	19. Mississippi River	29. Aransas Bay

30. Nueces Estuary
31. Upper Laguna Madre
32. Lower Laguna Madre
33. Laguna de Tamiahua
34. Laguna de Tuxapan
35. Barra de Tecolutla
36. Laguna de Alvardo
37. Grijalva-Usamacinta
38. Laguna de Terminos
39. Yucatan Estuaries

FIG. 1-1. Map showing the distribution of estuaries in the Gulf of Mexico.

5

rifting within the North American Plate as it began to drift away from the African and South American Plates. From the Late Jurassic to the early Cretaceous (approximately 145×10^6 yr BP) to the present, the basin has been a relatively stable geologic province. Since early to mid-Cretaceous time (140 to 100×10^6 yr BP), the formation of carbonates and evaporites on the submerged Florida and Yucatan platforms has dominated the eastern and southern regions, respectively. The present configuration of the northern and northwestern Gulf has evolved through persistent subsidence driven, in large part, by the deposition of clastic sediments carried by regional rivers into the Gulf, particularly during the Cenozoic (65×10^6 yr BP to the present). The Mississippi River, because of its vast drainage area, has been the major sediment source. In the west, the Sierra Madre Oriental mountain range of eastern Mexico was formed during the Laramide Orogeny from latest Cretaceous to early Tertiary time (70 to 55×10^6 yr BP).

The present-day physiography of the Gulf of Mexico has been strongly influenced by the extreme fluctuations in sea level that occurred from the end of the Pleistocene to the early Holocene (140,000 to 6000 yr BP). Today the Gulf is rimmed by a diverse combination of coastal plain-continental shelf sub-provinces. To the north and northwest, from Mobile Bay to the Rio Grande river valley, the coastal plain and continental shelf are widest and have a low slope angle, while to the west, the coastal plain and shelf adjacent to the mountains of eastern Mexico are narrower and steeper. In contrast, to the south and east, the coastal plains and continental shelves along the Yucatan and Florida Peninsulas, respectively, are both broad and flat. On a basin scale, the coastal plain-continental shelf region can be divided into two distinctively contrasting sedimentary provinces. The north and west, from the western side of the De Soto Canyon in the northern Gulf to the Campeche Canyon on the western side of the Yucatan shelf, are characterized by terrigenous clastic sediments and the influence of deltaic sedimentation processes. The south and east are composed primarily of carbonate deposits.

Geomorphology of the Coastal Plains

The lowlands bordering the western shoreline of Florida are low in elevation and include relict lagoons, coastal marine terraces, dune and beach ridges, relict shorelines, barrier islands, and coral reefs. The estuarine environments of the southwestern region are primarily small mangrove island-tidal creek complexes and extensive coastal wetlands areas, while rocky, drowned karst and wide shallows with extensive seagrass meadows and marshes typify the central and northern regions.

The northern coastal plains of the Gulf extend from the northwestern corner of the Florida Peninsula, around Apalachiocola Bay, to the Rio Grande river valley at the U.S.-Mexico border. The extensive floodplain of the Mississippi River separates an eastern Gulf coast province from a western Gulf coast province. In the east, along Mississippi, Alabama, and the Florida panhandle, estuarine environments vary from barrier island-back bay complexes with sandy beaches and dunes to well-developed drowned river valley systems. To the west, the eastern Louisiana coast is dominated by the Mississippi-Atchafalaya delta complex and the associated estuarine environments. The western Louisiana coast and upper Texas coast, over to Trinity-San Jacinto estuary (Galveston Bay), is predominantly strand plain and chenier plain systems with extensive marsh wetlands. A few small drowned river valley estuaries are also present.

The Texas and northern Mexico coast, from Trinity-San Jacinto estuary to the Sierra de Tamaulipas region of Mexico, is a nearly continuous series of long barrier islands that form at the mouths of extensive lagoon complexes and at the mouths of several drowned river estuaries. The coastal plain in the Tampico region extends south of Sierra de Tamaulipas, progressively narrowing, to where the Trans-Mexican Neovolcanic Belt approaches within a few kilometers of the Gulf. Barrier islands and lagoons are prominent in the north above Tuxpan (Bryant et al. 1991). The largest and most developed of these is Laguna Tamiahua, with marshes and dune fields as common features of the territory. A number of rivers discharge along this section of coast. The largest include the Tuxpan and Tecolutla Rivers, which form estuaries, and the Moctezuma-Panuco River, which forms a small delta.

To the south, in the Vera Cruz region, the coastal plain becomes steeper as the foothills of the Sierra Madre Oriental extend nearly to the shoreline (Bryant et al. 1991). The entire coast is more irregular but is still made up of barrier island-lagoon systems, the largest being Laguna de Alvarado at the mouth of the Papaloapan River. Along the Isthmus of Tehuantepec the coastal plain turns to the east, widens, and becomes nearly flat in some areas. Two major rivers, the Grijalva and Usumacinta, meander across the middle one-third of this territory, forming extensive wetlands composed of marshes, swamps, and lakes. They merge to form the third largest discharge of fresh water, behind the Mississippi and Atchafalaya Rivers, to flow into the Gulf of Mexico. The entire coastline is dotted with barrier island and lagoon systems, the most prominent being Laguna de Terminos. The Gulf coast of the Yucatan Peninsula is mostly a flat undissected plain. The exception is in the southwestern corner, which consists of low, hilly terrain (Bryant et al. 1991). Lagoons are common, and coral reef complexes occur throughout the region.

Sediment Regimes

Estuarine sediments are derived from fluvial, shoreline erosion, marine, eolian, and biological sources. The composition and distribution of sedimentary facies in an estuary are a function of the types and quantity of sedimentary material available for deposition, hydrodynamic processes, and the geometry of the basin. Sediments are transported to estuaries and moved around within estuaries both in suspension and as bed load. Suspended load is material distributed in the water column by the upward component of turbulence or by colloidal suspension and normally consists of clay- to silt-size particles. "Bed load" refers to larger materials, such as sands, gravels, and whole shells, that are moved by saltation or traction on or immediately above the sea bed. Sediment deposits composed of coarse material the size of sand grains or larger generally denote high-energy environments, whereas silt- and clay-sized material normally indicates low-energy environments (Chapter 3, this volume).

The coarsest sediments in most estuaries are ordinarily associated with (1) the barrier features fronting the adjacent open ocean; (2) the passes connecting estuaries with the ocean; (3) the terminus of inflowing rivers, normally located at the head of an estuary; and/or (4) the shallow edges along shorelines where erosion is occurring. Fine-grained deposits are generally confined to the deeper reaches of the estuary or in low-energy environments.

Sediment Distribution Patterns. Florida Bay is a broad, shallow region lying between the south Florida Peninsula and the Florida Keys. The bottom slopes gently from water depths of approximately 1 m in the northeast to between 2.5 and 3 m in the west. The interior of the bay is a complex of semi-isolated basins and banks. On the exposed edges of the banks, sediment is mostly sand; but most of the deep areas and the protected portions of the banks are dominated by silt and clay deposits (Folger 1972).

In Tampa Bay, sand-sized sediment is widespread throughout the bay, particularly along the flat, shallow (generally <1.8 m) margins. The coarsest material (e.g., very coarse sand and shell fragments) occurs in the deeper (5–6 m) channels where tidal currents are highest. Silt is abundant only in Hillsboro Bay (Goodell and Gorsline 1961). Apalachicola Bay is a coastal lagoon system with depths generally <2 m protected by extensive barrier islands. Surface sediments include a complex of sands, clays, and mixtures of the two; silts are conspicuously absent (Isphording et al. 1985). In contrast, the Pensacola Bay estuary has an average depth of just over 6 m, and bottom sediments are dominated by silts and sands, with very little clay (Isphording et al. 1985).

The Mobile Bay estuary is a classic submerged river valley with the vast Mobile River Delta at its head and Dauphin Island and the Fort Morgan peninsula enclosing it to the south. Water depths average 3 m. The basin of the bay contains high percentages of clay and silty clay mixtures. Sandy areas are present mostly near river mouths and along shorelines, and sand is the principal material building the coastal barrier features (Isphording et al. 1985). Mississippi Sound, a long, narrow, semi-protected lagoonal feature, similar in many respects to Apalachicola Bay, extends westward along the coasts of Alabama and Mississippi. It has an average water depth of 3 m. Sediments in the sound are predominantly sand-silt-clay mixtures in the east, silty clays and sands in the central region and along the northern shore and barrier islands, and clayey silts and sand-silt-clay mixtures in the west (Isphording et al. 1985).

The active birdfoot delta of the Mississippi River consists of a network of bifurcating distributaries protruding onto the adjacent shelf. Between the advancing passes are shallow embayments, some open to the Gulf of Mexico and others isolated by various barriers. To the north is the abandoned St. Bernard Delta, a complex marsh system consisting of a maze of small bays, interconnecting levees and channels, and islands. Silts and clays dominate the marsh and enclosed bay sediments, while open bay regions, exposed to wave action, become more sandy (Barrett et al. 1971b). Located west of the delta are the shallow and broad estuaries of Barataria, Timbalier, and Terrebonne Bays. The sediments within all three of the systems predominantly grade from silty sands on the shoreward sides of the barrier islands and in shoal areas to sandy silt and clayey silt in the central basins to silty clay adjacent to protected marsh wetlands (Barrett et al. 1971b).

Along the central Texas coast, the shallow estuaries of Matagorda, San Antonio, and Aransas Bays average water depths of about 2.9, 1.2, and 2.5 m, respectively. The central basins of these bays are covered by silty clays, while sands and mixtures of sand-silt-clay predominate along the shorelines and in shoal areas (Folger 1972). Laguna Madre, on the lower Texas coast, is one of the largest coastal lagoon systems in the Gulf of Mexico. Its bottom slopes evenly from exposed sand flats on the eastern side behind the barrier islands to a few deep (up to 2.5 m), elongated basins close to the mainland shore to the west. The 2.5- to 3.5-m-deep Intracoastal Waterway channel is also located on the western side. Bottom sediments are principally sand on the eastern side and silty to clayey sands, with some silty clay and mixtures of sand-silt-clay in the deeper sections of the western side (Rusnak 1960).

Clay Mineralogy. Clays are particles <2 μm in size, usually composed of common minerals such as quartz and feldspar and complexes of other distinctive clay minerals. Clay minerals are extremely variable in size, structure, and chemistry. The most common species belong to the kaolinite, montmorillonite, and illite groups (Brindley 1980; Degens 1989). The clay minerals that occur in coastal systems are derived principally from the weathering of rocks and soils within the watersheds that drain into them.

Knowledge of the presence and distribution patterns of the various clay mineral species in estuaries is important because their interactions, which occur between particulate and dissolved material, play key roles in geochemical and biogeochemical cycles (e.g., nutrients and heavy metals) (Burton and Liss 1976; Schlesinger 1997). Unfortunately, very little information on this subject is available for most of the estuaries in the Gulf of Mexico. However, for the northern Gulf, the following summary can be made based on the work of Griffin (1962), Folger (1972), and Isphording et al. (1985). For the estuaries on the lower west coast of Florida, kaolinite predominates in the deeper upper bay regions and in the mangrove swamps, while illite is most abundant near the bay mouths. In Tampa Bay, both kaolinite and montmorillonite are present but rare. On the upper west coast kaolinite is more abundant than either montmorillonite or illite. Along the northeastern rim of the Gulf from the Big Bend area over to the bays and bayous west of the Mississippi River, the dominant clay mineral shifts from nearly equaly amounts of kaolinite and montmorillonite in Apalachicola Bay to a westwardly increasing dominance of montmorillonite. Farther to the west, in the central coastal region of Texas, montmorillonite remains the dominant mineral, but both chlorite and illite often occur in moderate concentrations. Along the south Texas coast, illite becomes the predominant clay mineral.

HYDRODYNAMICS

Forcing Functions

It is generally accepted that estuarine circulation is forced by three dominant processes: net fresh-water delivery to the system, local momentum transfer from the wind, and variability at the estuary mouth caused by processes in the coastal ocean. Net fresh-water delivery to the system, runoff plus precipitation minus evaporation, is discussed extensively in Chapter 2 of this volume. Wind, and by inference wind stress, around the Gulf of Mexico coast has a strong annual signal, being influenced by the seasonal strength of the Bermuda high-pressure cell (Leipper 1954). Mean wind directions during fall and winter, when the high-pressure system is weakened and in a northeastern location, exhibit easterly and northeasterly components in the northern and eastern sections of the Gulf (Gutierrez de Velasco and Winant 1996). During spring and summer, when the high-pressure system is strengthened and in a south-western location, mean winds have southeasterly components. During all seasons, mean wind directions over the Yucatan shelf appear to be east-northeast-erly, while they exhibit northerly and north-northwesterly components in the Bay of Campeche.

The variance ellipses (Gutierrez de Velasco and Winant 1996) of the winds are generally extended in a north-south direction. They are more elliptical during the winter months (Gutierrez de Velasco and Winant 1996, Wiseman et al. in review). The dominant time scales of wind variability increase during summer months, and the strength of the wind variations decreases (Wiseman et al. in review). A land-sea breeze system is found along the Gulf coast throughout the year, with maximum development occurring during summer. Much of the synoptic variability in the wind field is due to passage of cold air outbreaks. These fronts pass across the northern Gulf coast with recurrence intervals of 3–10 d during winter. The strength and frequency of such events diminish during the summer (DiMego et al. 1976). Such outbreaks often extend as far south as the Yucatan Peninsula. Statistical analysis of these frontal passages over northern Gulf of Mexico waters (Fernandez-Partagas and Mooers 1975) indicates that they normally have a southwest-northeast orientation of the front. Pre-frontal winds are from the southeast, while post-frontal winds are from the northwest. The latter bring cold, dry air to the coast with massive associated heat fluxes from the coastal waters, at least along the northern Gulf coast. The most significant wave action associated with a frontal passage is largely dependent upon the coastline orientation and the associated fetch.

Exceptional perturbations to this pattern occur during the passage of tropical cyclones. Over the 101-year period of 1886–1986, an average of 3.72 tropical systems (tropical storms and hurricanes) affected the Gulf of Mexico (Florida A&M University 1988). While such events are infrequent, their impact on the morphology of estuarine systems can be extreme and long-lived. New inlets are opened across barrier islands, and others are closed (Hayes 1978). Bottom sediments are resuspended and redistributed (Schroeder et al. in press). The interior morphology of estuaries, particularly those in marsh environments, is dramatically modified (Jackson et al. 1995).

The third external forcing function is variability at the mouth of the estuary, associated with both local and far-field processes in the coastal zone occurring at tidal and sub-tidal periods. Tides in the Gulf are predominantly diurnal (Marmer 1954), although mixed tides are important along the west Florida shelf and where the shelf is particularly wide such as offshore of Atchafalaya Bay. The Gulf is classified as a microtidal environment, but tidal currents, forced by tidal water level variability on the open Gulf side of tidal inlets, remain the dominant motion in many estuaries for much of the year. The most important of these is Ekman set-up/set-down caused by local winds (Carter et al. 1979; Wiseman 1986). This process modifies the barotropic pressure gradient, forcing flow into the estuary. Other processes that affect this pressure gradient include shelf waves (Clarke and van Gorder 1986; Mitchum and Clarke 1986; Current 1996) and seasonal variations of coastal water level (Smith 1978) due to both wind forcing and steric adjustments (Blaha and Sturges 1981). While it is commonly assumed that it is the longshore component of wind stress that is responsible for barotropically forced exchange between the estuary and the coastal ocean, local variations in the nearshore bathymetry may result in an increased importance of cross-shore wind stress (Chuang and Wiseman 1983).

Although shelf-estuarine exchange due to variations in the barotropic pressure gradient has been extensively studied and described in the Gulf of Mexico (e.g., Schroeder and Wiseman 1986), there exists the potential for important exchange processes driven by the baroclinic pressure gradient and its variability. It has been noted (Barrett et al. 1971a) that high-salinity water is occasionally found between lower-salinity coastal water and low-salinity water in the upper reaches of Barataria Bay during the spring, when Mississippi River discharge is high. It is possible that river effluent rapidly lowers coastal salinities and establishes a reverse baroclinic pressure gradient between the estuary and the shelf, which alters flow regimes and exchange processes. In a similar fashion, upwelling and Loop Current eddies have the potential to significantly alter shelf water density and, consequently, baroclinic pressure gradients at the mouths of estuaries, thus modifying the intensity of the shelf-estuarine exchange.

Estuarine Dynamics

It is the baroclinic pressure gradients generated by fresh water derived from land drainage and the influence of this water on flow structure of estuarine boundaries that distinguish estuaries from bays (Pritchard 1967). Mid-latitude estuaries typically exhibit strong seasonal variations in discharge, with peak flows occurring in spring associated with melting of fresh water sequestered in snow and ice during winter. The Mississippi-Atchafalaya system clearly mimics this pattern in its discharge to the Gulf of Mexico. Many of the small subregional watersheds of remaining estuaries in the Gulf respond to local coastal patterns of rainfall, which are often highly variable throughout the year. Furthermore, the broad latitudinal extent of the Gulf suggests that rainfall and evaporation patterns will vary significantly across the region, modifying fresh-water discharge to the estuaries (Chapter 2 this volume).

The familiar drowned river geomorphology so typical of estuaries along the Middle Atlantic Bight is relatively absent from the Gulf of Mexico coast. Except for the important Mississippi and Atchafalaya Rivers, we find a predominance of broad, shallow, bar-built estuaries. The distance from the coastal ocean to the source of fresh water is typically very short compared to the tidal wavelength. Estuaries often are wide in the horizontal dimension orthogonal to the large-scale salinity gradient, and estuaries connect to the ocean through multiple inlets. Often, the best-studied estuaries have had their geomorphology significantly altered by dredging activities, such as Tampa Bay, Mobile Bay, Mississippi River, Trinity-San Jacinto Bay, and Laguna Madre. These deep ship channels provide a conduit for up-estuary movement of salt (e.g., Schroeder et al. 1996a), which can result in either the reinforcing of existing haline stratification or the development of stratification in regions where it previously did not occur.

Motions within an estuary, and between an estuary and coastal ocean, occur on a variety of time scales. More energetic motions occur at time scales of tides, synoptic weather systems (2–10 d), and seasons. The interaction of tidal currents with bathymetry can result in small-scale residual vorticities (Zimmermann 1978) that contribute to dispersion within an estuary. Given the relatively weak tidal flows in the Gulf, the second-order terms involved in the generation of residual vorticity are probably unimportant; dispersion due to topographic trapping associated with convoluted marsh environments found along the northern Gulf coast can be significantly more important (Inoue and Wiseman in review).

The sub-tidal, wind-driven flows occurring in estuaries of the northern Gulf of Mexico are particularly significant during the stormy winter season. These flows have been the subject of numerous site-specific studies and a few reviews (Schroeder and Wiseman 1986; Wiseman 1986). Coastal engineering literature on tidal inlets suggests that transport through a series of inlets, as water level changes at the coast, can be modeled as flow through a single equivalent inlet. This is clearly not the case for locally wind-driven flows. Coastal water levels change in response to an along-shore wind in the Gulf of Mexico. Coastal water levels increase when winds blow with the coast to the right of an observer looking downwind and decrease when winds blow with the coast to the left. This same wind stress, acting on waters of an estuary, will result in a downwind set-up of water levels within the estuary. This in turn, will result in hydraulic heads tending to force water out of the estuary at downwind inlets and into the estuary at upwind inlets. Net exchange will be greatly enhanced over that occurring in the absence of a tilting water surface within an estuary. Thus, the geomorphology of estuaries in the Gulf of Mexico, elongated in the long-shore direction with multiple tidal inlets, enhances shelf exchange in response to local along-shore wind stress events.

Mixing is affected by a variety of processes. Boundary shear stress, wind stress at the sea surface, and current stress at the sea bed generates turbulence, which contributes to estuarine mixing. As in most estuaries, shear associated with tidal currents is important in Gulf of Mexico estuaries. In shallow estuaries, which characterize so much of the Gulf, locally generated wind waves can interact effectively with the bottom and contribute significantly to local mixing and energy dissipation. The associated wave-current interactions (Grant and Madsen 1986) are expected to be significant. The ultimate importance of this interaction to mean circulation has yet to be determined. In regions where the diurnal sea breeze is important, a mean flow can be generated from this periodic forcing (Signell et al. 1990).

A summary for estuaries in the Gulf of Mexico is one of great variability in estuarine type and, consequently, patterns of circulation. Drowned river valleys are present, the most significant being the Mississippi River, with its multiple passes dominated by salt wedge intrusions. Yet salt wedges are not confined to the Mississippi, but exist in other estuaries with deep channels and high discharge, such as Mobile Bay. Other estuaries receiving significant fresh-water discharge and having a single entrance communicating with the sea behave like partially stratified or vertically homogeneous estuaries, such as Pensacola Bay or Fourleague Bay. Estuaries receiving extensive runoff but having multiple outlets to the sea, such as Mississippi Sound or Apalachicola Bay, present unique circulation patterns. When river discharge to such systems is greatly reduced, as in Terrebonne and Timbalier Bays, flow is strongly dominated by winds and tides. Finally, if evaporation exceeds runoff and precipitation, an inverse estuary or hypersaline lagoon situation results. Although not considered true estuaries in the sense of Pritchard (1967), these systems have many similarities with estuaries in the Gulf. In the following section, we present the circulation and hydrographic structure of selected examples of these environments.

Case Studies

Florida Bay. Florida Bay is a broad, V-shaped region lying between the south Florida Peninsula and the Florida Keys. Water depths are generally <3 m and shoal dramatically to the east. The inner regions of the Bay are separated into a series of semi-isolated basins by numerous shallow mudbanks. The region receives fresh-water drainage from the Everglades. This water source has been significantly modified by hydrologic changes during the last several decades (U.S. Department of Commerce 1993; Light and Dineen 1994). Exchange between Florida Bay and the coastal ocean takes place across a wide opening connected to the west Florida shelf and numerous passes connecting to the Straits of Florida. Although this system is described as an estuary (Wang et al. 1994), hypersaline conditions often prevail within it (Fourqurean et al. 1993; Zieman et al. 1994) (Fig. 1-2). The system may switch from hypersaline to hyposaline conditions during the course of a year (Ley et al. 1994). Vertical stratification is observed in the deeper channels of Florida Bay, but the water column is typically well mixed. The role of density stratification in forcing circulation is poorly understood. Strong lateral density gradients must contribute to long-term flow patterns, but the data necessary to determine the strength of this flow component are not yet available. These lateral density gradients can be further enhanced by winter cooling during cold air outbreaks (Walker et al. 1982, 1987; Roberts et al. 1983) and contribute to a gravitational exchange with waters of the Florida Reef tract when triggered by winds and tidal flows.

The effects of winds and tides on the shallow waters of Florida Bay seem to dominate the circulation. Winds over Florida Bay are predominantly southeasterly during summer months, but strong, cold, dry winds from the north cause major winter synoptic weather patterns (Walker et al. 1987; Wang et al. 1994). In shallow, well-mixed waters of Florida Bay proper, a downwind

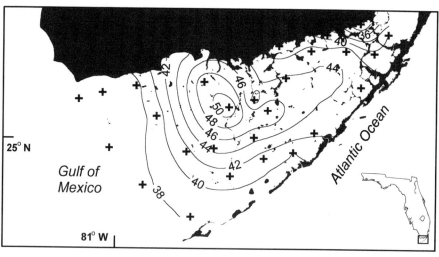

FIG. 1-2. *Average salinity values depicting hypersaline conditions in Florida Bay during the period June 1989–August 1990. Crosses indicate station locations (figure provided by J. W. Fourqurean, Florida International University, Miami, FL).*

flow may be expected to result from local wind stress. Water level variations within Florida Bay are also observed and are attributed to Ekman flow over the west Florida shelf (Wang et al. 1994). An additional set-up occurs in response to tidal stress when the energy in a tide dissipates as it propagates from west to east across Florida Bay from the west Florida shelf (Wang et al. 1994). Preliminary modeling efforts have been unable to reproduce this set-up using realistic bottom stress values. It was assumed that a significant portion of tidal dissipation was due to drag associated with islands and mud-banks. This set-up, and the set-down in sea level seaward of the Florida Keys, contribute to a tendency toward net outflow through passes among the Keys (Smith 1994). The strength of this outflow varies on a fortnightly scale due to tides and on shorter scales in response to wind-driven sea level variability.

Pensacola Bay and Tampa Bay. Pensacola Bay, on the western end of the Florida panhandle, is a typical drowned river valley estuary. The main stem of the bay divides into Escambia Bay and East Bay. These two bays are fed, respectively, by the Escambia River and the Blackwater and Yellow Rivers. Annual mean discharge of fresh water is only 324 m^3 s^{-1}, but the interannual variability in discharge can exceed a factor of 18. Little is known of the hydrography and circulation of this system. Flow patterns and salinity distributions (Fig. 1-3) appear to be consistent with a partially stratified estuary between types 2b and 3b on the Hansen and Rattray (1966) classification diagram (Ketchen and Staley 1979). While data for this study were of short duration, there is an indication that sub-tidal, wind-driven motion may be important.

While strong stratification due to fresh-water discharge is generally indicative of a partially stratified, vertically sheared estuarine circulation, weak vertical stratification does not necessarily imply barotropic circulation. The central axis of the lower reaches of Tampa Bay is relatively deep (5–6 m) and salinity is vertically well mixed most of the time (Galperin et al. 1991), but longitudinal salinity gradients are significant. The associated baroclinic torque has the potential to produce a significant mean flow (Weisberg and Williams 1985). Two-dimensional numerical models of bay circulation indicate a series of residual gyres, a salt balance dominated by horizontal diffusion, and underestimates of salinities in the upper bays (Galperin et al. 1991). Three-dimensional baroclinic models of the same system result in a salt balance dominated by horizontal advection, more realistic salinities in the upper bays than those produced by two-dimensional models, and mean profiles of tidal velocity shears similar to observed values (Galperin et al. 1985; Weisberg and Williams 1985).

Mobile Bay. Mobile Bay is a broad, shallow estuary of the Mobile River delta. It connects to the Gulf of Mexico through Main Pass and to Mississippi Sound through Pass aux Herons. The latter carries approximately 15% of the mean discharge from the estuary. While the mean depth of the estuary is only

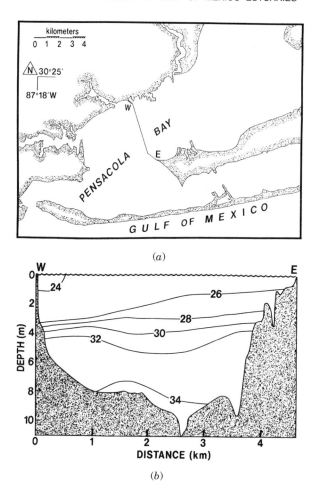

FIG. 1-3. (A) Map of lower Pensacola Bay, Florida, showing location of cross-section transect; (B) salinity distribution from the cross section in lower Pensacola Bay, Florida, at mid-flood tide on October 27, 1978 (adapted from Ketchen and Staley 1979).

3 m, a 12- to 15-m-deep ship channel, dredged in the middle of the 19th century, has important effects on the flow dynamics of Mobile Bay. Mobile River, which carries the combined flow of the Tombigbee and Alabama Rivers, delivers 95% of total fresh-water discharge to the estuary. It has a mean discharge of 1848 m^3 s^{-1}, but seasonal (Schroeder and Wiseman 1986) and intrannual variability (Schroeder et al. 1996b) is significant. The range of weekly mean discharge varies by a factor of 62 (Schroeder et al. 1980).

This significant variability in fresh-water discharge also implies a large variation in stratification (Schroeder et al. 1990, 1996b) and, consequently, in baroclinic pressure gradients in Mobile Bay. During a year, salinities at mid-

Bay stations may vary by more than 20 g liter^{-1}. Following a major river flood event, fresh water fills the entire shallow water regions of the Bay (Schroeder 1977; Schroeder et al. 1990); but this situation relaxes rapidly as river discharge diminishes and coastal waters move back into Mobile Bay (Fig. 1-4). In such shallow water, one might normally assume a vertically well-mixed condition; however, this is often not the case in Mobile Bay. Shallow bottom depressions and dredge spoil banks, in concert with normal estuarine dynamics of the system, allow vertical stratification in excess of 10 g liter^{-1} over 3 m (Schroeder et al. 1990). This stratification can be broken down through the combined

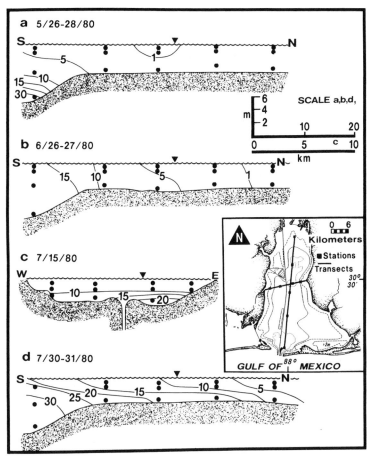

FIG. 1-4. *Longitudinal and lateral salinity sections in Mobile Bay, Alabama, May to July 1980. Note: Although the vertical scale remains the same in all sections, the vertical exaggeration of the longitudinal sections is twice that of the lateral section (adapted from Schroeder et al. 1990. Reprinted by permission of the Estuarine Research Federation. © Estuarine Research Federation).*

effects of wind and tides (Schroeder et al. 1990), with wind-generated waves enhancing bottom drag coefficients and consequent mixing (Grant and Madsen 1986). Reduced mixing under weak summer winds causes strong stratification, which can endure for extended periods of time. Warm water temperatures and consequent high metabolic rates result in the development of hypoxic to anoxic conditions in near-bottom waters. These hypoxic waters are advected by tidal and sub-tidal currents causing the "jubilee" phenomenon for which Mobile Bay is well known (May 1973; Schroeder and Wiseman 1988).

Wind-driven flows take two forms: wind-driven flows interior to Mobile Bay, which are locally driven, and exchanges with the coastal ocean, which are driven by winds over the adjacent continental shelf. The latter have been extensively discussed (Schroeder and Wiseman 1986). Ekman convergence and divergence at the coast, primarily driven by along-shore components of wind stress, force water in and out of Mobile Bay at periods within weather frequency (2–10 d). Interior to Mobile Bay, flow below the halocline over shallow reaches of the estuary is opposite wind stress (Noble et al. 1996). There are indications that upper-layer flows are downwind, but statistics are only marginally significant, presumably because the size of Mobile Bay allows the existence of other small-scale baroclinic motions, which make statistics of wind-driven flow noisy (Noble et al. 1996).

Surface layer flows are responsive to river forcings at low discharge levels (Noble et al. 1996), but at higher discharge levels (>3000 m^3 s^{-1}) the upper layer velocities no longer respond to increases in discharge. Lower-layer flows are unresponsive to low river discharge but respond at higher rates. Noble et al. (1996) suggests that this behavior can be related to an internal hydraulic control at Main Pass.

This description of velocities in Mobile Bay ignores the presence of a deep ship channel. This analysis is probably realistic only because the ship channel is narrow and the halocline of bay waters is so strong. This picture may well describe the situation that existed in Mobile Bay during the early years of the 19th century, prior to dredging of the channel from Main Pass to Port of Mobile. The channel, though, plays a significant role in maintaining the salt balance of the estuary. A well-developed salt wedge exists in the deeper waters of Mobile Bay. Under certain wind conditions, these saline waters upwell and overflow onto lateral shoals (Schroeder et al. 1996a). Vertical mixing may also bring waters entering Mobile Bay through the ship channel to the surface under specific wind conditions (Schroeder et al. 1996a).

Mississippi River. Passes of the Mississippi River Delta, through which the river debouches water from a drainage basin encompassing 43% of the contiguous United States and parts of two Canadian provinces, are the archetype for salt wedge estuaries. Oligohaline waters flow seaward in a surface layer above an intense halocline (Fig. 1-5). The pattern resulting from averaging over a tidal cycle to remove the large, oscillatory tidal currents shows that lower-layer waters creep slowly landward to maintain continuity as small amounts

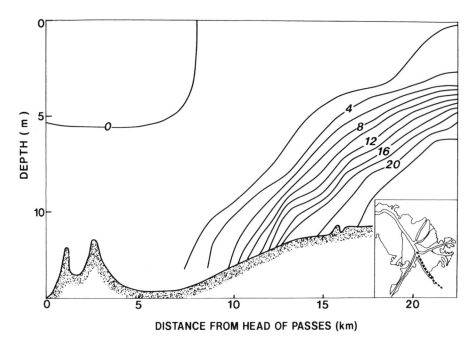

FIG. 1-5. *Longitudinal sigma-t section in South Pass channel of the Mississippi River, October 1, 1969, depicting a typical salt-wedge intrusion pattern during a flooding tide (adapted from Wright [1971]. Reprinted by permission of the American Society of Civil Engineers).*

of water are entrained into seaward-flowing surface layers. This lower-layer flow is modulated by changes in imposed pressure forces, which vary on tidal scales, weather band scales, and seasonal and longer time scales in response to river discharge and steric changes in coastal water level. During maximum flood, salt water can be completely flushed from the passes. At these times, there is no Mississippi River estuary sensu Pritchard.

Early work by Keulegan (1949) attributed the vertical entrainment of lower-layer fluid into the surface layers to breaking of internal waves, as observed in laboratory investigations. There is little field evidence to support this mechanism as the cause of entrainment. It seems more likely that processes at the toe of the wedge (Geyer and Farmer 1989) or mixing at the lateral boundaries and subsequent lateral advection along the halocline (Stigebrandt 1976) are responsible for the observed effective entrainment.

Terrebonne and Timbalier Bays. Terrebonne-Timbalier Bay is a broad, shallow estuary on the northern Gulf of Mexico coast. While associated with an abandoned distributary of the Mississippi River, it no longer receives significant runoff from any source other than a small, local drainage basin (Prager 1992). Except during periods of freshets, the water column salinity, and hence density, are nearly homogeneous vertically. However, strong lateral salinity gradients still exist (Fig. 1-6). Model studies (Elliott and Reid 1976) suggest that such lateral stratification is relatively ineffectual in controlling dispersion throughout a shallow estuarine system on any but the very longest time scales.

Although tides are weak throughout the Gulf of Mexico (Marmer 1954), current meter measurements from the Bay system indicate that tidal flows are the dominant pattern (McKee et al. 1994). The orientation of these flows is robust. In contrast, the sub-tidal flows are highly variable in direction and, consequently, mean flows are weak. These motions are coherent with the wind stress within the weather band. Salinity variability within the estuary is controlled by tidal advection at diurnal time scales (Wiseman and Inoue 1993a; McKee et al. 1994). However, there is significant variability over longer time

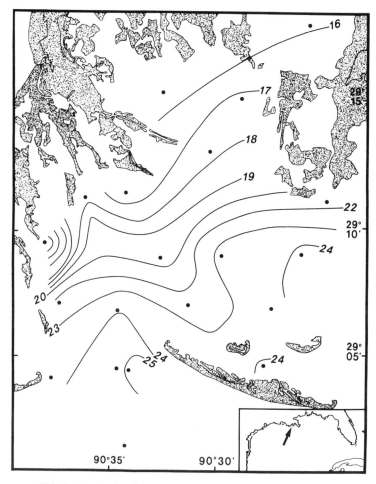

FIG. 1-6. Lateral salinity gradients in Terrebonne Bay, Louisiana.

periods. Within the weather band (periods on the order of 2–10 d), salt exchange between the estuary and the coastal ocean is important, and the processes responsible for it are many. A moderately deep channel (5–6 m) extends northward from Cat Island Pass. Cross-shore wind stress events result in important estuarine-shelf exchange processes, as have been observed elsewhere (Weisberg and Sturges 1976). Along-shore wind events also contribute to the intensity of the exchange processes. In discussing the role of such winds, focus is normally on the coast Ekman convergence/divergence driven by the wind stress (e.g., Schroeder and Wiseman 1986). These same winds drive a local downwind set-up within the estuary. This creates a hydraulic head, which drives outflow through downwind tidal passes and accelerates inflow through upwind tidal passes (Wiseman and Inoue 1993b). At the longest periods measured, salinity variability in the estuary is a response to salinity variability within the coastal ocean (Wiseman et al. 1990; Wiseman and Inoue 1993a).

A variety of processes contribute to the dispersion characteristics of shallow estuaries (Ridderinkhof and Zimmerman 1990). Although tidal currents dominate the energy spectrum of the Terrebonne-Timbalier Bay system, velocities are not sufficiently strong that residual vorticity generation appears to be an important process in the region. Rather, topographic trapping (Okubo 1973) seems to be particularly important (Inoue and Wiseman in review).

Laguna Madre. Laguna Madre, including Baffin Bay, is a bar-built estuary along the south Texas coast. Local runoff and precipitation are much lower than evaporation. During the early decades of the 20th century, salinities were generally >50 g liter^{-1} and salinities of 100 g liter^{-1} were not unusual (Collier and Hedgepeth 1950). These waters had a restricted exchange with those of Corpus Christi Bay, and a tidal mean circulation akin to turning a partially stratified estuary upside down was suggested (Collier and Hedgepeth 1950). Completion of the intracoastal waterway is likely to have greatly increased flushing of the system and exchange with sources of lower-salinity water. Laguna Madre is no longer hypersaline, although hypersaline conditions are still observed in some of the peripheral regions such as upper Baffin Bay (T. Whitledge, personal communication).

Tides in Laguna Madre are minimal, and water level is usually reflective of mean water levels outside the lagoon, the initially small (0–30 cm) Gulf of Mexico tide having been further dampened by passage through restricted tidal passes (Collier and Hedgepeth 1950). Wind tides are significant and can result in prolonged exchange between Laguna Madre and its adjacent water bodies, Corpus Christi Bay and the Gulf of Mexico.

Other Systems. Little is known of the circulation and dynamics of the many other estuaries of the Gulf of Mexico. Two bar-built estuaries receiving large discharges of fresh water, Apalachicola Bay and Mississippi Sound, are particularly interesting and deserve further study. The high river discharge suggests strong baroclinic pressure gradients; the shallow depths indicate important wind-forcing effects; the relatively small size implies important tidal flushing; and the multiple connections to the ocean allow important forcing interactions between local and far-field wind-driven set-up. Which processes will control the long-term, low-frequency dynamics are presently unclear.

Barataria Bay, Atchafalaya Bay and its tributary bays, Calcasieu Lake, Sabine River, Galveston Bay, and Corpus Christi Bay are similar to Mobile Bay and Terrebonne-Timbalier Bays. The relative size of fresh-water discharge, connection to the coastal ocean, and number of tidal passes will modify the anticipated dynamics but are not expected to modify the general character of the resultant flows; Sabine Lake may be exceptional. A ship channel dredged along the eastern shore may allow salt to penetrate aperiodically to the upper part of the estuary without passing through the broad mid-reaches of the system, thus generating anomalous baroclinic gradients at particular times (Baskaran et al. 1997; Bianchi et al. 1997).

SUMMARY

Our concepts of estuarine circulation are dominated by experiences on the east and west coasts of the United States and the west coast of Europe. This experience is heavily influenced by the presence of coastal plain (drowned river valley) and fjord estuaries with moderate to high tidal energy along these coastlines. Thus, our traditional picture of an estuary includes a narrow, moderately deep to deep channel with a localized upstream fresh-water source and tidal mixing that is often sufficiently strong to significantly control the distribution of baroclinic pressure gradients within the estuary. While such estuarine systems occur in the Gulf of Mexico (e.g., Pensacola Bay), they are not the dominant type. The more typical Gulf coast estuary is a bar-built system or a combination bar-built and coastal plain. These estuarine systems are characterized by broad regions of relatively flat topography and, consequently, a lack of topographic steering of the low-frequency flow. Further, they are often as large in their along-shore direction as in their dimension perpendicular to the coast. The internal Rossby radius of deformation, the length scale appropriate to baroclinic motions, may become commensurate with the width of the estuary.

Tides in the Gulf are small (tidal range is 0–30 cm) and occur as predominantly diurnal or mixed types. Under low wind forcing, the most energetic estuarine motions remain barotropic tidal currents directed by topography. However, over longer time periods or under strong wind forcing, the systems are responsive to wind forcing and orientations of currents are highly variable. Along portions of the east and west coasts of the United States and the west coast of Europe, local landscape topography is often sufficient to direct wind stress on estuarine waters. The generally low-lying coastal terrain, characteristic of the Gulf of Mexico coastal plain provinces, frequently precludes such topographic steering of winds, resulting in wind-driven estuarine currents.

This allows development of a much more variable flow field in Gulf coast estuaries than might be expected from experiences elsewhere in the world.

Finally, we reiterate the important role that multiple inlets appear to play in the circulation patterns within Gulf coast estuaries. Not only do phase lags in tidal waves entering different channels have potentially important effects, as seen in other estuaries (e.g., the Chesapeake and Delaware canal), the interaction of local and far-field, wind-driven sea level slopes can drive long-period flows through the estuaries. These circulation patterns drive very efficient renewal of estuarine waters. Thus, the shallow Gulf of Mexico estuaries offer an interesting range of circulation, with variable scales that are only beginning to be appreciated and understood. The very long period baroclinic circulations are probably the least well studied of such motions.

ACKNOWLEDGMENTS

We thank L. Sanford and R. R. Twilley for their critical review of the manuscript and J. W. Fourqurean for providing the figure of salinity values for Florida Bay. WWS was supported in part by NOAA Office of Sea Grant, Department of Commerce under Grant No. NA56RG0129 (Project No. R/ ER-35), the Mississippi-Alabama Sea Grant Consortium, The University of Alabama, and the Dauphin Island Sea Lab. WJW was supported by the U.S. Minerals Management Service and Louisiana State University through the LSU Coastal Marine Institute. This publication is Contribution No. 247 from the Aquatic Biology Program, University of Alabama, and Contribution No. 293 from the Dauphin Island Sea Lab, Dauphin Island, Alabama.

REFERENCES

Barrett, B. B., J. W. Tarver, G. B. Adkins, W. R. Latapie, W. J. Gaidry, J. F. Pollard, C. J. White, W. R. Mock, and J. S. Mathis. 1971a. Cooperative Gulf of Mexico Estuarine Inventory and Study, Louisiana. Phase II, Hydrology, p. 8–130. *In:* B. B. Barrett [ed.], Cooperative Gulf of Mexico estuarine inventory and study. Louisiana Wildlife and Fisheries Commission.

———, J. W. Tarver, G. B. Adkins, W. R. Latapie, W. J. Gaidry, J. F. Pollard, C. J. White, and W. R. Mock. 1971b. Cooperative Gulf of Mexico Estuarine Inventory and Study, Louisiana. Phase III, Sedimentology, p. 131–187. *In:* B. B. Barrett [ed.], Cooperative Gulf of Mexico estuarine inventory and study. Louisiana Wildlife and Fisheries Commission.

Baskaran, M., M. Ravichandran, and T. S. Bianchi. 1997. Cycling of Be-7 and Pb-210 in a high DOC, shallow, turbid estuary of southeast Texas. Estuar. Coast. Shelf Sci. **45:** 165–176.

Bianchi, T. S., M. Baskaran, J. DeLord, and M. Ravichandran. 1997. Carbon cycling in a shallow turbid estuary of southeast Texas: The use of plant pigment biomarkers and water quality parameters. Estuaries **20:** 404–415.

Blaha, J., and W. Sturges. 1981. Evidence for wind-forced circulation in the Gulf of Mexico. J. Mar. Res. **39:** 711–734.

Brindley, G. W. 1980. Crystal structures of clay minerals and their X-ray identification. Mineral Society.

Bryant, W. R., J. Lugo, C. Cordova, and A. Salvador. 1991. Physiography and bathymetry, p. 13–30. *In:* A. Salvador [ed.], The Gulf of Mexico Basin, the geology of North America, Vol. J. Geological Society of America.

Burton, J. D., and P. S. Liss. 1976. Estuarine chemistry. Academic Press.

Carter, H. H., T. O. Najarian, D. W. Pritchard, and R. E. Wilson. 1979. The dynamics of motion in estuaries and other coastal water bodies. Rev. Geophys. **17:** 1585–1590.

Chuang, W. S., and W. J. Wiseman, Jr. 1983. Coastal sea level response to frontal passages on the Louisiana-Texas coast. J. Geophys. Res. **88:** 2615–2620.

Clarke, A. J., and S. Van Gorder. 1986. A method for estimating wind-driven frictional, time-dependent, stratified shelf and slope water flow. J. Phys. Oceanogr. **16:** 1013–1028.

Collier, A., and J. W. Hedgpeth. 1950. An introduction to the hydrography of tidal waters of Texas. Publ. Inst. Mar. Sci. **1:** 125–194.

Current, C. L. 1996. Spectral model stimulation of wind driven subinertial circulation on the inner Texas-Louisiana shelf. Ph.D. dissertation, Texas A&M University.

Degens, E. T. 1989. Perspectives on biogeochemistry. Springer-Verlag.

DiMego, G. J., L. F. Bosart, and G. W. Endersen. 1976. An examination of the frequency and mean conditions surrounding frontal incursions into the Gulf of Mexico and Caribbean Sea. Monthly Weather Rev. **104:** 709–718.

Elliot, B. A., and R. O. Reid. 1976. Salinity induced horizontal circulation. pp. 425–442. ASCE Journal of the Waterways, Harbors and Coastal Engineering Division.

Emery, K. O., and M. Uchupi. 1972. Western North Atlantic Ocean: Topography, rocks, structure, water, life and sediments. American Association of Petroleum Geologists Memoir. 17.

Fairbridge, R. W. 1980. The estuary: Its definition and geodynamic cycle, p. 1–36. *In:* E. Olausson and I. Cato [eds.], Chemistry and biogeochemistry of estuaries. Wiley.

Fernandez-Partagas, J., and C. N. K. Mooers. 1975. A subsynoptic study of winter cold fronts in Florida. Monthly Weather Rev. **103:** 742–744.

Florida A&M University. 1988. Meteorological database and synthesis for the Gulf of Mexico. OCS Study/MMS-88-0064. U.S. Department of Interior, Minerals Management Service, Gulf of Mexico OCS Regional Office.

Folger, D. W. 1972. Characteristics of estuarine sediments of the United States. Geol. Surv. Professional Paper 724. U.S. Government Printing Office.

Fourqurean, J. W., R. D. Jones, and J. C. Zieman. 1993. Processes influencing water column nutrient characteristics and phosphorus limitation of phytoplankton biomass in Florida Bay, FL, USA: Inferences from spatial distributions. Estuar. Coast. Shelf Sci. **36:** 295–314.

Galperin, B., A. F. Blumberg, and R. H. Weisberg. 1985. A time-dependent three-dimensional model of circulation in Tampa Bay, p. 77–97. *In:* S. F. Treat and P A. Clark [eds.], Proceedings, Tampa Bay Area Scientific Information Symposium 2.

————, A. F. Blumberg, and R. H. Weisberg. 1991. The importance of density driven circulation in well mixed estuaries: The Tamp Bay experience, p. 332–343. *In:* Estuarine and Coastal Modeling, Second International Conference WW Division ASCE. American Society of Civil Engineers.

Geyer, W. R., and D. M. Farmer. 1989. Tide-induced variation of the dynamics of a salt wedge estuary. J. Phys. Oceanogr. **19:** 1060–1072.

Goodell, H. G., and D. S. Gorsline. 1961. A sedimentologic study of Tampa Bay, Florida. Proceedings of Twenty-first International Geological Congress, Copenhagen, 1960, Rept, pt. **23:** 75–88.

Grant, W. D., and O. S. Madsen. 1986. The continental-shelf bottom boundary layer. *In:* M. Van Dyke, J. V. Wehausen, and J. L. Lumley [eds.], Annual reviews of fluid mechanics **18:** 265–305.

Griffin, G. M. 1962. Regional clay-mineral facies—products of weathering intensity and current distribution in the northeastern Gulf of Mexico. Geol. Soc. Amer. Bull. **73:** 737–768.

Gutierrez de Velasco, G., and C. D. Winant. 1996. Seasonal patterns of wind stress and wind stress curl over the Gulf of Mexico. J. Geophys. Res. **101:** 18127–18140.

Hansen, D. V., and M. Rattray. 1966. New dimensions in estuary classification. Limnol. Oceanogr. **11:** 319–326.

Hayes, M. O. 1978. Impact of hurricanes on sedimentation in estuaries, bays, and lagoons, p. 323–346. *In:* M. L. Wiley [ed.], Estuarine interactions. Academic Press.

Hudson, J. D. 1963a. The recognition of salinity controlled molluscan assemblages in the Great Estuarine Series (Middle Jurassic) of the Inner Hebrides. Paleontology **6:** 318–326.

————. 1963b. The ecology and stratigraphical distribution of invertebrate fauna of the Great Estuarine Series. Paleontology **6:** 327–348.

Inoue, M., and W. J. Wiseman, Jr. (in review) Transport, mixing and stirring processes in a Louisiana estuary: A model study. J. Geophys. Res.

Isphording, W. C., J. A. Stringfellow, and G. C. Flowers. 1985. Sedimentary and geochemical systems in transitional marine sediments in the northeastern Gulf of Mexico. Transact. Gulf Coast Assoc. Geol. Soc. **XXXV:** 397–408.

Jackson, L. L., A. L. Foote, and L. S. Blistrieri. 1995. Hydrological, geomorphological and chemical effects of Hurricane Andrew on coastal marshes of Louisiana. J. Coast. Res. Special Issue **21:** 306–323.

Ketchen, H. G., and R. C. Staley. 1979. A hydrographic survey in Pensacola Bay. Technical Report, Department of Oceanography, Florida State University.

Keulegan, G. H. 1949. Interfacial stability and mixing in stratified flows. J. Res. Nat. Bur. Stds. **43:** 487–500.

Leipper, D. F. 1954. Marine meteorology of the Gulf of Mexico, a brief review, p. 89–98. *In:* Gulf of Mexico, its origin, waters, and marine life. Bull. Fish. Wild. Ser. 55.

Ley, J. A., C. L. Montague, and C. C. McIvor. 1994. Food habits of mangrove fishes: A comparison along estuarine gradients in northeastern Florida Bay. Bull. Mar. Sci. **54:** 881–899.

Light, S. S., and J. W. Dineen. 1994. Water control in the Everglades: A historical perspective, p. 47–67. *In:* S. M. Davis and J. C. Ogden [eds.], Everglades—the ecosystem and its restoration. St. Lucie Press.

Marmer, H. A. 1954. Tides and sea level in the Gulf of Mexico, p. 101–118. *In:* Gulf of Mexico, its origin, waters, and marine life. Bull. Fish. Wild. Ser. 55.

May, E. B. 1973. Extensive oxygen depletion in Mobile Bay, Alabama. Limnol. Oceanogr. **18:** 353–366.

McKee, B. A., W. J. Wiseman, Jr., and M. Inoue. 1994. Salt-water intrusion and sediment dynamics in a bar-built estuary, p. 13–16. *In:* K. R. Dyer and R. J. Orth [eds.], Changes in fluxes in estuaries. Olsen and Olsen.

Mitchum, G. T., and A. J. Clarke. 1986. Evaluation of frictional, wind-forced, long-wave theory on the west Florida shelf. J. Phys. Oceanogr. **16:** 1029–1037.

Nichols, M. M., and R. B. Biggs. 1985. Estuaries, p. 77–186. *In:* R. A. Davis, Jr. [ed.], Coastal sedimentary environments. Springer-Verlag.

Noble, M., W. W. Schroeder, W. J. Wiseman, Jr., H. F. Ryan, and G. Gelfenbaum. 1996. Subtidal circulation patterns in a shallow, highly-stratified estuary: Mobile Bay, Alabama. J. Geophys. Res. **101:** 25689–25703.

Okubo, A. 1973. Effect of shoreline irregularities on streamwise dispersion in estuaries and other embayments. Neth. J. Sea Res. **6:** 213–224.

Prager, E. J. 1992. Modeling of circulation and transports in Terrebonne Bay, Louisiana and its implications for oyster harvesting management. Ph.D. dissertation, Louisiana State University.

Pritchard, D. W. 1955. Estuarine circulation patterns. Proc. Amer. Soc. Civil Engr. **81:** 717-1 to 717-11.

———. 1967. What is an estuary: Physical viewpoint, p. 3–5. *In:* G. H. Lauff [ed.], Estuaries. American Association Advancement Science, Pub. 83.

Ridderinkhof, H., and J. T. F. Zimmerman. 1990. Mixing processes in a numerical model of the Western Dutch Wadden Sea, p. 194–209. *In:* R. T. Cheng [ed.], Residual currents and long-term transport processes. Coastal and Estuarine Studies 38. Springer-Verlag.

Roberts, H. H., L. J. Rouse, Jr., and N. D. Walker. 1983. Evolution of cold-water stress conditions in high-latitude reef systems: Florida reef tract and the Bahama Banks. Carib. J. Sci. **19:** 55–60.

Rusnak, G. A. 1960. Sediments of Laguna Madre, Texas, p. 153–196. *In:* F. P. Shepard, F. B. Phleger, and T. H. Van Andel [eds.], Recent sediments, northwestern Gulf of Mexico—a symposium summarizing the results of work carried on in Project 51 of the American Petroleum Institute. American Association of Petroleum Geologists.

Schlesinger, W. H. 1997. Biogeochemistry—an analysis of global change. Academic Press.

Schroeder, W. W. 1977. The impact of the 1973 flooding of the Mobile River system on the hydrography of Mobile Bay and east Mississippi Sound. Northeast Gulf Sci. **1:** 68–76.

———, J. L. W. Cowan, J. R. Pennock, S. A. Luker, and W. J. Wiseman, Jr. (in press). Natural restoration of resource excavations in a coastal plain estuary. Estuaries.

———, S. P. Dinnel, and W. J. Wiseman, Jr. 1990. Salinity stratification in a river-dominated estuary. Estuaries **13:** 145–154.

———, J. R. Pennock, and W. J. Wiseman, Jr. 1996a. A note on the influence of a deep ship channel on estuarine-shelf exchange in a broad, shallow estuary, p. 159–170. *In:*

C. Pattiaratchi [ed.], Mixing in estuaries and coastal seas., Coastal and Estuarine Studies 50. American Geophysical Union.

————, and W. J. Wiseman, Jr. 1986. Low-frequency shelf-estuarine exchange processes in Mobile Bay and other estuarine systems on the northern Gulf of Mexico, p. 355–367. *In:* D. Wolfe [ed.], Estuarine variability. Academic Press.

————, and W. J. Wiseman, Jr. 1988. The Mobile Bay estuary: stratification, oxygen depletion and jubilees, p. 41–52. *In:* B. Kjerfve [ed.], Hydrodynamics of estuaries, volume 2. CRC Press.

————, W. J. Wiseman, Jr., J. R. Pennock, and M. Noble. 1996b. A note on very low-frequency salinity variability in a broad, shallow estuary, p. 255–263. *In:* D. Aubrey and C. Friedrichs [eds.], Buoyancy effects on coastal and estuarine dynamics. Coastal and Estuarine Studies 53. American Geophysical Union.

————, W. J. Wiseman, Jr., A. Williams, Jr., D. C. Raney, and G. C. April. 1980. Climate and oceanography, p. 27–51. *In:* Mobile Bay: Issues, resources, status, and management. NOAA Estuary of the Month Seminar Series Number 15.

Schubel, J. R., and D. J. Hirschberg. 1978. Estuarine graveyard, climate change, and the importance of the estuarine environment, p. 285–303. *In:* M. Wiley [ed.], Estuarine interactions. Academic Press.

Signell, R. P., R. C. Beardsley, H. C. Graber, and A. Capotondi. 1990. Effect of wave-current interaction on wind-driven circulation in narrow, shallow embayments. J. Geophys. Res. **95:** 9671–9678.

Smith, N. P. 1978. Long-period estuarine-shelf exchanges in response to meteorological forcing, p. 147–159. *In:* J. C. J. Nihoul [ed.], Hydrodynamics of estuaries and fjords. Elsevier.

————. 1994. Long-term Gulf-to-Atlantic transport through tidal channels in the Florida Keys. Bull. Mar. Sci. **54:** 602–609.

Stigebrandt, A. 1976. Vertical diffusion driven by internal waves in a sill fjord. J. Phys. Oceanogr. **6:** 486–495.

U.S. Department of Commerce. 1993. NOAA Workshop on the restoration of Florida Bay. P. B. Ortner, D. E. Hoss and J. A. Browder [eds.]. Miami, Florida, July 14–16, 1993.

Walker, N. D., H. H. Roberts, L. J. Rouse, Jr., and O. K. Huh. 1982. Thermal history of reef-associated environments during a recent cold-air outbreak event. Coral Reefs **1:** 82–87.

————, L. J. Rouse, Jr., and O. K. Huh. 1987. Response of subtropical shallow-water environments to cold-air outbreak events: Satellite radiometry and heat flux modeling. Cont. Shelf Res. **7:** 735–757.

Wang, J. D., J. Van de Kreeke, N. Krishnan, and D. Smith. 1994. Wind and tide response in Florida Bay. Bull. Mar. Sci. **54:** 579–601.

Weisberg, R. H., and W. Sturges. 1976. Velocity observations in the West Passage of Narragansett Bay: A partially mixed estuary. J. Phys. Oceanogr. **6:** 345–354.

————, and R. G. Williams. 1985. Initial findings on the circulation of Tampa Bay, p. 49–66. *In:* S. F. Treat and P. A. Clark [eds.], Proceedings, Tampa Bay Area Scientific Information Symposium 2.

Williams, E. G. 1960. Marine and fresh water folliliferous beds in the Pottsville and Allegheny Group of western Pennsylvania. J. Paleontol. **34:** 905–922.

Wiseman, W. J., Jr. 1986. Estuarine-shelf interactions, p. 109–115. *In:* C. N. K. Mooers [ed.], Baroclinic processes on continental shelves. Coastal and Estuarine Sciences 3. American Geophysical Union.

———, J. M. Grymes III, and W. W. Schroeder. (In press). Coastal wind stress variability along the northern Gulf of Mexico. *In:* J. Dronkers [ed.], Estuarine and coastal morphology. Balkema.

———, and M. Inoue. 1993a. Salinity variations in two Louisiana estuaries, p. 1230–1242. *In:* O. T. Magoon, W. S. Wilson, H. Converse, and T. Tobin [eds.], Coastal Zone '93. Volume 1. Proceedings, Eighth Symposium on Coastal and Ocean Management. American Shore and Beach Preservation Association/ASCE.

———, and M. Inoue. 1993b. Local wind effects on multiple inlet estuaries. The Science of Management of Coastal Estuarine Systems (Abstracts). 12th Biennial Estuarine Research Federation meeting, pp. 137–138.

———, E. M. Swenson, and F. J. Kelly. 1990. Control of estuarine salinities by coastal ocean salinity, p. 184–193. *In:* R. T. Cheng [ed.], Residual currents and long-term transport processes. Coastal and Estuarine Studies 38. Springer-Verlag.

Wright, L. D. 1971. Hydrography of South Pass, Mississippi River. J. Waterways, Harbors and Coastal Engineering Div., ASCE, **97(WW3):** 491–504.

Zieman, J. C., R. Davis, J. W. Fourqurean, and M. B. Robblee. 1994. The role of climate in the Florida Bay seagrass dieoff. Bull. Mar. Sci. **54:** 1088.

Zimmerman, J. T. F. 1978. Dispersion by tide-induced residual current vortices, p. 207–216. *In:* J. C. J. Nihoul [ed.], Hydrodynamics of estuaries and fjords. Elsevier.

Chapter 2

Hydrography, Mixing Characteristics, and Residence Times of Gulf of Mexico Estuaries

Ruben S. Solis and Gary L. Powell

INTRODUCTION

Gulf of Mexico estuaries (Fig. 2-1) serve a wide variety of purposes. They encompass vast areas protected as national wildlife refuges, support large near-shore fisheries, and harbor the most heavily utilized ports in the United States and Mexico. The recreational, commercial, environmental, and aesthetic values of these estuaries depend on their ability to dilute, process, and export anthropogenic nutrients and wastes. The mechanisms that control these processes include tidal, meteorological, and hydrological forcings, density currents, and Coriolis forces, along with tidal dispersion and other types of mixing. Proper management of these valuable estuarine resources depends on understanding these geophysical processes. The physics governing estuarine hydrodynamics is described in detail in Officer (1976), Ippen (1966), Dyer (1973), and Fischer (1976). Comprehensive and descriptive reviews of the

Biogeochemistry of Gulf of Mexico Estuaries, Edited by Thomas S. Bianchi, Jonathan R. Pennock, and Robert R. Twilley.
ISBN 0-471-16174-8 © 1999 John Wiley & Sons, Inc.

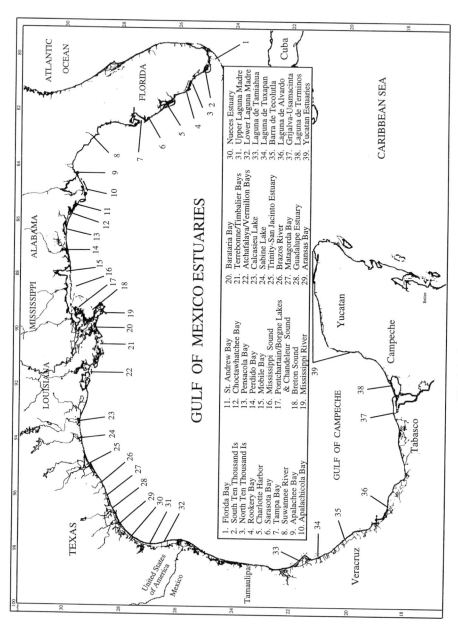

FIG. 2-1. Gulf of Mexico estuaries.

GULF OF MEXICO ESTUARIES

1. Florida Bay
2. South Ten Thousand Is
3. North Ten Thousand Is
4. Rookery Bay
5. Charlotte Harbor
6. Sarasota Bay
7. Tampa Bay
8. Suwannee River
9. Apalachee Bay
10. Apalachicola Bay
11. St. Andrew Bay
12. Choctawhatchee Bay
13. Pensacola Bay
14. Perdido Bay
15. Mobile Bay
16. Mississippi Sound
17. Pontchartrain/Borgne Lakes
    & Chandeleur Sound
18. Breton Sound
19. Mississippi River
20. Barataria Bay
21. Terrebonne/Timbalier Bays
22. Atchafalaya/Vermilion Bays
23. Calcasieu Lake
24. Sabine Lake
25. Trinity-San Jacinto Estuary
26. Brazos River
27. Matagorda Bay
28. Guadalupe Estuary
29. Aransas Bay
30. Nueces Estuary
31. Upper Laguna Madre
32. Lower Laguna Madre
33. Laguna de Tamiahua
34. Laguna de Tuxpan
35. Barra de Tecolutla
36. Laguna de Alvardo
37. Grijalva-Usamacinta
38. Laguna de Terminos
39. Yucatan Estuaries

geomorphology, bathymetry, circulation, salinity, tides, climatology, meteorology, and fresh-water inflow characteristics of U.S. estuaries on the Gulf coast are provided in a series of publications by the National Oceanic and Atmospheric Administration (NOAA 1985b, 1990) and Ward (1980). In this chapter we will summarize the more important physical processes affecting estuarine mixing and provide examples to illustrate the complexity and variability of Gulf of Mexico estuaries. We will also use the common indices of "residence time" and "mixing efficiency" to characterize mixing processes in estuaries and to compare processes among Gulf estuaries.

FACTORS AFFECTING CIRCULATION AND MIXING

Circulation and dispersion processes play a role in determining the effectiveness of biological and chemical cycling and the dilution of pollutants in estuaries. These processes, in turn, are driven by tidal, meteorological, and hydrological forcing and are often affected by human activities.

Tides

Astronomical tides are the periodic variations in water level influenced primarily by the relative positions of the earth, moon, and sun. Tidal currents induced by the spatial gradient in water levels are strongest near tidal passes and diminish as one moves farther into the estuary. These currents are effective mechanisms in advecting and dispersing dissolved and particulate matter contained in the water column.

Astronomical tides are the most highly predictable component of estuarine variability due to the periodic movements of the sun, moon, and earth. Gulf coast tides are classified as micro-tidal, with ranges (difference in elevation between consecutive high and low tides) typically around 1 m along the lower Florida coast and between 0.5 and 0.7 m throughout the rest of the Gulf of Mexico (NOAA 1991). Tides throughout the Gulf are chiefly diurnal, with one high and one low tide per day. Exceptions include the Florida Peninsula from Apalachee Bay to just north of Tampa Bay and the area south of Rookery Bay, where tides normally have two highs and two lows per day and are classified as mixed. Tidal amplitudes are modulated throughout the month in association with the moon's declination (the celestial equivalent of latitude), the principal factor influencing the amplitude of diurnal tides (Marmer 1954). Amplitudes are largest when the moon is at extreme declinations and smallest when it is over the equator.

As a tide progresses through an estuary, its amplitude is attenuated and its phase becomes lagged in comparison to the Gulf signal. This diminution in amplitude, and of currents generated by tidal action, has the effect of creating a spatial variation in tidal dispersion within the estuary, being greatest near the passes and decreasing farther into the estuary. However, in rare cases, as in Mobile Bay, tidal amplitudes will increase in limited regions within the estuary due to rapid convergence of flow cross section (Ward 1980).

Diurnal Gulf tides are superimposed upon a lower-frequency, seasonal scale variation in sea level that occurs in the Gulf of Mexico. This low-frequency variation is thought to be related more to meteorological influences and Gulf currents than to astronomical effects (Marmer 1954). Along the Louisiana and Texas coasts, distinct peaks in this low-frequency variation are found in the spring and fall seasons. At most other locations, a single peak is observed in the fall. Minima are observed in the winter and summer seasons. The range of this seasonal variation is about 0.3 m. This slow change in sea level affects circulation and mixing processes at short time scales (hours and days) by varying the tidal prism (i.e., the volume of water entering the estuary during flood tide) and thus affecting tidal currents and dispersion. The effect at longer time scales is simply a slow flooding and emptying of the estuary, with seasonal changes in total volume.

Discernible astronomical tides are found throughout most Gulf of Mexico estuaries except for Laguna Madre. Laguna Madre is shallow (<1 m in most areas) and has direct contact with Gulf tides only through Brazos Santiago Pass at its extreme south end and through Port Mansfield Channel midway up the Lower Laguna. To the north, tides enter through Aransas Pass and must cross Corpus Christi Bay before entering the Upper Laguna Madre through Humble Channel and the Intracoastal Waterway. As a result of frictional effects and limited tidal access, astronomical tides in Laguna Madre are highly attenuated. Analysis of harmonic constituents and spectral analysis of estuary water levels indicated that tides here are more strongly influenced by meteorological forcing than by true astronomical forcing (NOAA 1995).

Meteorological Forcing

Wind-generated stress has four effects on the hydrography of estuaries: (1) it alters water levels in estuaries through the mechanism of wind setup; (2) it induces circulation patterns by directly forcing surface waters; (3) it increases vertical mixing through the actions of wind-generated waves; and (4) it creates lower-frequency variations in water levels that lead to exchange with the Gulf of Mexico. Because Gulf estuaries are shallow, they are highly susceptible to forcing by wind stresses and, as a result, are considered meteorologically dominated (Ward 1980). Meteorological forcing ranges from daily variations in sea breeze, to weekly passage of weather fronts, and to less frequent but higher-energy gales associated with tropical storms and hurricanes that may occur from mid-summer to early fall.

The effect of wind on the water surface is most easily seen in tidal records, where the meteorologically induced tide often overwhelms astronomical variations. Ward (1980) presents examples of the evacuation of Sabine Lake and Galveston Bay in response to the late fall passage of weather fronts. Roughly half the volume of the bay, or over three times the volume of the tidal prism, was reported to have been discharged from Galveston Bay. Clearly, these events can play a significant role in estuarine mixing simply through their effectiveness in bulk replacement of estuarine water with Gulf waters.

Studies of exchange between estuaries and Gulf waters in response to low-frequency meteorological forcing are presented by Wiseman et al. (1988), Smith (1988), and others (see references in Schroeder and Wiseman 1986). Wiseman et al. (1988) suggest that low-frequency exchange is on the same order as tidal diffusion and density currents in Mobile Bay. In numerical experiments of flushing characteristics of Laguna Madre above Baffin Bay, an estuary with perhaps the weakest astronomical tides on the Gulf coast, virtually none of the flushing of this system was found due to astronomical tides, but rather to low-frequency, meteorologically driven variations in water level (Smith 1988).

Despite the importance of wind as a driving force in estuarine circulation, relatively few data exist that establish the magnitude and direction of wind-generated circulation currents. Currents are commonly monitored at tidal passes and in channels, where flows are concentrated and strongest, rather than in open areas of estuaries, where currents are weakest. However, long-term measurements in open areas are needed for circulation patterns to be established, as shown in a study of residual circulation in Sarasota Bay and Tampa Bay (Sheng and Peene 1993).

Precipitation and Evaporation

Precipitation provides fresh water to estuaries in the form of surface runoff and ground water inflows and, more directly, in the form of rain falling on the estuary. In most estuaries, including those of the Gulf of Mexico, surface runoff is the most significant source of fresh water. However, in some Gulf estuaries, the dominant source of fresh water is ground water. In other Gulf estuaries, direct rainfall on the estuary, though not dominant, contributes significantly to the total fresh-water budget. Finally, in a few Gulf estuaries, evaporation from the surface exceeds all fresh-water inflows.

Heaviest rainfall on the Gulf coast is concentrated in two areas—in Mexico on the Tabasco plain along the Isthmus of Tehuantepec, and in the United States from the Mississippi Delta to the Florida panhandle. Precipitation in some areas of the Tabasco plain, the heaviest anywhere on the Gulf of Mexico coast (West 1964), reaches nearly 500 cm annually. Rainfall rates decrease eastward to less than 50 cm yr^{-1} on the northwestern corner of the Yucatan Peninsula. Northward and westward from the maximum, rainfall decreases uniformly to nearly 60 cm yr^{-1} in south Texas. On the U.S. Gulf coast in the area from the Mississippi Delta to the western panhandle of Florida, average annual precipitation exceeds 160 cm. Precipitation decreases eastwardly to a minimum of about 120 cm yr^{-1} near Tallahassee and then increases again along the Florida peninsula to roughly 140 cm yr^{-1} in the Florida Keys.

Westward from the Mississippi Delta, average annual precipitation decreases uniformly through Louisiana and Texas to a minimum of about 60 cm yr^{-1}.

Heaviest rainfall on the Mexican Gulf coast generally occurs during the hot months from June through October. The minimum generally occurs in the cooler months from December through April. September is typically the wettest month and March the driest. Extreme variation exists in the seasonal hydrographs. For example, average monthly rainfall fluctuates from nearly 1 cm in March to over 36 cm in September in Veracruz and from nearly 5 cm in April to over 52 cm mo^{-1} in September, October, and November in Coatzacoalcos (Arbingast et al. 1975).

Seasonal frontal passages, combined with convective thunderstorms during the summer, account for the seasonal distribution of rainfall on the U.S. Gulf coast. Seasonal peaks in precipitation are typically found during the spring and either the summer or early fall. During the spring and fall, frontal passages are frequent, and atmospheric moisture fed by the onshore flow from the Gulf is abundant. This combination of circumstances leads to more frequent and heavier precipitation during these months. On most of the Texas Coast, maximum monthly precipitation occurs in September, with a secondary peak in May or June. From northeast Texas through Louisiana, the maximum is found in July, with a secondary peak in May. Along the northeastern Gulf coast, the maximum occurs from July through September, with a secondary minimum in March (Cross 1974). The secondary March peak diminishes on the lower half of the Florida peninsula, and the peak shifts from July through September, similar in shape to the seasonal hydrograph for the northern Yucatan Peninsula in Mexico.

Rainfall on the U.S. Gulf coast is more uniformly distributed seasonally than in Mexico. For example, average monthly rainfall near Lake Pontchartrain varies from about 8 cm in October to less than 18 cm in May (Newton 1972). Greatest seasonal variability on the U.S. Gulf coast occurs in southern Florida, where rainfall varies from about 3 cm in December to over 24 cm in June and September (Fernald 1981).

In addition to precipitation associated with frontal passages and convective activity, tropical storms and hurricanes contribute significant rainfall throughout the Gulf coast on a less frequent basis from June to September.

Evaporation rates are more uniform around the Gulf coast than rainfall rates, varying only from 120 to 160 cm yr^{-1} and generally increasing with decreasing latitude. Minimum evaporation rates on the Gulf coast occur from the Florida panhandle to Mississippi, where gross evaporation is just over 120 cm yr^{-1}. Evaporation increases southwardly through Florida and Texas to just over 135 cm yr^{-1} in the Florida Keys and about 140 cm yr^{-1} in south Texas (Linsley et al. 1975). In Mexico, evaporation rates remain nearly uniform between 130 and 140 cm yr^{-1} on the Gulf coastal plain, increase to approximately 160 cm yr^{-1} near Laguna Terminos, and decrease again to about 150 cm yr^{-1} in the Yucatan Platform (Arbingast et al. 1975; Deegan et al. 1986).

Fresh-water Inflow

Fresh-water inflow into the Gulf of Mexico from Florida to the Yucatan Peninsula totals approximately 1110×10^9 m^3 yr^{-1}. Of this, 866×10^9 m^3 yr^{-1} comes from U.S. watersheds except the Rio Grande (NOAA 1985b), 229×10^9 m^3 yr^{-1} comes from Mexico drainage (West 1964), and 12×10^9 m^3 yr^{-1} comes from the Rio Grande (West 1964). The Mississippi and Atchafalaya Rivers contribute about 55% of the total, followed by the Rio Grijalva-Usumacinta system, which contributes just under 10%. Average inflow is highly variable around the Gulf of Mexico perimeter (Fig. 2-2). Discharge generally decreases in magnitude eastward from the Mississippi River toward the Florida peninsula and westward around the Gulf coast toward Mexico, reaching another peak near the Grijalva-Usumacinta system. Variation in discharge arises from a combination of factors, namely, the distribution of rainfall around the

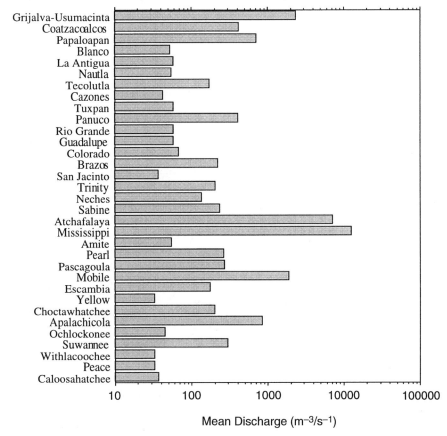

FIG. 2-2. *Distribution of average fresh-water discharge to the Gulf of Mexico perimeter (data from NOAA 1985a).*

Gulf, the size of drainage areas associated with each river, and anthropogenic consumption of fresh water from some watersheds.

Drainage areas of Florida estuaries from Tampa Bay south are among the smallest on the Gulf of Mexico coast (NOAA 1985a), most of them covering <10,000 km^2 and limited by the width of the Florida peninsula. Estuarine drainage basins from the Suwannee River to Mobile Bay are larger, limited to the north by the southern extension of the Appalachian Mountains. The Mississippi River basin covers 2.93×10^6 km^2. This is by far the largest of all Gulf basins and represents 61% of the total Gulf drainage area. Headwaters for the Mississippi River and its tributaries rise as far north as Canada and as far west as the continental divide. Estuaries between the Atchafalaya River and Laguna Madre drain regional watersheds lying largely within the states of Louisiana and Texas. These watersheds vary in area from nearly 7000 km^2 (Aransas Bay) to over 130,000 km^2 (Matagorda Bay). South Texas watersheds and those of northern Mexico are among the largest and most arid of the Gulf of Mexico. The Rio Grande drainage basin covers 472,000 km^2 (West 1964), the second largest on the Gulf coast. Despite its large drainage area, the Rio Grande has one of the smallest runoff to watershed-area ratios for the entire Gulf, approximately 0.03×10^6 m^3 km^{-2} yr^{-1}. This results from the aridity of the Rio Grande drainage basin and from heavy agricultural withdrawals in the lower basin. By contrast, runoff to watershed ratios exceed 1.0×10^6 m^3 km^{-2} yr^{-1} for the Rio Nautla and Rio Coatzacoalcos.

Ground water and springs provide steady low flows of fresh water into a number of estuaries along the Florida peninsula and, notably, virtually the entire discharge to some estuaries along the Yucatan Peninsula. For example, Celestun Lagoon, a shallow coastal lagoon in an area of the Yucatan Peninsula with few rivers due to the underlying karst geology, receives fresh water only from direct rainfall and ground water discharge (Herrera-Silveira 1995).

Seasonal variability in inflows tends to follow local rainfall patterns except for some estuaries with large drainage basins that are subject to inland climatological patterns such as the Mississippi-Atchafalaya systems, the Suwannee River, and Apalachicola Bay (Orlando et al. 1993). The largest relative difference between average maximum and minimum monthly inflows occurs in the estuaries of south Texas and Florida. Average maximum monthly inflows are nearly 10 times the average monthly minimum flows in south Texas estuaries, and in Tampa Bay they are 7 times the average monthly minimum (NOAA 1985b).

Channelization and Dredging

Most Gulf of Mexico estuaries in the United States are highly channelized. In total they contain over 7000 km of dredged navigational channels (NOAA 1990). The deepest of these, such as the nearly 15-m-deep Houston Ship Channel and the Matagorda Ship Channel, link inland ports to the Gulf. These channels have increased the volume of tidal prism, altered circulation, and increased salt flux into the estuary through the mechanism of density currents. Most deep navigation channels have been dredged to greater depths over time in response to competition among ports to allow access to deep-draft vessels. The longest continuous channel in the Gulf is the approximately 2100-km Gulf Intracoastal Waterway (GIWW), which parallels the Gulf Coast continuously from Brownsville, Texas, to St. Marks, Florida, and continues again in Florida from Tampa to Fort Myers. This shallow, approximately 4-m-deep by 30-m-wide channel has increased circulation within estuaries (e.g., Laguna Madre in Texas) and has created direct connections among estuaries (e.g., Sabine Lake in Texas and Calcasieu Lake in Louisiana). Both deep and shallow channels have created regions where vertical stratification and density currents are commonly found in what are otherwise classified as vertically well-mixed estuaries.

Placement of material associated with channel dredging has partitioned some estuaries into discrete regions between which there is limited exchange. An extreme example of this is Sabine Lake, where dredge material from the Sabine-Neches and Port Arthur Canals, which run along the western edge of Sabine Lake, was placed on the eastern bank of the canal. This resulted in the creation of a continuous island that extends almost the entire length of Sabine Lake. Exchange between the canal and Sabine Lake is now restricted to the extreme north and south ends of the estuary.

Water Impoundments, Diversions, Watershed Changes

Changes in the quantity and timing of fresh-water inflows are a major concern in estuarine management. Dams, reservoirs, and other control structures have been built for the purpose of managing inland water resources. Such structures control the release of river flows to coastal systems. Consumptive use of water diverted from natural streams is generally a small fraction of the average streamflow, though diversions can be a significant fraction of flow during droughts. Provisions for minimum flow to estuaries have been agreed to, such as the Agreed Order between the Texas Natural Resource Conservation Commission and the City of Corpus Christi. This agreement stipulates multi-stage operating rules for the Choke Canyon Reservoir/Lake Corpus Christi water supply system that provides critical fresh-water inflows from these impoundments to maintain the ecological health of Corpus Christi Bay and Nueces River estuary. Impoundment structures and operations do not generally cause significant changes in average streamflows, but they do attenuate flood peaks. Low flow conditions may be alleviated through more steady releases of water from impoundments and wastewater return flows; in some cases, the timing of releases has affected seasonal inflow patterns.

While flood peaks are generally reduced downstream of impoundments, they are amplified in streams whose watersheds have become increasingly urbanized. Streets, parking lots, and buildings prevent rainfall from infiltrating into soil, causing accelerated runoff into drainage structures, creeks, and rivers and creating more intense floods than those occurring in non-urbanized water-sheds.

Major human activities include redirecting the entire flow of major rivers. Discharge from the Colorado River in Texas prior to 1992 was directly into the Gulf of Mexico. During that time, only a small fraction of the river's flow entered Matagorda Bay through a series of small cuts connecting the river and the eastern arm of the bay during high flow conditions. However, levees constructed in 1992 have rerouted the entire river flow into the eastern arm of the bay. A large fraction of Mississippi River discharge prior to the 20th century reached the Gulf of Mexico through an extensive wetland flood plain, including the region of Fourleague Bay (Madden et al. 1988; Day et al. 1994). Presently, more of the river water (65%) reaches the Gulf directly through the main river channel due to construction of a levee system that confines most of the flow to the main river channel.

In Florida, drainage of fresh-water inflows into Florida Bay has been altered in timing and volume due to over 100 yr of construction in the Everglades (Light and Dineen 1994). While the original intent of construction there was primarily to make this area of southern Florida habitable, more recent objectives have shifted from those of flood control and water supply to that of solving environmental problems.

SALINITY AND MIXING EFFICIENCY

Salinity

Salinity distributions in an estuary reflect the integrated effects of mixing processes occurring within the estuary. Salt is introduced from its primary source, the Gulf of Mexico, through tidal inlets, and is mixed with fresh water in the estuary, dispersed, and eventually exchanged with Gulf of Mexico waters. The utility of using salinity as a tracer, and its significance to the physiology of marine organisms, are widely recognized (Orlando et al. 1993).

Typical salinities in Gulf estuaries can be evaluated based on data compiled by NOAA (NOAA 1985b, 1990, 1997). NOAA (1990) contains physical characteristics for 36 Gulf estuaries in the United States including watershed drainage area, estuarine surface area, average depth and volume, and average fresh-water inflows. Tidal prism volumes are presented in NOAA (1985b). NOAA (1997) contains estimates for volumes within each estuary corresponding to the following salinity distributions: "tidal fresh" (0.0 to 0.5 g liter^{-1}), "mixing zone" (0.5 to 25.0 g liter^{-1}), and "sea water" (>25.0 g liter^{-1}). While the NOAA data sets are fairly comprehensive and convenient to use, data used in these compilations were obtained from a variety of programs with differing purposes, data collection methodologies, and periods of record. As a result, care must be taken in making comparisons among estuaries given the spatial and temporal biases of these data. Finally, no single data parameter in these data sets was complete across estuaries; in other words, it was necessary to develop estimation procedures for each parameter because of inconsistent or missing data.

With the above caveats in mind, volume-weighted average salinities for Gulf estuaries were computed from the NOAA datasets (Fig. 2-3). Significant variation in average salinity is found across the Gulf, from <9 g liter^{-1} in the Atchafalaya-Vermilion Bay and Mississippi River systems to >20 g liter^{-1} on the lower coasts of Texas and Florida. Salinity in Laguna Madre, one of only four major estuaries in the world that regularly exhibit hypersalinity, is thought to be higher than presented in the NOAA data sets but was limited in the data set by the upper salinity, 33.0 g liter^{-1}, assumed for sea water. The

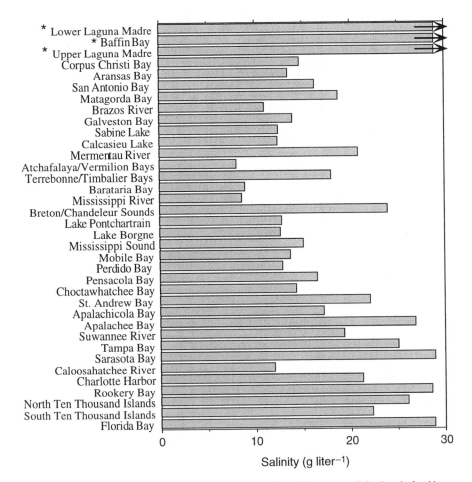

FIG. 2-3. Average salinities for Gulf of Mexico estuaries. *Average salinity levels for Upper Laguna Madre, Lower Laguna Madre, and Baffin Bay were assumed to be at sea-water concentration in NOAA data sets, although they are probably higher (U.S. data from NOAA 1997).

frequency and severity of hypersalinity in Laguna Madre of Texas have decreased since the mid-1940s because of increased tidal flushing brought about by dredging of the GIWW longitudinally through the lagoon and to the opening of a direct connection to the Gulf of Mexico at Port Mansfield. Except for smaller river-mouth estuaries (Mermentau River, Caloosahatchee River), average salinity inversely reflects inflow trends around the Gulf; lower salinities are found toward, and increase away from, the Atchafalaya-Mississippi River system. The intent of presenting average salinities here is to identify general Gulf-wide trends, not to define a single characteristic salinity for each estuary. Salinities presented in Fig. 2-3 are close to the values found elsewhere (Orlando et al. 1993), but due to data limitations and differences in data sets, they can differ from others.

"Estuaries are inherently variable in time and space" (Wolfe and Kjerfve 1986). Spatial variability is manifested as horizontal salinity gradients across an estuary and as vertical gradients in the water column. Temporal variability is evident in time traces of salinity at any location in an estuary. Spatial and temporal variability is demonstrated in Sabine Lake, Texas, by changes in vertical salinity structure over two 3-day study periods that occurred under significantly different hydrological conditions (Fig. 2-4; Texas Water Development Board 1997). Strong vertical stratification that varies hourly in response to tidal forcing appears under high inflow conditions (Fig. 2-4a). Under low flow conditions (Fig. 2-4b) and similar tidal forcing, vertical stratification never occurs, even over several tidal cycles.

Time-space variations of salinity along the axis of an estuary were found in upper Laguna Madre, Texas, in a 2-year study (Fig. 2-4c; Smith 1988). Here spatial variability over scales of kilometers corresponds to distance from primary sources of fresh water, while temporal variability corresponds to low-frequency climatological forcing. These examples demonstrate that the time scale of variability in an estuary depends on the forcing mechanism, and ranges from hours in response to tidal forcing, to days, weeks, and years in response to meteorological and hydrological forcing. Spatial scales of variability depend on the governing physics and on the location of sources and sinks of fresh and salt water.

Gulf estuaries whose salinity regimes are dominated by large fresh-water inflows (Atchafalaya-Vermilion Bay, Lake Pontchartrain-Borgne, Chandeleur Sound, Mississippi Sound) and those with feeble inflows whose salinity is determined largely by Gulf salinities (Tampa Bay, Corpus Christi Bay, Sarasota Bay, Laguna Madre) experience little intra-annual salinity variability and are categorized as stable in one classification scheme (Orlando et al. 1993). Intra-annual salinity in the remaining estuaries is significantly variable and is influenced by several physical factors, such as tidal exchange, wind mixing and so on.

Fresh-water Inflows, Precipitation, Evaporation, and Tidal Prism

Comparison of gross mixing characteristics of estuaries begins by assuming that estuaries can be treated as simple mixing chambers, that is, that complete

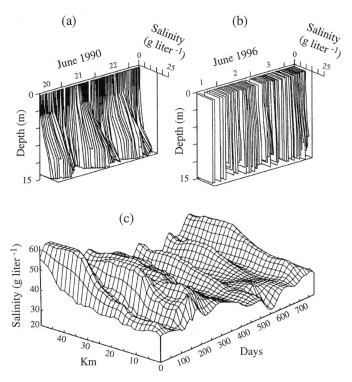

FIG. 2-4. Space-time plots of salinity showing (a) salinity stratification in Sabine Lake, June 1990, under high inflow conditions; (b) in June 1996 under low flow conditions, upper Sabine Lake at Highway 87 "Rainbow" Bridge (Texas Water Development Board 1997); and (c) estuary-wide variability in upper Laguna Madre south of Corpus Christi Bay. Salinity in all figures is in g liter⁻¹. (From Smith 1988; reprinted with permission from "Hydrodynamics of Estuaries, Volume II, Estuarine Case Studies." Copyright CRC Press, Boca Raton, Florida. © 1988).

mixing occurs within the estuary. In this simplification, assume that salt water and fresh water enter the estuary on the flood tide and mix to yield water of average salinity s, and that mixed water of salinity s is discharged on the ebb tide. Let inflows and outflows be limited to (1) salt-water influx through the tidal prism with salinity σ, (2) fresh-water influx with salinity of 0 through riverine inflows, ground water discharge, and precipitation, (3) tidal outflow of salinity s, and (4) fresh-water losses by evaporation. The total mass of salt entering the estuary is given by $\sigma \cdot V_t$, where V_t is the average tidal prism volume that enters during the flood cycle. The total volume of water entering during this period is given by $V_{fr} + V_t$, where V_{fr} represents the total volume of inflowing fresh water. For a stationary system and with complete mixing, the incoming water mixes completely to yield salinity s:

$$\frac{\sigma \cdot V_t}{V_t + V_{fr}} = s \qquad (2\text{-}1)$$

Re-writing for normalized (by σ) average salinity for the entire estuary:

$$\frac{s}{\sigma} = \frac{1}{1 + (V_{fr}/V_t)} = \frac{1}{1 + (V_r + V_p - V_e)/V_t} \tag{2-2}$$

Here V_r is average volume of riverine and ground water inflow, V_p is the average precipitation volume falling on the estuary, and V_e is the average volume lost to evaporation. [In principle, NOAA (1985b) provides intertidal volume, which includes tidal influx, riverine inflows, and evaporation losses. We assume here that this is nearly equal to the tidal prism.] The quantities V_r, V_p, and V_e are accumulated over one tidal cycle.

Equation 2-2 shows that the ratio of total fresh-water inflow volume to the tidal prism volume, V_{fr}/V_t, alone determines average estuarine salinity. The relative importance of river inflows to precipitation and evaporation is found by separating V_{fr} into its components and normalizing by V_r, yielding the net rainfall:inflow ratio $(V_p - V_e)/V_r$. For $|(V_p - V_e)|V_r \gg 1$, rainfall or evaporation is significant in computing water and salt balances, while $|(V_p - V_e)|/V_r \ll 1$ indicates that precipitation and evaporation can be neglected. The net rainfall:inflow ratio is insignificant in most estuaries (Fig. 2-5). However, this ratio is significant for Texas estuaries south of Galveston Bay. Here the ratio is negative (evaporation exceeds precipitation) and falls as low as -2.5 in the lower Laguna Madre. The ratio remains positive (precipitation exceeds evaporation) from Sabine Lake to the Suwannee River, reaching its greatest positive value (0.18) on the U.S. Gulf coast in the Breton/Chandeleur Sounds. $(V_p - V_e)/V_t$ becomes negative again south of the Suwannee River but is <0.015 and thus insignificant for Florida estuaries. Rainfall and evaporation are usually neglected in computing estuarine water balances, but this omission can clearly lead to misleading comparisons among Gulf estuaries.

The ratio V_{fr}/V_t is recognized as a general descriptor of estuarine character-istics. Bowden (1980) suggests that V_{fr}/V_t ratios on the order of 1, 0.1, and 0.01 apply to salt-wedge, partially mixed, and well-mixed estuaries, respec-tively. Based on this classification, most Gulf of Mexico estuaries are partially or well mixed (Fig. 2-6), the exceptions being the Brazos River, Sabine Lake, and the Atchafalaya-Vermilion Bays. (The Mississippi River is also considered a salt wedge estuary but was not considered separately in the NOAA datasets.)

Mixing Efficiency

The above expression for s/σ (Eq. 2-2) assumes that salt water introduced during the flood cycle mixes completely with estuarine water. In reality, tidal waters are only partially entrained with estuarine waters during the flood cycle, and portions of essentially unmixed tidal water escape back to the ocean during the ebb cycle. In a gross sense, the process can be quantified in terms of a bulk mixing efficiency, in which efficiency is defined as that fraction of

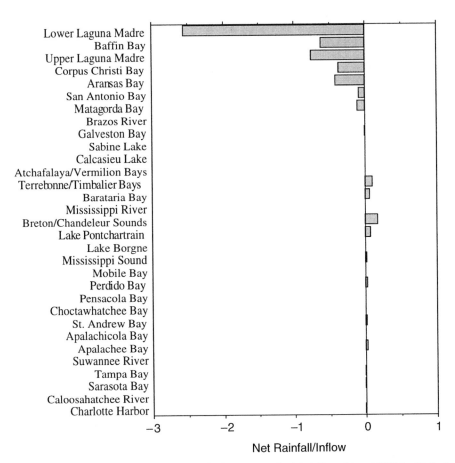

FIG. 2-5. Ratio of net precipitation (precipitation-evaporation) to inflow, $(V_p - V_e)/V_r$, for Gulf of Mexico estuaries.

the tidal prism that is available for mixing with estuarine water. The assumption of fractional as opposed to perfect mixing is used by Dyer and Taylor (1973), Smith (1985), and van de Kreeke (1988). However, van de Kreeke (1988), instead of assuming fractional retention of flood tide waters, assumes fractional loss of estuarine water entrained in the ebb tide, followed by the return of the remaining estuarine water combined with additional "new" sea water on the flood cycle.

Assuming imperfect mixing of the inflowing tidal water, the expression for s/σ becomes

$$\frac{s}{\sigma} = \frac{e}{e + (V_{fr}/V_t)} = \frac{e}{e + (V_r + V_p - V_e)/V_t} \tag{2-3}$$

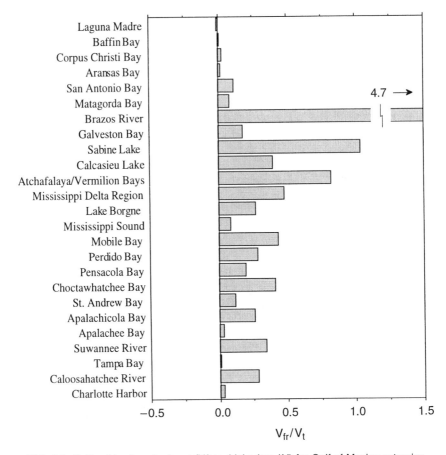

FIG. 2-6. *Ratio of fresh-water input (V_{fr}) to tidal prism (V_t) for Gulf of Mexico estuaries.*

where *e* is the bulk mixing efficiency. Solving for *e*, we get

$$e = \frac{V_{fr}}{V_t} \cdot \frac{s}{\sigma - s} = \frac{V_{fr}}{V_t} \cdot \frac{s/\sigma}{1 - s/\sigma} \tag{2-4}$$

Equation 31 in van de Kreeke (1988), $e = (V_{fr}/V_t) \cdot \sigma/(\sigma - s)$, differs from Eq. 2-4 due to the incomplete expression for fresh-water balance in Eq. 30 of van de Kreeke (1988). Equation 2-4 shows that bulk mixing efficiency can be calculated if one knows the volumes for riverine inflow and tidal prism, average salinity for the entire estuary, and salinity of the receiving water body, which we assume here to be sea water. Applying the above expression to the NOAA datasets, the mixing efficiency for Gulf estuaries can be computed (Fig. 2-7). With the exception of the Suwannee River ($e = 0.42$) and Sabine Lake ($e = 0.57$), $e < 0.3$ for all U.S. Gulf of Mexico estuaries and falls as low as 0.03 in

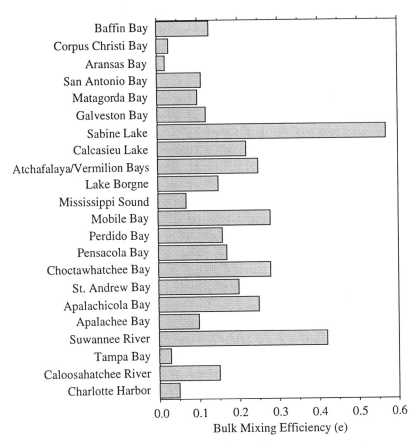

FIG. 2-7. Bulk mixing efficiency (e) for Gulf of Mexico estuaries.

Aransas Bay, Tampa Bay, and Corpus Christi Bay. These results agree with magnitudes of e published elsewhere (van de Kreeke 1988). As discussed above concerning calculated average salinities, this exercise is not intended to put precise values on mixing efficiency, but rather to put reasonable bounds on its magnitude.

One might expect that mixing efficiency would depend upon several factors in addition to V_{fr} / V_t including (1) the morphological characteristics of the estuary; (2) the presence of circulation patterns within the estuary; and (3) the presence or lack of meteorological mixing. Also, assuming that some of the estuarine water in the outgoing tide returns on the next flood tide, mixing efficiency also depends on the strength of long shore currents and mixing at the mouth of the estuary. Although none of these factors are explored in the above equations, the simple relationship between e and V_{fr} / V_t is shown in Fig. 2-8. Evidently, mixing efficiency increases with V_{fr} / V_t, although scatter

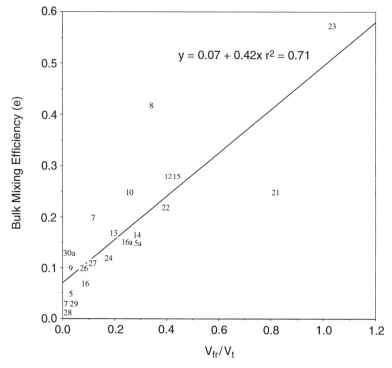

FIG. 2-8. *Mixing efficiency (e) versus ratio of freshwater input (V_{fr}) to tidal prism (V_t) for Gulf of Mexico estuaries. Data point numbers correspond to estuarine identification numbers in Fig. 2-1.*

about the regression line ($r^2 = 0.71$) indicates the importance of the above factors. The regression equation $e = 0.07 + 0.42 (V_{fr} / V_t)$ was applied to Eq. 2-3 for U.S. Gulf estuaries (Fig. 2-9). The equation (dashed line, Fig. 2-9) and data points representing U.S. Gulf estuaries are shown in comparison to the "perfect" mixing case (solid line).

More encompassing schemes for classifying and comparing estuarine salinity and mixing are presented by Hansen and Rattray (1966), Fischer (1992), and Orlando et al. (1993). The stratification-circulation classification of Hansen and Rattray (1966) uses salinity stratification and the average surface velocity : discharge velocity ratio to differentiate between well-mixed, partially mixed, and salt-wedge estuaries. The estuarine Richardson number is used to quantify buoyancy effects on stratification (Fischer 1972). Another scheme classifies the stability of salinity regimes for Gulf estuaries in terms of their temporal salinity variability with a view to developing strategies for estuarine management (Orlando et al. 1993). In this scheme, estuaries dominated by high inflows and those dominated by tidal fluxes due to low river inflows are

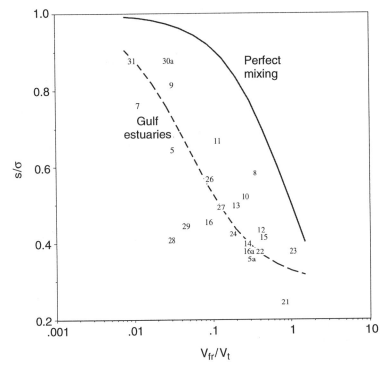

FIG. 2-9. *Normalized salinity (s/σ) versus ratio of freshwater input (V_{fr}) to tidal prism (V_t) for Gulf of Mexico estuaries, where s is the average estuarine salinity and σ is sea-water salinity. Solid line represents perfect mixing (Eq. 2-2), s/σ = 1/(1 + V_{fr}/V_t). Dotted line represents Gulf of Mexico estuaries, s/σ = e/(e + V_{fr}/V_t), where e is given by the best fit equation (e = 0.07 + 0.42 V_{fr}/V_t, r² = 0.71) from Fig. 2-8. Data point numbers correspond to estuarine identification numbers in Fig. 2-1.*

considered to have the most stable intra-annual salinity regimes, while those in intermediate riverine inflows and tidal ranges are considered less stable.

ESTUARINE DISPLACEMENT RATE

Many indices are used to compare the physical characteristics of estuaries. Perhaps the most basic is estuary volume: area ratio, which provides an estimate of average estuary depth. Gulf of Mexico bays and estuaries typically have volume: area ratios ranging from <1 m, characteristic of shallow, wind-driven flats and lagoons, to >3 m, indicative of estuaries with deeper primary bays (Armstrong 1982). However, a related and more useful measure of hydrography is the fresh-water inflow displacement rate, calculated as the average annual fresh-water inflow to the estuary divided by the estuary's average volume, which gives a crude approximation of the frequency of water turnover in an estuary.

Large displacement rates (>12 yr^{-1}) indicate that substantial amounts of allochthonous materials, as well as in situ biological production, particularly plankton blooms, will be washed out of the estuary to coastal waters. An example is the Sabine-Neches estuary (Sabine Lake), which has an in-flow: volume ratio of 53 and exhibits the lowest average salinity and production of marine organisms among major Texas estuaries (Texas Department of Water Resources 1981; Solis 1994). Such estuaries may exhibit "outwelling" (Odum 1971), the export of much of their nutrient load and planktonic produc-tion into the neritic waters of the Gulf of Mexico. This export can enhance productivity of the near-shore continental shelf. Positive estuaries of this type occur predominantly from eastern Texas across Louisiana, Mississippi, and Alabama to the Florida panhandle.

Habitat type and area may also be related to fresh-water inflow displace-ment rate in Gulf of Mexico estuaries. For example, the amount of open water habitat in seven major Texas estuaries is maximum at an inflow: volume ratio of about 2 (Longley 1995), whereas the amount of oyster reef habitat is maximum at an inflow: volume ratio of approximately 6 (Fig. 2-10). For both habitat types, the habitat area decreases rapidly in moving away from the optimum inflow: volume ratios. One possible explanation for the oyster reef pattern is that the oyster's physiology does not tolerate extremes very well, exhibiting better growth and survival at intermediate levels of salinity and depth. In another case, the area of salt-brackish marshes tends to in-crease while seagrass areas decrease with increasing inflow: volume ratios (Fig. 2-11). The larger amounts of nutrients and sediments transported by increasing inflows may favor salt-brackish marshes, while increased turbidity can limit seagrass abundance due to reduced light.

The Texas coastal region can be used to illustrate the relationship between relative habitat area and inflow: volume ratios because of the wide variation in net precipitation across the state. High inflow estuaries occur on the upper Texas coast near Louisiana, with inflows decreasing down the coastline to the extremely dry bays and lagoons of the lower coast near Mexico. Inflow: volume ratios for other regions in the Gulf of Mexico also exhibit wide variations, but the clinal transition may not be as smooth as it is in Texas. The relationships in Figs. 2-10 and 2-11 can be used to predict changes in habitat area and type from such events as global climate change and eustatic sea level rise that could alter fresh-water inflows, estuary water volume, and consequently the inflow: volume ratio of an estuary. For example, a long-term temperature increase of 2°C, combined with a 20% decrease in net precipitation and a sea level rise of 0.46 m, would substantially reduce the inflow: volume ratio of Texas bays and estuaries. These levels would be expected to decrease salt-brackish marshes by $>50\%$ while increasing the relative amount of seagrass habitat by a similar margin (Longley 1995).

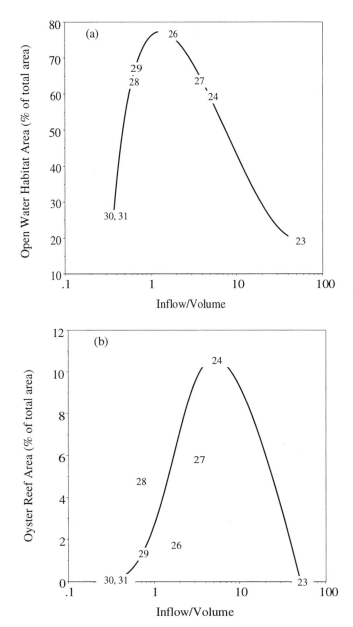

FIG. 2-10. *Relative habitat areas of open water (a) and oyster reef (b) in seven major Texas estuaries as a function of estuary inflow : volume ratios (adapted from Longley 1995). Data point numbers correspond to estuarine identification numbers in Fig. 2-1.*

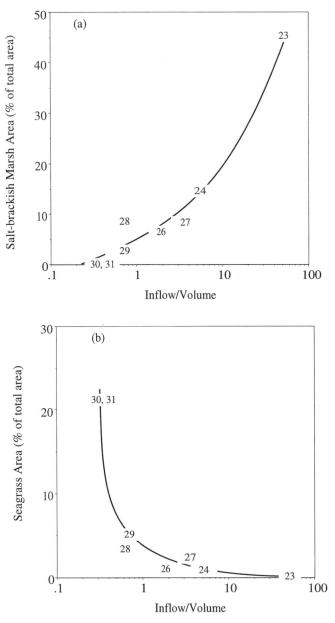

FIG. 2-11. *Relative habitat areas of salt-brackish marsh (a) and seagrass (b) in seven major Texas estuaries as a function of estuary inflow : volume ratios (adapted from Longley 1995). Data point numbers correspond to estuarine identification numbers in Fig. 2-1.*

ESTUARINE RESIDENCE TIMES

Definition of and Method for Computing Residence Time

Another parameter used to characterize the mixing characteristics of an estuary is residence time. Care is required in distinguishing between the various definitions of residence time found in the literature and in the many methods used to calculate residence time. Residence time, or fresh-water flushing time, as used here is the average amount of time required to replace the equivalent fresh water in the estuary by fresh-water inputs (Bowden 1967; Officer 1976). Other definitions that have been used for residence time include the time required to replace the entire volume of the estuary, the average age of particles in an estuary (Officer 1980), and the time required to replace a given fraction of particles originally found in the estuary (Miller and McPherson 1991). The definitions of Officer (1980) and Miller and McPherson (1991) make no assumptions regarding the source of particles in the estuary, in contrast to the more widely used definition based on particles originating at the fresh-water inflow point. Time scales used in reservoir theory that are similar to residence time include transit time, average age, and flushing time, which for the case of the well-mixed box model (large internal diffusivity) all approach fresh-water flushing time (Zimmerman 1976, 1988). The value most often reported in the literature corresponds to the definition of fresh-water flushing time.

Residence time can be computed as an integral quantity to characterize an entire estuary, for sub-areas (compartments or boxes) of an estuary, and as a function of time, although the effort required to compute residence times for sub-areas or as a function of time can approach that required to develop and apply a numerical transport model (Ward 1996). Residence times can be computed for different flow conditions and as a function of the location of the source of tracer of interest. Residence time is useful in classifying salinity mixing processes and fluxes in estuaries because it is based on conservation of mass principles (Jay et al. 1997).

Methods for computing residence time include the fresh-water fraction method, which is used here, as well as the tidal prism method (Officer 1976), two-layer method for stratified estuaries (Bowden 1967), segmentation methods (Ketchum 1951; Dyer and Taylor 1973), and box-model methods (Officer 1980; Miller and McPherson 1991). Again, care is required in comparing the results from different computation methods. For example, the box-model method of Officer (1980) is based on the average age of particles in the estuary, while that of Miller and McPherson (1991) is based on the time of removal of particles. Each of these differences in definition and computation methods can yield significantly different residence times.

More recently, methods based on hydrodynamic and conservative transport models (Dimou and Adams 1990; Sheng et al. 1993) have been used to compute residence times. While numerical models are capable of handling highly complex flows, multiple input and output points, and unsteady conditions, Geyer and Signell (1992) warn that accurately simulating dispersive processes in complex estuaries using the simplified dispersion formulations commonly assumed in these models poses many difficulties. While the use of improved hydrodynamic transport models for computing residence times holds much promise for the future, residence times are now most commonly found using simple salinity or tidal-prism measurements.

The interpretation of residence time can become muddled in estuaries with stratification, with multiple inflow or outflow points, with significant channelization where evaporation exceeds rainfall and river inflows, and with significant horizontal variation in salinity or flows (i.e., in most real estuaries). Nonetheless, residence times can in a gross sense characterize estuaries and are useful in making cross-system comparisons. Residence time is also useful as a correlative index when considering physical, chemical, and biological processes, which rely on the time of contact with the overlying water.

Limiting values for residence time, t, are given by (Zimmerman 1976):

$$\frac{V_f}{Q} \approx \frac{T \cdot V}{V_{fr} + V_t} \leq t \leq \frac{V}{Q} \qquad (2\text{-}5)$$

where

$$V = V_{lt} + \frac{V_{fr} + V_t}{2} \qquad (2\text{-}6)$$

Here V_f represents the fresh-water content of the estuary, Q is the fresh-water inflow rate (equal to fresh-water inflow volume per tidal period, V_{fr}/T), T is the tidal period, V is the estuary volume, V_{lt} is the estuary volume at low tide, V_{fr} is the volume of fresh-water input over time T, and V_t is the tidal influx in time T. (Residence time, t, varies depending on whether V is defined as the estuary volume at high tide, mean tide, or low tide. The magnitude of t if V is defined as the volume at high tide is $T/2$ more than if V is defined as the volume at mean tide, and T more than if V is defined as the volume at low tide.) The two leftmost expressions in Eq. 2-5 are generally applied definitions for residence time, while the expression on the extreme right, the displacement time, is the inverse of the displacement rate discussed above. The expressions to the left of the first inequality assume perfect mixing of tidal and fresh water, while the right-hand expression assumes that no mixing occurs. Neither expression is completely accurate for estuaries, so residence times fall somewhere between the values of both expressions. For estuaries in which fresh-water inflow volume relative to tidal influx is large, $V_{fr}/V_t \gg 1$, residence time and displacement time approach one another. For estuaries with large tidal volume relative to river inflow, $V_{fr}/V_t \ll 1$, displacement time can exceed residence time by orders of magnitude.

Other limits on residence time for conditions of low fresh-water inflows and large evaporation rates are found by considering the expression for residence time based on the tidal prism method (Officer 1976):

$$\frac{t}{T} = \frac{V}{V_{iv}} = \frac{V}{V_t + V_{fr}} = \frac{V}{V_t} \cdot \frac{1}{1 + V_{fr}/V_t} \tag{2-7}$$

Here V_{iv} represents intertidal volume, which is the sum of V_t, the tidal prism volume that enters the estuary over a tidal cycle, and V_{fr}, the volume of fresh water entering over the same period. As $V_{fr} \to 0$, that is, as evaporation approaches combined precipitation and river inflow, $t/T \to V/V_t$, and (from Eq. 2-3) $s/\sigma \to 1$. Flushing in this case is strictly due to tidal mixing, and estuarine salinity approaches sea water salinity. If evaporation exceeds river inflows plus precipitation, that is, $V_{fr} < 0$, then both t/T in Eq. 2-7 and s/σ in Eq. 2-3 increase, that is, residence time and average salinity tend to increase. Conditions leading to large residence times and hypersalinity are tempered by large tidal fluxes. As tidal flux increases, $|V_{fr}|/V_t \ll 1$, and again $t/T \to V/V_t$ and $s/\sigma \to 1$.

Fresh-Water Fraction Method

The term V_f in Eq. 2-5 represents fresh-water content found in the estuary. Fresh-water content is the product of estuary volume and the equivalent fresh-water fraction, f, where f is given by

$$f = \frac{\sigma - s}{\sigma} \tag{2-8}$$

where s represents the average salinity of the estuary and σ represents the salinity of the receiving waters, here again assumed equal to sea-water salinity. This can be used to compute residence time, t, as follows:

$$t = \frac{f \cdot V}{Q} = \frac{V_f}{Q} \tag{2-9}$$

where V is volume of the estuary, Q is fresh-water inflow rate, and V_f is volume of fresh water in the estuary. Fresh-water salinity is assumed equal to zero.

Residence Time Estimates

As shown above, residence time is influenced by fresh-water inflow rates, precipitation and evaporation rates from the estuarine surface, and the rate of influx of saline water due to the tides. These factors are used below to estimate Gulf residence times.

Sources of Data. Sources of physical and hydrological data compiled in the NOAA data sets include Diener (1975), Barrett (1971), Louisiana Wildlife and Fisheries Commission (1971), Bault (1972), Christmas (1973), and McNulty et al. (1972), as well as many unpublished data bases. Deegan et al. (1986) present physical characteristics including open water area, mean depth, mean riverine discharge, precipitation, evaporation, and mean tide for 64 estuaries in the Gulf of Mexico from Florida to the Yucatan Peninsula. However, data for many smaller estuaries on the Mexican coast are not presented.

Residence Times for Gulf Estuaries. Residence times for several U.S. Gulf of Mexico estuaries were calculated by applying the fresh-water fraction method, Eq. 2-9, to the NOAA (1985b, 1997) data sets (Fig. 2-12). Like the calculations for average salinity presented earlier, these estimates for residence times were made to allow comparison among estuaries. Published values for residence times for any particular estuary can differ significantly. For example, Austin (1954) estimated the flushing time for Mobile Bay to be about 50 d, Wiseman et al. (1988) suggested time scales on the order of 1 mo, and the value computed here with the NOAA data is 10 d. There is, however, reasonable agreement for other estuaries. Armstrong (1982) reports residence times of 7 d (0.02 yr), 40 d (0.11 yr), and 77 d (0.21 yr) for Sabine Lake, Galveston Bay, and Matagorda Bay, respectively, while the values presented here are 9 d (0.025 yr), 40 d (0.11 yr), and 67 d (0.18 yr). Miller and McPherson (1991) report a fresh-water replacement time for Charlotte Harbor under average flow conditions of 50 d (0.14 yr), close to the value of 59 d (0.16 yr) reported here. Inconsistencies likely arise from differences in the conditions under which data sets were collected, in the methods used to estimate parameters, and in the definition of estuarine boundaries. For example, estuarine residence time for Charlotte Harbor under average flow conditions using the method of Miller and McPherson (1991), that is, the time required to remove 95% of the particles in the estuary, is 130 d (0.36 yr). Results presented here are meant to represent "average" conditions. Residence times were not computed for all estuaries due to missing or inadequate information.

The longest residence times computed here are for south Texas estuaries and Lake Pontchartrain in Louisiana, all over 200 d. Large residence times for south Texas estuaries result from the relatively small inflows these systems receive. Lake Pontchartrain's large residence time arises from the relatively small inflows it receives with respect to its volume and to its virtual lack of a tidal prism to assist in flushing. The smallest computed residence times, all

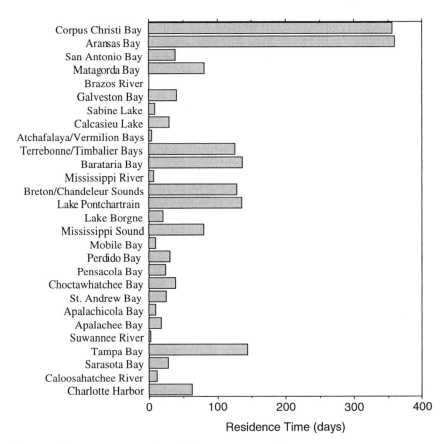

FIG. 2-12. *Residence times for Gulf of Mexico estuaries, NOAA (1985b, 1997) data, calculated using the fresh-water fraction method (Eq. 2-7).*

<10 d, are found in systems with large inflow:estuary volume ratios. These include the Suwannee River (3 d), Apalachicola Bay (10 d), Mobile Bay (10 d), Mississippi River (7 d), Atchafalaya-Vermilion Bay (5 d), Sabine Lake (9 d), and Brazos River (1 d).

Spatial and temporal variability in residence times results from the variability discussed above for tides, river inflows, and salinity. Herrera-Silveira (1995) presents temporal and spatial variation in residence times of a small coastal lagoon in the Yucatan Peninsula. Residence times varied by up to a factor of 7 from the head to the mouth of the estuary, and by a factor of 2 for a given site over a period of only 1 year. This variation is likely typical for Gulf estuaries.

Relationship of Residence Time to Nutrients and Fisheries. Residence time may be used to explore a number of ecological features, such as the relationship between estuarine nutrient loading and fisheries production. How-ever, to explore the underlying relationship, it is important to consider the amount of time the estuary has to process the nutrients. Multiplying an estu-ary's total nutrient loading by residence time results in an estimate of the effective loading rate. For example, Brock et al. (1996) compared the effective nitrogen loadings of five Texas estuaries to their average fishery harvests (Fig. 2-13). This comparison illustrates the general agreement in trends between nutrient levels and fishery production in these western Gulf of Mexico estuar-ies. The trends match best for the Trinity-San Jacinto estuary (Galveston Bay), the Guadalupe estuary (San Antonio Bay), and the Lavaca-Colorado estuary (Matagorda Bay). However, the Mission-Aransas estuary (Aransas Bay) appears more productive than anticipated, while the Nueces estuary (Corpus Christi Bay) seems less productive than expected.

Although the relationships between estuarine nutrient loading and the harvest of seafood species are not completely known, the explanation of anomalies may lie in special circumstances affecting the estuaries. For instance, the Mission-Aransas estuary may be more productive than anticipated because it is significantly influenced by that portion of the fresh-water inflows and nutrients from neighboring Guadalupe estuary that flow south toward Aransas

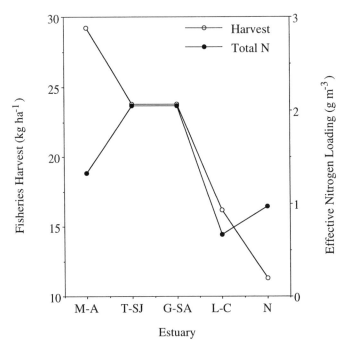

FIG. 2-13. *Effective nitrogen loadings versus fishery harvests in five major Texas estuaries (adapted from Brock et al. 1996). M-A: Mission-Aransas estuary; T-SJ: Trinity-San Jacinto estuary; G-SA: Guadalupe-San Antonio estuary; L-C: Lavaca-Colorado estuary; N: Nueces es-tuary.*

Pass and the Gulf of Mexico. Indeed, the Guadalupe estuary lacks its own Gulf inlet, except for the small and intermittently flowing cedar Bayou, and up to half of this estuary's seaward flow passes through the Mission-Aransas estuary on its way to the Gulf (Matsumoto 1994). The Nueces Estuary exhibits the opposite result, being less productive of fisheries than expected based upon its nutrient loadings. In this case, urbanization has caused a loss of fishery habitats such as oyster reefs, seagrass beds, and intertidal marshes (Tunnell and Dokken 1996).

SUMMARY

Physical factors that govern mixing and transport processes in Gulf of Mexico estuaries, including tides, wind, rainfall and evaporation, fresh-water inflows, and anthropogenic alterations to estuaries and their watersheds, vary throughout the Gulf. The most significant variation is found in rainfall, which varies by nearly an order of magnitude around the Gulf, and riverine runoff, which spans over two orders of magnitude. Heaviest rainfall and greatest runoff volumes are found in the Tabasco Plain in Mexico and in the area surrounding the Mississippi Delta in the United States, while minimum rainfall and runoff volumes are found in south Texas. Regional variations in average salinity reflect the rainfall/runoff patterns in Gulf estuaries. In contrast, tidal amplitudes and evaporation rates are more uniform Gulf-wide. Inflows, tidal prisms, and salinity were used to compute bulk mixing efficiencies, which were found to range from 0.02 to 0.57 in U.S. Gulf estuaries. Useful relationships between inflow : volume ratios and selected ecological factors were presented. Estimates for estuarine residence times were found to vary from less than 5 d to over 300 d.

ACKNOWLEDGMENTS

The authors would like to thank Paul Orlando of NOAA, Strategic Assessments Branch, Ocean Assessments Division, for generously supplying and providing an explanation for much of the data utilized in this report. We would also like to thank David A. Brock, Texas Water Development Board, for his useful comments and insights during the preparation of this manuscript. Finally, thanks go to the reviewers of this chapter for their helpful comments and suggestions, which led to a much improved final product.

REFERENCES

Arbingast, S. A., C. P. Blair, J. R. Buchanan, C. C. Gill, R. K. Holz, C. A. Marin, R. H. Ryan, M. E. Bonine, and J. P. Weiler. 1975. Atlas of Mexico (2nd ed.). Bureau of Business Research. The University of Texas at Austin.

Armstrong, N. E. 1982. Responses of Texas estuaries to freshwater inflows, p. 103–120. *In:* V. S. Kennedy [ed.], Estuarine comparisons. Academic Press.

Austin, G. B. 1954. On the circulation and tidal flushing of Mobile Bay, Alabama. Part 1. Texas A&M College Research Foundation Project 24, Technical Report 12.

Barrett, B. B. 1971. Cooperative Gulf of Mexico estuarine inventory and study, Louisiana, phase II, hydrology and phase III, sedimentology. Louisiana Wildlife and Fisheries Commission.

Bault, E. I. 1972. Hydrology of Alabama estuarine areas—cooperative Gulf of Mexico estuarine inventory. Alabama Marine Resources Bulletin No. 7. Alabama Marine Resources Laboratory.

Bowden, K. F. 1967. Circulation and diffusion, p. 15–36. *In:* G. H. Lauff [ed.], Estuaries. Publication No. 83. American Association for the Advancement of Science.

———. 1980. Physical factors: Salinity, temperature, circulation, and mixing processes, p. 37–70. *In:* E. Alausson and I. Cato [eds.], Chemistry and biogeochemistry of estuaries. Wiley.

Brock, D. A., R. S. Solis, and W. L. Longley, 1996. Guidelines for water resources permitting: Nutrient requirements for maintenance of Galveston Bay productivity. Final report to U.S. Environmental Protection Agency by the Texas Water Development Board.

Christmas, J. Y. 1973. Cooperative Gulf of Mexico estuarine inventory and study, Mississippi: Phase I, area description, phase II, hydrology. Gulf Coast Research Laboratory.

Cross, R. D. 1974. Atlas of Mississippi. University Press of Mississippi.

Day, J. W., Jr., C. J. Madden, R. R. Twilley, R. F. Shaw, B. A. McKee, M. J. Dagg, D. L. Childers, R. C. Raynie, and L. J. Rouse. 1994. The influence of Atchafalaya River discharge on Fourleague Bay, Louisiana (USA), p. 151–160. *In:* K. R. Dyer and R. J. Orth [eds.], Changes in fluxes in estuaries. Olsen and Olsen.

Deegan, L. A., J. W. Day, Jr., J. G. Gosselink, A. Yanez-Arancibia, G. Soberon Chavez, and P. Sanchez-Gil. 1986. Relationships among physical characteristics, vegetation distribution and fisheries yield in Gulf of Mexico estuaries, p. 83–100. *In:* D. A. Wolfe [ed.], Estuarine variability. Academic Press.

Diener, R. A. 1975. Cooperative Gulf of Mexico estuarine inventory and study—Texas: Area description. National Marine Fishery Service Circular 393.

Dimou, K. N., and E. E. Adams. 1990. 2-D particle tracking model for estuary mixing, p. 472–481. *In:* M. L. Spaulding [ed.], Estuarine and coastal modeling. ASCE.

Dyer, K. R. 1973. Estuaries: A physical introduction. Wiley.

———, and P. A. Taylor. 1973. A simple, segmented prism model of tidal mixing in well-mixed estuaries. Estuar. Coast. Shelf Sci. **1:** 411–418.

Fernald, E. A. 1981. Atlas of Florida. Florida State University Foundation.

Fischer, H. B. 1972. Mass transport mechanisms in partially stratified estuaries. J. Fluid Mech. **53:** 671–687.

———. 1976. Mixing and dispersion in estuaries, p. 107–133. *In:* M. Van Dyke, W. G. Vincenti, and J. V. Wehausen [eds.], Annual review of fluid mechanics, volume 8. Annual Reviews.

Geyer, W. R., and R. P. Signell. 1992. A reassessment of the role of tidal dispersion in estuaries and bays. Estuaries **15:** 97–108.

Hansen, D. V., and M. Rattray. 1966. New dimensions in estuary classification. Limnol. Oceanogr. **11:** 319–326.

Herrera-Silveira, J. A. 1995. Salinity and nutrients in a tropical coastal lagoon with groundwater discharges to the Gulf of Mexico. Hydrobiologia **321:** 165–176.

Ippen, A. T. 1966. Estuary and coastline hydrodynamics. McGraw-Hill.

Jay, D. A., R. J. Uncles, J. Largier, W. R. Geyer, J. Vallino, and W. R. Boynton. 1997. A review of recent developments in estuarine scalar flux estimation. Estuaries **20:** 262–280.

Ketchum, B. H. 1951. The exchange of fresh and salt waters in tidal estuaries. J. Mar. Res. **10:** 18–37.

Light, S. S., and J. W. Dineen. 1994. Water control in the Everglades: A historical perspective, p. 47–84. *In:* S. M. Davis and J. C. Ogden [eds.], Everglades—the ecosystem and its restoration. St. Lucie Press.

Linsley, R. K., M. A. Kohler, and J. L. H. Paulhus. 1975. Hydrology for engineers (2nd ed.). McGraw-Hill.

Longley, W. L. 1995. Estuaries, p. 80–118. *In:* G. R. North, J. Schmandt, and J. Clarkson [eds.], The impact of global warming on Texas. University of Texas Press.

Louisiana Wildlife and Fisheries Commission (LWFC). 1971. Cooperative Gulf of Mexico estuarine inventory and study, Louisiana. Phase I, area description and phase IV, biology. LWFC.

Madden, C., J. Day, and J. Randall. 1988. Coupling of freshwater and marine systems in the Mississippi deltaic plain. Limnol. Oceanogr. **33:** 982–1004.

Marmer, H. A. 1954. Tides and sea level in the Gulf of Mexico, p. 101–118. *In:* Fishery Bulletin 89. U.S. Government Printing Office.

Matsumoto, J. 1994. Guadalupe estuary example analysis, p. 277–323. *In:* W. L. Longley [ed.], Freshwater inflows to Texas bays and estuaries: Ecological relationships and methods for determination of needs. Texas Parks and Wildlife Press.

McNulty, J. K., W. N. Lindall, Jr., and J. E. Sykes. 1972. Cooperative Gulf of Mexico estuarine inventory and study, Florida: Phase 1, area description. NOAA Technical Report NMFS Circ-368. U.S. Department of Commerce.

Miller, R. L., and B. F. McPherson. 1991. Estimating estuarine flushing and residence times in Charlotte Harbor, Florida, via salt balance and a box model. Limnol. Oceanogr. **36:** 602–612.

National Oceanic and Atmospheric Administration (NOAA). 1985a. The Gulf of Mexico coastal and ocean zone strategic assessment: Data atlas. Strategic Assessment Branch, Ocean Assessments Division.

National Oceanic and Atmospheric Administration (NOAA). 1985b. National estuarine inventory: Data atlas, volume I: Physical and hydrologic characteristics. Strategic Assessment Branch, Ocean Assessments Division.

National Oceanic and Atmospheric Administration (NOAA). 1990. Estuaries of the United States, vital statistics of a national resource base. Strategic Assessment Branch, Ocean Assessments Division.

National Oceanic and Atmospheric Administration (NOAA). 1991. Tide tables 1991, East Coast of North and South America. U.S. Department of Commerce.

National Oceanic and Atmospheric Administration (NOAA). 1995. Tidal characteristics and datums of Laguna Madre, Texas. NOAA Technical Memorandum NOS OES 008. NOAA.

National Oceanic and Atmospheric Administration (NOAA). 1997. Unpublished data. Strategic Assessment Branch, Ocean Assessments Division.

Newton, M. B., Jr. 1972. Atlas of Louisiana; a guide for students. Louisiana State University School of Geosciences.

Odum, E. P. 1971. Fundamentals of ecology. W.B. Saunders.

Officer, C. B. 1976. Physical oceanography of estuaries (and associated coastal waters). Wiley.

———. 1980. Box models revisited, p. 65–114. *In:* P. Hamilton and K. B. MacDonald [eds.], Estuarine and wetland processes with emphasis on modeling. Plenum Press.

Orlando, S. P., Jr., L. P. Rozas, G. H. Ward, and C. J. Klein. 1993. Salinity characteristics of Gulf of Mexico estuaries. National Oceanic and Atmospheric Administration, Office of Ocean Resources Conservation and Assessment.

Schroeder, W. W., and W. J. Wiseman, Jr. 1986. Low-frequency shelf-estuarine exchange processes in Mobile Bay and other estuarine systems on the northern Gulf of Mexico, p. 355–567. *In:* D. A. Wolfe [ed.], Estuarine variability. Academic Press.

Sheng, Y. P., H. K. Lee, and C. E. Demas. 1993. Simulation of flushing in Indian River Lagoon using 1-D and 3-D models, p. 366–380. *In:* M. L. Spaulding et al. [eds.], Estuarine and coastal modeling III. ASCE.

Sheng, Y. P., and S. J. Peene. 1993. A field and modeling study of residual circulation in Sarasota Bay and Tampa Bay, Florida, p. 641–655. *In:* M. L. Spaulding et al. [eds.], Estuarine and coastal modeling III. ASCE.

Smith, N. P. 1985. Numerical simulation of bay-shelf exchanges with a one-dimensional model. Contr. Mar. Sci. **28:** 1–13.

———. 1988. The Laguna Madre of Texas: Hydrography of a hypersaline lagoon, p. 31–40. *In:* B. Kjerfve [ed.], Hydrodynamics of estuaries, volume II, estuarine case studies. CRC Press.

Solis, R. S. 1994. Coastal hydrology and the relationships among inflow, salinity, nutrients, and sediments, p. 23–71. *In:* W. L. Longley [ed.], Freshwater inflows to Texas bays and estuaries: Ecological relationships and methods for determination of needs. Texas Parks and Wildlife Press.

Texas Department of Water Resources. 1981. Sabine-Neches Estuary: A study of the influence of freshwater inflows. LP-116.

Texas Water Development Board. 1997. Unpublished data, Sabine Lake 1990 and 1996 studies. TWDB Environmental Section.

Tunnell, J. W., Jr., and Q. R. Dokken. 1996. Current status and historical trends of the estuarine living resources within the Corpus Christi Bay National Estuary Program study area. Final report in volumes 1–4. Texas A&M University at Corpus Christi.

van de Kreeke, J. 1988. Dispersion in shallow estuaries, p. 27–39. *In:* B. Kjerfve [ed.], Hydrodynamics of estuaries, volume I, estuarine physics. CRC Press.

Ward, G. H., Jr. 1980. Hydrography and circulation processes of Gulf estuaries, p. 183–215. *In:* P. Hamilton and K. B. MacDonald [eds.], Estuarine and wetland processes with emphasis on modeling. Plenum Press.

———. 1996. Estuaries, p. 12.1–12.114. *In:* L. W. Mays [ed.], Water resources handbook. McGraw-Hill.

West, R. C. 1964. Handbook of Middle American Indians. Natural environment and cultures. University of Texas Press.

Wiseman, W. J., Jr., W. W. Schroeder, and S. P. Dinnel. 1988. Shelf-estuarine water exchanges between the Gulf of Mexico and Mobile Bay, Alabama, p. 1–8. *In:* M. P. Weinstein [ed.], Larval fish and shellfish transport through inlets. American Fisheries Society Symposium 3.

Wolfe, D. A., and B. Kjerfve. 1986. Estuarine variability: An overview, p. 3–17. *In:* D. A. Wolfe [ed.], Estuarine variability. Academic Press.

Zimmerman, J. T. F. 1976. Mixing and flushing of tidal embayments in the western Dutch Wadden Sea I: Distribution of salinity and calculation of mixing time scales. Neth. J. Sea Res. **10:** 149–191.

———. 1988. Estuarine residence times, p. 75–84. *In:* B. Kjerfve [ed.], Hydrodynamics of estuaries, volume I, estuarine physics. CRC Press.

Chapter *3*

Sedimentary Processes of Gulf of Mexico Estuaries

Brent A. McKee and M. Baskaran

INTRODUCTION

Sediments that are produced from weathering of both igneous and sedimentary rocks are transported to the ocean primarily by rivers. These sediments, both as suspended loads and as bed loads, are discharged into either open oceans, marginal seas, or isolated bodies of coastal waters. In coastal waters, the terrigenous sedimentary debris forms a major component of the sediments, compared to the biogenic (derived from the hard parts of organisms) and cosmogenic (extraterrestrial debris) fractions. The purpose of this chapter is to review the sedimentary processes that control the distribution of terrigenous sedimentary materials in Gulf of Mexico estuaries.

Water discharge data are of sufficient quality and availability (usually in the gray literature) worldwide to allow calculation of global input fluxes. In contrast, high-quality sediment discharge data are relatively rare, and the fates of fluvially derived sediments in estuarine environments are poorly understood (Milliman 1991). Sediment discharge estimates for North American rivers (although incomplete and of varying quality) are arguably the best available for any continent (Milliman and Meade 1983). Approximately 90% of sediments discharged by rivers in the conterminous United States enter estuarine envi-

Biogeochemistry of Gulf of Mexico Estuaries, Edited by Thomas S. Bianchi, Jonathan R. Pennock, and Robert R. Twilley.
ISBN 0-471-16174-8 © 1999 John Wiley & Sons, Inc.

ronments bordering the Gulf of Mexico (Milliman and Meade 1983). In fact, the abundance of fine-grained sediments and their influence on other important processes is one of the salient features of Gulf estuaries. Understanding sediment inputs and sedimentary processes in estuaries is central to understanding many other important phenomena. For example, turbidity induced by particle suspension can influence processes such as primary and secondary productivity, particulate material flux, and pollutant dispersal in coastal environments. Despite their importance, little is known about sediment dynamics and sedimentary processes in Gulf of Mexico estuaries.

The combined watershed area of all Gulf of Mexico estuaries is 4.4 × 10^6 km^2 (Fig. 3-1). This is 80% of the fresh-water drainage from the conterminous United States into coastal margins (Wilson and Iseri 1969). Over 70% of the fresh-water inflow to the Gulf of Mexico is supplied by the Mississippi and Atchafalaya Rivers. Nevertheless, estuaries in the Gulf receive much greater input of fresh water than other regions in the conterminous United States (North Atlantic, Middle Atlantic, South Atlantic, and Pacific) even when the Mississippi River is not considered (NOAA 1990). On average, Gulf of Mexico estuaries are the most shallow (mean: less than 2.5 m) and have the largest water surface area among estuarine regions in the United States (NOAA 1990). Three unique characteristics of Gulf of Mexico estuaries result from the combined effects of these physical factors: (1) flushing times are typically very short (Chapter 2, this volume); (2) estuarine waters are intimately coupled with bottom sediments; and (3) wind-driven re-suspension of estuarine sediments is prevalent (Chapter 1, this volume).

The physical characteristics of Gulf of Mexico estuaries define sediment dynamics in these waters. The large water surface areas (providing a large fetch for prevailing winds) and the small tidal range (<0.5 m) typical of Gulf of Mexico estuaries usually result in systems where the physical dynamics are dominated by winds rather than tidal forcing. High rates of sediment discharge are associated with fresh-water inputs to many of these estuaries. Frequent re-suspension of bottom sediments results from the shallow depths and the turbulent nature of these estuaries.

SEDIMENTARY PROCESSES

The nature and distribution of sedimentary deposits in estuaries are highly variable and are controlled by several factors, including (1) rate of sediment supply; (2) type of source material; (3) grain-size distribution; and (4) energy conditions at the sediment-water interface. In this section, we will discuss sediment inputs, estuarine turbidity, benthic sediment processes (sediment deposition, mixing, and accumulation), and the export of sediments from Gulf of Mexico estuaries. A schematic diagram showing the various sedimentary processes discussed in this chapter is presented in Fig. 3-2.

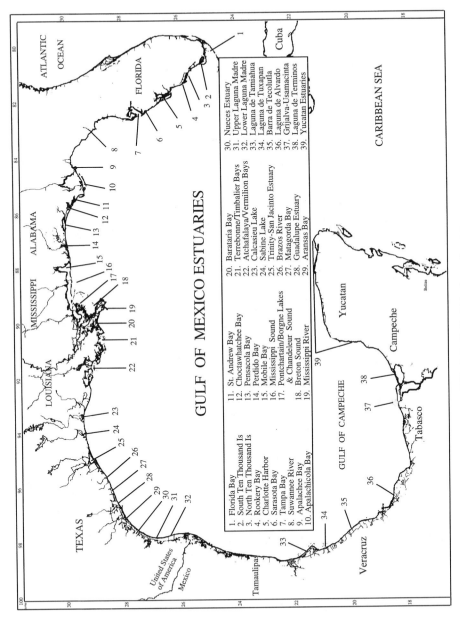

GULF OF MEXICO ESTUARIES

1. Florida Bay
2. South Ten Thousand Is
3. North Ten Thousand Is
4. Perdido Bay
5. Charlotte Harbor
6. Sarasota Bay
7. Tampa Bay
8. Suwannee River
9. Apalachee Bay
10. Apalachicola Bay

11. St. Andrew Bay
12. Choctawhatchee Bay
13. Pensacola Bay
14. Perdido Bay
15. Mobile Bay
16. Mississippi Sound
17. Pontchartrain/Borgne Lakes
    & Chandeleur Sound
18. Breton Sound
19. Mississippi River

20. Barataria Bay
21. Terrebonne/Timbalier Bays
22. Atchafalaya/Vermilion Bays
23. Calcasieu Lake
24. Sabine Lake
25. Trinity-San Jacinto Estuary
26. Brazos River
27. Matagorda Bay
28. Guadalupe Estuary
29. Aransas Bay

30. Nueces Estuary
31. Upper Laguna Madre
32. Lower Laguna Madre
33. Laguna de Tamiahua
34. Laguna de Tuxapan
35. Barra de Tecolutla
36. Laguna de Alvardo
37. Grijalva-Usamacinta
38. Laguna de Terminos
39. Yucatan Estuaries

FIG. 3-1. Gulf of Mexico estuaries.

65

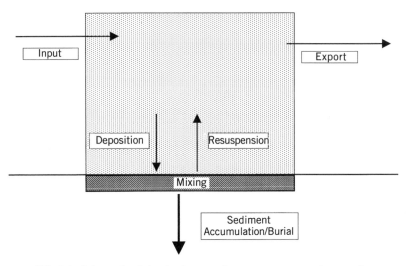

FIG. 3-2. Schematic of the dominant sedimentary processes in estuaries.

Riverine Sediment Inputs

As a result of the spatial and temporal variability in sediment inputs, there are inherent difficulties in measuring the discharge-weighted flux of sediments from rivers (Meade et al. 1990). Globally, information regarding sediment discharge rates for rivers is inadequate (Milliman and Meade 1983). The reason for this is twofold. (1) A protocol for quantitative assessment of riverine sediment discharge rates has been in existence for less than 30 yr and has been in general use for only a decade (Guy and Norman 1970; Nordin et al. 1983; Meade and Stevens 1990). Estimates of sediment discharge rates derived from other (less quantitative) techniques are not directly comparable to the rates calculated using quantitative measurements. (2) Sediment discharge rates are temporally variable in most river systems; therefore, good estimates of annual rates require an intensive sampling regime throughout the year and over many years. One study of North American rivers observed that in many cases >50% of annual sediment discharge takes place in <1% of the annual cycle (Meade et al. 1990). Globally, there is a paucity of temporally intensive sediment discharge measurements, especially those using quantitative techniques (Milliman and Meade 1983; Meade 1996). This scarcity of quantitative sediment discharge rate measurements also applies to rivers entering the Gulf of Mexico.

Estimating Sediment Discharge. Streamflow has been monitored for most Gulf of Mexico rivers since the turn of the 20th century, with significant continuous records for the past 50 yr. However, little is known about riverine inputs of sediment to Gulf of Mexico estuaries with the exception of the

Mississippi/Atchafalaya system. Sediment discharge measurements have been recorded since 1950 for the Mississippi and since 1973 for the Atchafalaya. The Mississippi River ranks seventh in the world in terms of sediment discharge, now transporting an average of 210×10^6 mt yr^{-1} of suspended sediment to Gulf of Mexico estuaries (Meade 1996).

In a compilation of inflow data for U.S. Gulf of Mexico rivers, Slade (1992) reported the existence of multi-year sediment discharge records for the Mississippi and Atchafalaya Rivers and only five other Gulf of Mexico rivers: Rio Grande (1966–1983), Colorado (1957–1973), Brazos (1966–1986), Trinity (1955–1971), and Pearl (1967–1987). The Brazos River ranks third (after the Mississippi and Atchafalaya) among U.S. Gulf of Mexico rivers for sediment discharge (11×10^6 mt yr^{-1}; Meade and Parker 1985), while the Alabama and Tombigbee Rivers entering the Mobile Bay system combine to rank fourth (4×10^6 mt yr^{-1}; Ryan and Goodell 1972). Presently, the Rio Grande discharges only 0.8×10^6 mt yr^{-1}, although it was a major source of sediments to the Gulf of Mexico until the beginning of the 20th century (Meade et al. 1990). The Sabine-Neches river system has the largest fresh-water inflow to bay volume ratio of all Texas estuaries (Ward, 1980) and has a sediment discharge rate of 0.75×10^6 mt yr^{-1} (Jansen et al. 1979). Judson and Ritter (1964) reported a sediment discharge rate of 0.17×10^6 mt yr^{-1} for the Apalachicola River, the largest fresh-water input to the Gulf of Mexico from Florida. Little is known about the sediment discharge of the Grijalva-Usumacinta River system in Mexico, which ranks third in terms of water discharge to the Gulf of Mexico (Chapter 2, this volume).

Meade et al. (1990) discussed the importance and difficulty of measuring riverine sediment discharge and pointed out that the wash load (suspended silts and clays) may be relatively uniform from surface to bottom within a cross section of a river; however, it may be quite non-uniform laterally across the river. Collecting daily records of sediment discharge is expensive and often impractical. One possible way of extending our knowledge of sediment discharge rates is via sediment rating curves. These are empirical relationships between sediment discharge rates and water discharge rates, which are much better documented. This relationship is often very complex (Mossa 1996) but, if applied with care (especially in systems where the relationship is relatively simple), sediment rating curves can be effective (Meade et al. 1990). However, sediment discharge rates based on water discharge data and sediment rating curves can introduce as much as 50% error (Yorke and Ward 1986). To construct a good sediment rating curve, sediment discharge data must be collected over a range of water discharges. These data are used to construct a relationship between water discharge and sediment discharge or sediment concentration (usually a simple power function) specific to the site from which the data are collected and not readily transferable to another site. Observed relationships between suspended sediment concentration and water discharge usually display a clockwise loop—indicative of exhaustion or depletion effects. For example, Everett (1971) noted a clockwise distribution of suspended sediment concentrations with water discharge—peaking at ~1200 mg liter^{-1} and ~19,800 m^3 s^{-1} during the March–April 1964 period, and ~1600 mg liter^{-1} and ~28,300 m^3 s^{-1} during January–March 1960. Sediments stored on the bed during low water are in plentiful supply as waters rise, but stored materials are soon resuspended and become depleted as the river reaches maximum water discharge (e.g., Mississippi and Rio Grande Rivers; Nordin and Beverage 1965; Robbins 1977; Mossa 1996). The sediment discharge peak in the Mississippi River at Tarbert Landing precedes the water discharge peak during high discharge years and is coincident with the water discharge peak during average or low discharge years. At Belle Chasse, water and sediment discharge peaks for the Mississippi River correspond very closely in time (Mossa 1996).

The nature of sediment discharge to Gulf of Mexico estuaries varies greatly longitudinally due to differences in drainage basin geology (from very low sediment discharge rates for the carbonate-rich, groundwater-dominated rivers of the western Florida coast to high sediment discharge rates for the sediment-rich, river-dominated middle Gulf of Mexico estuaries). Drainage basin size and land use (percent urban vs. agricultural) within each basin are important factors in differentiating the sediment load of Gulf of Mexico rivers. Climatology is also an important factor for water and sediment discharge from the smaller basins within the Gulf. Comparing precipitation and evaporation rates for Gulf drainage basins reveals a net evaporative deficit for western Gulf of Mexico (Texas) estuaries and southern Florida systems (Chapter 2, this volume). Most of the middle Gulf of Mexico estuaries (Louisiana, Mississippi, Alabama) have a large net surplus of precipitation, and local runoff dominates discharge. For example, Sabine Lake and Mobile Bay have among the highest fresh-water inflow per unit of estuarine volume in the United States (Ward 1980). In contrast, Denes and Bayley (1983) demonstrated that for a major river such as the Atchafalaya, local runoff is a trivial part of the discharge.

Temporal Variability in Sediment Discharge. The ratio of sediment discharge during flood events (due to storms, continuous heavy rains, and hurricanes) to that of average low-discharge conditions is likely to be high for many of the rivers that drain into the Gulf of Mexico. For example, during the period 1963–1987, sediment discharge for the Red River (which empties into the Atchafalaya River) varied by a factor of over 4 between the lowest and highest reported values (Mossa 1996). In contrast, sediment discharge for the lower Mississippi River varied by a factor of 7 between the lowest and highest values (Mossa 1996). The Mississippi River's greatest measured flow (64,800 m^3 s^{-1}) is less than seven times the lowest annual recorded flow (9660 m^3 s^{-1}) because of its large, diverse drainage basin (Gunter 1979). A 56-month record of sediment and water discharge for the Mississippi River at Tarbert Landing (Fig. 3-3) demonstrates the intra-annual and inter-annual variability of this system.

Historical Changes in Sediment Discharge. The suspended sediment load of Gulf of Mexico rivers has changed considerably over the past 200 yr due

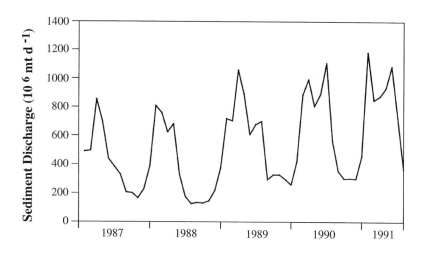

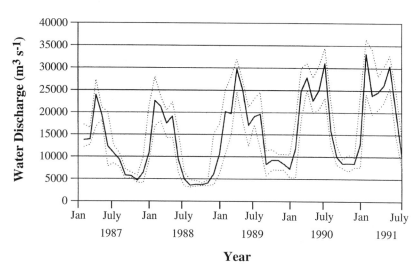

FIG. 3-3. *A 56-mo record of monthly sediment and water discharge for the Mississippi River at Tarbert Landing.*

to several anthropogenic factors. These include (1) deforestation and agriculture; (2) changes in the land management strategy; and (3) construction of dams, diversions, and levees. This anthropogenic alteration of sediment supply is a general phenomenon for Gulf of Mexico estuaries and is exemplified by the Mississippi River system. The suspended sediment load carried by the Mississippi River to the Gulf of Mexico has decreased by approximately one-half over the past 200 yr (Meade et al. 1990). Most of this decrease has occurred relatively recently (1950s) as a result of the construction of dams and reservoirs on the Missouri and Arkansas Rivers (Meade and Parker 1985). The source and composition of particulate material discharged by the Mississippi River have also changed during the past century. Increased sediment discharge by the Ohio River (resulting from increased erosion from poor land management) has partially compensated for the decreased sediment discharge from the western tributaries (Keown et al. 1986).

Historically, major floods (and efforts to control them) have played a major role in sediment and water discharges from the Mississippi River. As early as March 1543, a flood was noted by De La Vega (DeSoto's chronicler) that lasted for about 80 d, with water covering approximately 150 km of the valley on either side. Other major floods were noted in 1664 and in 1717–1718 in New Orleans. The four largest floods in recent times were in 1828, 1882, 1927, and 1973 (~50 yr apart). Records of annual peak discharges have been recorded intermittently since 1897 and continuously since 1927. The largest flood during that period was in May 1927 (64,800 m^3 s^{-1}), causing extensive damage along the lower Mississippi River Valley.

Levee construction along the Mississippi began ca. 1717 with the founding of New Orleans and increased gradually until the 1880s, when the building rate was accelerated. In the 1930s the whole river system was greatly extended and stabilized after the disastrous 1927 flood. Subsequent to that flood period, the U.S. Army Corps of Engineers began an extensive flood-control program. Channelization (preventing overbank flooding and eliminating 185 km of river channel), levee construction, and construction of storage reservoirs on the tributaries were three flood-control measures taken. Storage reservoirs enable the tributaries to store more water during high flow, thus decreasing the peak discharge. As a result, there has been a noticeable decrease in peak discharges since the 1920s. Everett (1971) noted that between 1890 and 1970, the number of annual peaks decreased and the accumulated storage of water in the basin increased. Prior to the 1940s, an average of four water discharge maxima were noted each year (taking the mean of 10-yr intervals). Since 1950, there has been about one discharge maximum per year. Water storage has increased dramatically (by a factor of 6) since the early 1950s.

The amount of suspended sediment carried by the river depends upon streamflow, turbulence, particle size, and water temperature. Suspended sediment concentrations in the Mississippi River decreased from >900 mg liter^{-1} in early 1950s to <200 mg liter^{-1} in the 1990s, a decrease of over 400%. Everett (1971) noted that in 1964 the suspended sediment concentration in the Baton

Rouge area of the Mississippi exceeded 1500 mg liter^{-1} during peak sediment discharge. During the period 1949–1964, suspended sediment concentrations were <100 mg liter^{-1} about 7% of the time and >1000 mg liter^{-1} 8% of the time in the Baton Rouge area of the river. Kesel (1988) also noted an historical change in the amount of sand in suspension (comparing Quinn's 1894 data to present data) for the Mississippi River. The percentage of sand in suspension at Tarbert Landing during 1950–1983 was about one-half of that noted during 1879–1893. At Belle Chasse, this difference was approximately 72% less.

Bed material composition also exhibited an historical change (1933 vs. the post-1965 period), decreasing in grain size from 0.2 to 0.1 mm in the zone 320–160 km upriver and from 0.17 to 0.05 mm in the zone 0–160 km upriver (Keown et al. 1986). The grain size of bed sediment exhibits a distinct fining below the Old River control structure (Keown et al. 1986; Nordin and Queen 1992). At Tarbert Landing, the bed sediments are 1% silty clay and 96% fine sand; at Belle Chasse they are 70% silty clay and 30% fine sand; and at Venice they are 80% silty clay and 20% fine sand (Nordin and Queen 1992). At intermediate locations, the riverbed may be as much as 98% silty clay.

Sediment Storage in River Systems. Approximately 90% of the sediment currently being eroded off the land surface of the conterminous United States is being stored somewhere between the river and the sea (Meade et al. 1990). Understanding the delivery of sediments to estuaries must include some understanding of this storage (and subsequent remobilization) process. The Mississippi River system exemplifies this phenomenon.

Seasonal sediment storage and remobilization is well documented for the lower Mississippi River (Everett 1971; Wells 1980; Meade and Parker 1985). Because of the extensive levee system, there are no inputs or outlets in the lower 500-km stretch of the river from Tarbert Landing (just below the control structure and diversion of the Atchafalya River) to Venice (near the mouth of the river). This fact has permitted researchers to make upriver-downriver comparisons and calculate mass balances for particulate materials in this portion of the river. Streamflow in the lower Mississippi River is affected by a diversion of approximately 30% of the total flow to the Atchafalaya River through the Old River control structure just above Tarbert Landing. The target for the percentage of Mississippi River flow that is diverted to the Atchafalaya is 30%; however, data show that this fraction fluctuates from year to year—ranging from 15% to 29% (Mossa 1996). Below Tarbert Landing, the river channel's cross-sectional area increases with downstream distance, generally resulting in decreased stream velocities. The mean depth of flow varies from ~6 to 60 m.

The hydrologic control of sediment storage and remobilization has been characterized, and a Lagrangian transport model has been formulated that reasonably reproduces the rate and timing of storage and remobilization in the lower Mississippi River (Demas and Curwick 1988; Mossa 1996). At low discharge there are large differences in cross-sectional area between upriver and downriver locations, and the resultant deepening of the channel results in decreased velocities downstream, promoting sediment deposition. During higher discharge stages the differences in upriver-downriver cross-sectional area are much smaller, and increased current velocities are accompanied by a steeper downstream surface-water gradient resulting in bottom shear stresses that surpass threshold values for resuspension (Mossa 1996). Wells (1980) examined suspended sediment concentrations at Head of Passes (near the river mouth) and St. Francisville (700 km upstream) and noted that a threshold discharge rate of about 17,000 m^3 s^{-1} separated trends of increasing and decreasing sediment concentrations downstream. At a water discharge rate of 10,000 m^3 s^{-1}, 43% and 25% of the material in suspension at St. Francisville remained in suspension at New Orleans and Venice, respectively. From these data, Wells (1980) inferred a net storage (sedimentation) of sediments in the lower river during low discharge periods. Kesel (1988) also observed a trend toward lower suspended sediment concentrations at New Orleans relative to Tarbert Landing and attributed these differences (35% decrease from 1930 to 1952; 48% decrease from 1953 to 1962; 42% decrease from 1963 to 1982) to deposition in the lower river. This study used data collected with different measuring techniques; therefore, a direct comparison is not possible. Kesel (1988) extended the database of the above studies to include the periods from 1851 to 1852 (Humphreys and Abbott 1861), from 1879 to 1893 (Quinn 1894), and from 1930 to the present (New Orleans Water and Sewage Board).

Using a quantitative transport model, periods of net storage and remobilization of sediments in the lower Mississippi can be predicted based on daily water discharge rates measured at Tarbert Landing (Demas and Curwick 1988). Significant deposition and net storage of river sediments occurs below Tarbert Landing during periods of river discharge less than 14,000 m^3 s^{-1}. For example, Demas and Curwick (1988) observed that suspended sediment concentrations decreased by 80% downriver of Tarbert Landing during flows between 7300 and 7600 m^3 s^{-1} and that 60,000 to 90,000 mt d^{-1} of sediment was deposited as riverbed sediments. Significant re-suspension and net re-mobilization of sediments occurred during periods of river discharge greater than 20,000 m^3 s^{-1}. A 30% increase in suspended sediment concentrations downriver of Tarbert Landing was noted during flows between 25,000 and 28,000 m^3 s^{-1} as a result of sediment re-suspension rates of 140,000 to 180,000 mt d^{-1} (Demas and Curwick 1988). Periods of sediment storage are typically 4–8 mo in the lower Mississippi River, as illustrated for the years 1987–1992 (Fig. 3-4). The months in which conditions for sediment storage are met (after discharge drops below 14,000 m^3 s^{-1} and before it exceeds 20,000 m^3 s^{-1}) are indicated in Fig. 3-4 by shaded bars. From these studies, it appears that temporary storage and re-mobilization of sediments in the lower river can greatly influence the flux of sediments that reaches the estuary.

Estuarine Turbidity

Shallow river-dominated estuarine environments are usually very turbid. High turbidity conditions are often observed during high flow conditions as a result

Mississippi River @ Tarbert Landing

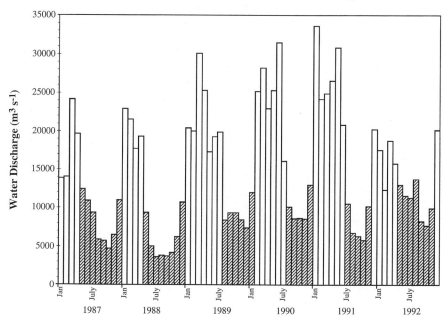

FIG. 3-4. *Monthly water discharge rates for the Mississippi River at Tarbert Landing (1987–1992). Shaded bars indicate months for which net sediment storage is predicted in the lower river, based on a Lagrangian transport model.*

of large riverine inputs of sediments. In shallow water estuaries, tides and winds can easily re-suspend surficial sediments, resulting in high turbidity. It has been reported that river discharge and wind conditions control the distribution of suspended sediments in Atchafalaya Bay (Caffrey and Day 1986; Denes and Caffrey 1988). During the winter season, cold front passages are the most important forcing function for sediment dynamics in central Gulf of Mexico estuaries (Roberts et al. 1988; Moeller et al. 1993). Salinity and turbidity fields in the Sabine Lake estuary (which has the greatest fresh-water inflow to bay volume ratio among all the Texas estuaries; Ward 1980) is strongly influenced by wind forcing, especially when associated with frontal passages (Orlando et al. 1993). McKee et al. (1995) reported weekly mapping of turbidity and salinity in Terrebonne Bay for a 2-yr period. Their observations revealed that most sediments were redistributed within the bay and not exported out. There was no evidence of sediments from coastal waters being imported into Terrebonne Bay or of bay sediments being exported to the inner shelf (McKee et al. 1995; see also Fig. 3-5). In contrast, tidal forcing was found to be important in controlling the turbidity plume in outer Mobile Bay and the export of sediments to the inner shelf (Stumpf et al. 1993; see also Fig. 3-6).

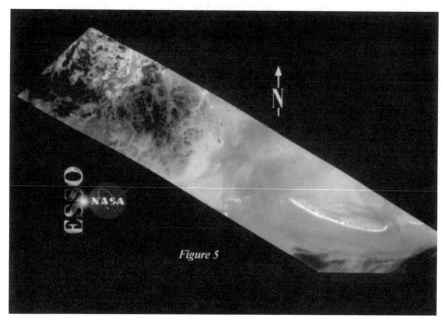

FIG. 3-5. Surface reflectance image (CAMS) of Terrebonne Bay indicating the distribution and relative concentrations of suspended sediments.

Turbidity fields within estuaries influence many other processes. For example, water column turbidity reduces irradiance and inhibits primary productivity. Randall and Day (1987) observed that turbidity controls primary productivity in Atchafalaya Bay. Since turbidity patterns in shallow Gulf of Mexico estuaries are spatially and temporally variable, the best way to understand variations in turbidity is via remote sensing (Miller et al. 1994). Remotely sensed turbidity/reflectance data from Terrebonne and Mobile Bays are given in Figs. 3-5 and 3-6, respectively. These images demonstrate the complex spatial variability in suspended sediment concentrations that is typical of Gulf of Mexico estuaries. The turbidity field for Mobile Bay demonstrates the export of particulate material from the estuary. In Terrebonne Bay, locally isolated high turbidity values reflect the importance of re-suspension with little or no export to the coastal ocean.

Benthic Sediment Processes

Temporary versus permanent storage of particulate material in bottom sediments, and the difference in time scales that separate these processes, are important factors in understanding sedimentological and biogeochemical processes. The development of radiochemical techniques has provided tools for identifying areas of active sedimentation, for examining rates of sedimentary

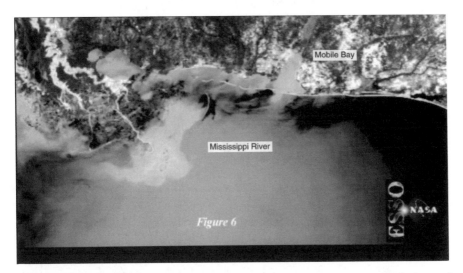

FIG. 3-6. *Surface reflectance image (AVHRR) of Mobile Bay indicating the distribution and relative concentrations of suspended sediments.*

processes, and for establishing geochronologies within bedded sediments. The terms used to describe sedimentation on various time scales (deposition, accumulation) can be distinguished quantitatively when radiochemical techniques are used. We define deposition as temporary emplacement of particulate material on a sediment surface during a specified period of time. Accumulation is the sum of deposition and removal over a longer time scale (McKee et al. 1983).

Sediment Deposition. Information on recent sediment deposition can be obtained using short-lived, naturally occurring, and anthropogenic radionuclides. In sediment-dominated estuaries, particle-reactive radiochemical tracers such as ^{234}Th and ^{7}Be can often be used to provide insight into monthly and seasonal deposition rates and patterns (Canuel et al. 1990). In sediment-starved or dynamic estuaries, these tracers can be utilized to delineate spatial and temporal patterns of net deposition. For example, in the Galveston Bay system, tracer studies indicate no deposition in areas of the middle bay. Only in isolated pockets (such as East Bay, Offats Bayou, and selected places in Trinity Delta) is there ongoing deposition. In another sediment-starved estuary, Terrebonne Bay, the distributions of ^{234}Th$_{xs}$ and ^{210}Pb$_{xs}$ suggest that the dominant time scale for sediment deposition and re-distribution is greater than 1 yr and less than 10 yr (Fig. 3-7; McKee et al. 1995). Observed variations in seasonal inventories of ^{234}Th were attributed to the influence of local runoff events that control the supply of sediments to Terrebonne Bay (McKee et al. 1995). The radioisotope data suggest that sediments are temporarily stored

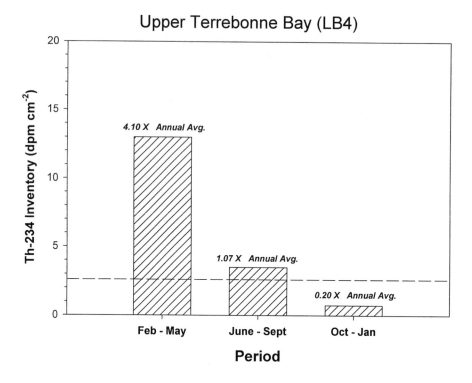

Upper Terrebonne Bay (LB4)

Th-234 Inventory (dpm cm^{-2})

4.10 X   Annual Avg.

1.07 X   Annual Avg.

0.20 X   Annual Avg.

Feb - May   June - Sept   Oct - Jan

Period

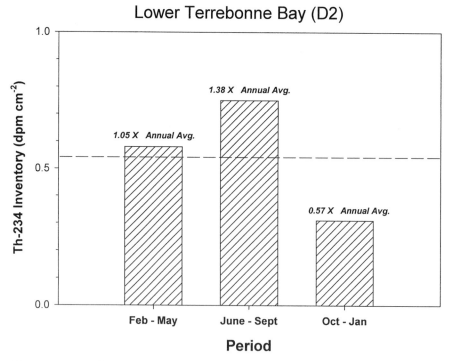

Lower Terrebonne Bay (D2)

Th-234 Inventory (dpm cm^{-2})

1.38 X   Annual Avg.

1.05 X   Annual Avg.

0.57 X   Annual Avg.

Feb - May   June - Sept   Oct - Jan

Period

FIG. 3-7. Seasonal [234]Th sediment inventories for two sites in Terrebonne Bay indicating the relative importance of sediment deposition (from McKee et al. 1995).

in Terrebonne Bay for a few years because typical storm and tidal energy is not sufficient to disperse sediments out of the bay. This study suggested that only catastrophic events such as hurricanes (which impact the area approximately every 5 yr) provide sufficient energy to export sediments.

Analyses of ^{7}Be and excess ^{234}Th can be utilized to obtain information on sediment deposition rates in estuaries directly influenced by riverine sediment discharge. Much of the sediments from the Atchafalaya River that enter Fourleague Bay are deposited in the upper bay during high flow regimes but are redistributed to the lower bay after winter storms begin (Day et al. 1995). This was revealed by a comparison of deposition rates determined by excess ^{234}Th profiles in upper and lower Fourleague Bay (Fig. 3-8). The seasonal rates of sediment deposition are high relative to long-term accumulation rates (determined by ^{210}Pb; Fig. 3-8), indicating that some sediments are stored for a short time before being exported from the bay.

Sediment Accumulation. Yearly to decadal sediment accumulation rates are commonly determined using a set of particle-reactive radionuclides (^{210}Pb, ^{239,240}Pu, and ^{137}Cs). Based on the ^{239,240}Pu profiles for four sediment cores in the Sabine-Neches Estuary, Ravichandran et al. (1995) estimated the sedimentation rate to vary from 0.4 to 0.5 cm yr^{-1}. The sedimentation rate in the Trinity Delta area of Galveston Bay is 0.25 cm yr^{-1} (Baskaran et al. 1997). High rates of sediment accumulation (2 to 5 cm yr^{-1}) were reported in Fourleague Bay of the Atchafalaya system (Fig. 3-8; Day et al. 1995). In contrast, lower rates of sediment accumulation (0.24 to 1.69 cm yr^{-1}) were reported in the sediment-starved estuaries: Terrebonne and Barataria Bays (McKee et al. 1995; Booth and McKee submitted).

In addition to providing information on sediment accumulation rates, these tracers provide some insight into sedimentary processes such as sediment focusing or sediment erosion by comparing the input of these nuclides to their inventories in the sediment. For example, the inputs of ^{210}Pb, ^{239,240}Pu, and ^{137}Cs to any given site are fairly well known. The ratio of the measured inventory of these nuclides to their fluxes is a measure of sediment focusing/accumulation/erosion.

In the Sabine-Neches estuarine system, fewer than 30% of the sediment coring locations were found to have undergone net sedimentation over a time scale of 100 yr (Ravichandran, 1994). In many estuarine areas of the Trinity-San Jacinto estuary there is no excess ^{210}Pb, suggesting that there has been no net sedimentation in the last 100 yr (Baskaran et al. submitted).

In areas where physical and/or biological mixing is dominant, a combination of radiochemical tracers (e.g., ^{234}Th, ^{7}Be, ^{210}Pb, ^{239,240}Pu, and ^{137}Cs isotopes) can be used to understand the mixing of surface sediments. ^{239,240}Pu and ^{137}Cs are by-products of atmospheric testing of nuclear weapons, introduced to the environment around 1952, with maximum fallout in 1963. In areas where physical and/or biological mixing is negligible, the peak fallout retained in the sedimentary record corresponds to the year 1963. Any deviation from the

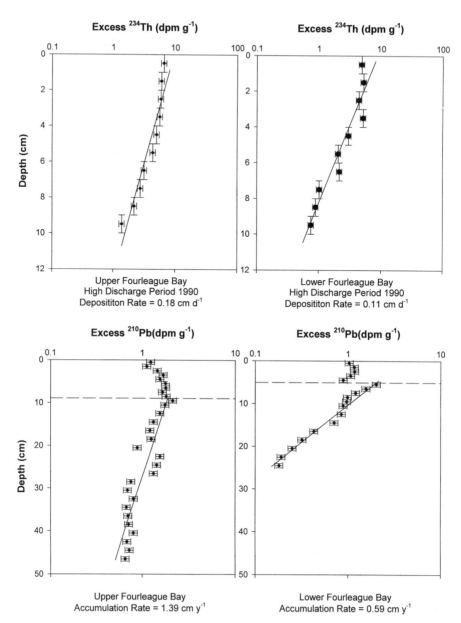

FIG. 3-8. Sediment profiles of ^{234}Th and ^{210}Pb for upper and lower Fourleague Bay. These profiles were used to quantify the rates of sediment deposition and accumulation (from Day et al. 1995).

expected atmospheric fallout in sediments is attributed to mixing, and thus the rates of mixing and sedimentation can be delineated (e.g., Ravichandran et al. 1995). In one sediment core collected from Galveston Bay, a very broad ^{239,240}Pu peak was attributed to the dominance of biological mixing with low sedimentation (Baskaran et al. submitted). In contrast, the vertical distributions of ^{239,240}Pu concentrations in sediment cores from the Sabine-Neches estuary suggest that Pu fallout is clearly tracked in sediments (Fig. 3-9; Ravichandran et al. 1995) and is preserved as a sharp sub-surface peak. The minimum and maximum sedimentation rates calculated using the ^{210}Pb profiles from these cores bracket the rate predicted based on a sedimentation-mixing model utilizing the ^{239,240}Pu data. A similar observation of a distinctive ^{239,240}Pu profile was observed in a sediment core collected from the Mississippi Delta (Baskaran et al. submitted). In this case, mixing was negligible. The apparent sedimentation rate (uncorrected for mixing) derived from the ^{210}Pb profile was identical to the net sedimentation rate derived from the ^{239,240}Pu profile.

Sediment accumulation rates within an estuary are generally not constant due to spatial variability in sediment inputs and redistribution. In the Sabine-Neches Estuary, two sampling sites a few meters apart yielded very different net sedimentation values: zero (no excess ^{210}Pb in the surficial layer) and 0.5 cm yr^{-1}. Thus, a large number of cores is required before the average sedimentation rate for an estuary can be determined.

Sediment Export Onto/Across the Continental Shelf. It is difficult to quantify directly or indirectly the magnitude of sediment export from estuaries. Direct measurements of sediment exported from estuaries are difficult because inlets are often large and the deployment of an adequate number of instruments is prohibitively expensive. The net difference between sediment inputs to an estuary and sediment accumulation within an estuary is an indirect measure of the sediment exported from an estuary. However, high temporal and spatial variability in input and accumulation rates within an estuary makes sediment budgeting of this kind very difficult.

Remote sensing techniques provide an alternative method for evaluating the temporal and spatial patterns of sediment concentrations and potentially for quantifying sediment export rates (Miller et al. 1994). Qualitatively, it has been demonstrated that the discharge of fine-grained sediments from Atchafalaya Bay onto the continental shelf is taking place in the form of a mud stream (Wells and Roberts 1980; Adams et al. 1982). Stumpf et al. (1993) demonstrated the export of sediment from Mobile Bay using remote sensing techniques. The use of such techniques to quantify sediment export from estuaries is an area of active research that holds great promise for the future.

In some cases, changes in the export flux of sediment from an estuary to the shelf can be evaluated via sediment geochronologies. For example, a log raft in the upper Atchafalaya began to grow between the 1500s and the late 1700s, blocking most of the river's flow (Comeaux 1970). The Atchafalaya was virtually bypassed until 1839, when raft removal was first attempted and

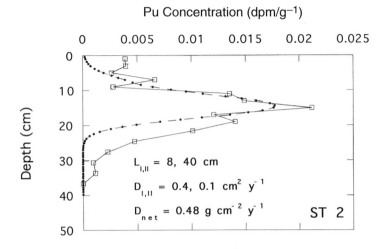

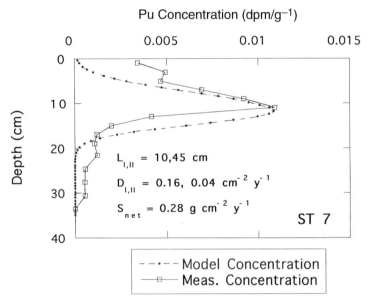

FIG. 3-9. *Sediment profiles of* ^{239,240}*Pu from the Sabine-Neches Estuary. The measured profiles are compared to profiles generated using a mixing-sedimentation model with parameter values indicated in Fig. 3-8 (from Ravichandran et al. 1995).*

flow increased (Gunter 1979). The Atchafalaya River carried approximately 10% of the Mississippi flow in 1858 (Elliot 1932). Removal of the raft was completed in 1861 and was followed by a dramatic growth in the Atchafalaya Delta. Eventually, the inland lakes filled with sediments and fine-grained materials escaped Atchafalaya Bay onto the adjacent inner shelf. This is demonstrated in a $^{210}Pb_{xs}$ profile from a station on the inner shelf adjacent to Atchafalaya Bay (Fig. 3-10; McKee et al. submitted). The profile exhibits a distinct break in slope and therefore in sedimentation rate. The increase in the sediment accumulation rate since the early 1970s suggests that this is when fine-grained sediments began to escape Atchafalaya Bay and accumulate on the inner shelf.

SUMMARY

The abundance of fine-grained sediments and the profound influence that these sediments have on estuarine processes is one of the salient features of Gulf of Mexico estuaries. However, little is known (quantitatively) about sedimentary processes in these estuarine environments. With the exception

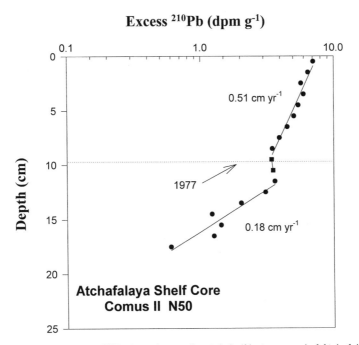

Excess ^{210}Pb (dpm g^{-1})

FIG. 3-10. Sediment profile of ^{210}Pb from the continental shelf just seaward of Atchafalaya Bay. The change in slope (and in the associated sedimentation rate) indicates that the sediment supply to this site increased significantly after ~1977 (from McKee et al. submitted).

of the Mississippi and Atchafalaya Rivers, riverine sediment inputs are poorly characterized for Gulf of Mexico estuaries. The large surface area and shallow water depths of Gulf of Mexico estuaries result in very dynamic sedimentary processes governed primarily by wind forcing. Sediment distributions in the water column and in bed sediments are highly variable (temporally and spatially) due to the dominance of wind-driven resuspension. Short-term (seasonal) depositional rates can be locally high (centimeters per month), but net (decadal) sediment accumulation rates are significantly lower. Winter storms and hurricanes provide sufficient energy to redistribute sediments within the estuarine system and may be a driving force in exporting sediments to the coastal ocean.

The major challenge to understanding sedimentary processes in these environments is to develop instrumentation and techniques adequate to measure processes occurring on short-duration, high-frequency time scales and in physically demanding conditions. Based on what is known from the sediment record, the sedimentary processes associated with episodic events (such as storms and hurricanes) and their impacts on estuaries are very important. However, little is presently known about these event-scale phenomena on a mechanistic level. Because such episodic events are unpredictable and hazardous for equipment and personnel, field observations during such events will continue to be a challenge.

REFERENCES

Adams, C. E., Jr., J. T. Wells, and J. M. Coleman. 1982. Sediment transport on the central Louisiana continental shelf: Implications for the developing Atchafalaya River Delta. Contrib. Mar. Sci. **25**: 133–148.

Baskaran, M., B. J. Presley, S. Asbill, P. H. Santschi, and R. Taylor. (Submitted). Reconstruction of historical contamination of trace metals in Mississippi River Delta, Tampa Bay and Galveston Bay sediments. Environ. Sci. Tech.

Booth, J. G., and B. A. McKee. (Submitted). Deposition and burial of particulate material and organic carbon in subaqueous bottom sediments of the Barataria Basin, LA. Geochim. Cosmocim. Acta.

Caffrey, J., and J. Day. 1986. Control of the variability of nutrients and suspended sediments in a Gulf Coast estuary by climatic forcing of spring river discharge. Estuaries **9**: 295–300.

Canuel, E., C. Martens, and L. Benninger. 1990. Seasonal variations in ^{7}Be activity in the sediments of Cape Lookout Bight, North Carolina. Geochim. Cosmochim. Acta **54**: 237–245.

Comeaux, M. L. 1970. The Atchafalaya River raft. Louisiana Studies. Winter, pp. 217–227.

Day, J., Jr, C. Madden, R. Twilley, R. Shaw, B. McKee, M. Dagg, D. Childers, R. Raynie, and L. Rouse. 1995. The influence of Atchafalaya River discharge on Fourleague Bay, Louisiana (USA), p. 151–160. *In:* K. Dyer and R. Orth [eds.], Changes in fluxes in estuaries. Olsen and Olsen.

Demas, C., and P. Curwick. 1988. Suspended sediment and associated chemical transport characteristics of the lower Mississippi River, Louisiana. Louisiana Department of Transportation, Water Resources Tech. Report 45.

Denes, T. A., and S. E. Bayley. 1983. Long-term rainfall and discharge in the Atchafalaya River basin, Louisiana. Louisiana Acad. Sci. **46:** 114–121.

——, and J. M. Caffrey. 1988. Changes in seasonal water transport in a Louisiana estuary, Fourleague Bay, Louisiana. Estuaries **11:** 184–191.

Elliot, D. C. 1932. The improvement of the lower Mississippi River for flood control and navigation. U.S. Waterways Experiment Station, U.S. Army Corps of Engineers, War Department.

Everett, D. 1971. Hydrologic and quality characteristics of the lower Mississippi River. Louisiana Department of Public Works Tech. Report 5.

Gunter, G. 1979. The annual flows of the Mississippi River. Gulf Res. Report **6:** 283–290.

Guy, H. P., and V. W. Norman. 1970. Field methods for measurement of fluvial sediment: U.S. Geological Survey techniques of water-resources investigations. book 3, chapter C2. U.S. Geological Survey.

Humphreys, A. A., and H. L. Abbott. 1861. Report upon the physics and hydraulics of the Mississippi River upon the protection of the alluvial region against overflow and upon the deepening of the mouths. J. B. Lippincott.

Jansen, P. P., L. Van Bendegom, J. Van Den Berg, M. DeVries, and A. Zanen. 1979. Principles of river engineering. Pitman.

Judson, S., and D. F. Ritter. 1964. Rates of regional denudation in the U.S. J. Geophys. Res. **6:** 3395–3401.

Keown, M. P., E. A. Dardeau, Jr., and E. M. Causey. 1986. Historic trends in the sediment flow regime of the Mississippi River. Water Resour. Res. **22:** 1555–1564.

Kesel, R. H. 1988. The decline in the suspended load of the lower Mississippi River and its influence on adjacent wetlands. Environ. Geol. Water Sci. **11:** 271–281.

McKee, B. A., J. G. Booth, and P. W. Swarzenski. (Submitted). Sediment deposition, redistribution and accumulation in the Mississippi River Bight. Cont. Shelf Res.

——, C. A. Nittrouer, and D. J. DeMaster. 1983. Concepts of sediment deposition and accumulation applied to the continental shelf near the mouth of the Yangtze River. Geol **11:** 631–633.

——, W. Wiseman, and M. Inoue. 1995. Salt water intrusion and sediment dynamics in a bar-built estuary: Terrebonne Bay, LA, p. 13–16. *In:* K. Dyer and R. Orth [eds.], Changes in fluxes in estuaries. Olsen and Olsen.

Meade, R. H. 1996. River-sediment inputs to major deltas, p. 63–83. *In:* J. D. Milliman and B. U. Haq [eds.], Sea-level rise and coastal subsidence. Kluwer.

——, and R. Parker. 1985. Sediment in rivers of the United States. *In:* National water summary 1984: U.S. Geological Survey Water Supply Paper 2275, p. 49–60.

——, and H. H. Stevens. 1990. Strategies and equipment for sampling suspended sediment and associated toxic chemicals in large rivers—with emphasis on the Mississippi River. Sci. Total Environ. **97/98:** 125–135.

——, T. Yuzyk, and T. Day. 1990. Movement and storage of sediment in rivers of the United States and Canada, p. 255–280. *In:* W. H. Riggs [ed.], The geology of North America. Geological Society of America.

Miller, R. L., J. F. Cruise, E. Otero, and J. M. Lopez. 1994. Monitoring suspended particulate matter in Puerto Rico: Field measurements and remote sensing. Wat. Res. Bull. **30:** 271–282.

Milliman, J. 1991. Flux and fate of fluvial sediment and water in coastal seas, p. 69–90. *In:* R. Mantoura, J. Martin, and R. Wollast [eds.], Ocean margin processes in global change. Wiley.

———, and R. H. Meade. 1983. World-wide delivery of river sediment to the oceans. J. Geol. **91:** 1–21.

Moeller, C. C., O. K. Huh, H. H. Roberts, L. E. Gumley, and W. P. Menzel. 1993. Response of Louisiana coastal environments to a cold front passage. J. Coast. Res. **9:** 434–447.

Mossa, J. 1996. Sediment dynamics in the lowermost Mississippi River. Eng. Geol. **45:** 457–479.

National Oceanic and Atmospheric Administration (NOAA). 1990. Estuaries of the United States, vital statistics of a national resource base. Strategic Assessment Branch, Ocean Assessments Division.

Nordin, C. F., Jr., and J. P. Beverage. 1965. Sediment transport in the Rio Grande, New Mexico: U.S. Geological Survey Professional Paper 462-F, p. F1–F35.

———, C. Cranston, and A. Mejia. 1983. New technology for measuring water and suspended sediment discharge of large rivers, p. 1145–1158. *In:* Proceeding of the Second International Symposium in River Sedimentation. Water Resources and Electric Power Press.

———, and B. Queen. 1992. Particle size distributions of bed sediments along the thalweg of the Mississippi River, Cairo, Illinois, to Head of Passes, September 1989. U.S. Corps of Engineering Potamology Program Report 7.

Orlando, S. P., Jr., L. P. Rozas, G. H. Ward, and C. J. Klein. 1993. Salinity characteristics of Gulf of Mexico estuaries. National Oceanic and Atmospheric Administration, Office of Ocean Resources Conservation and Assessment.

Quinn, J. B. 1894. Chief of Engineers Report, in House of Representatives Executive Documents 1, Pt. 2, 53rd Congress 3rd Session, volume 2, part 3, p. 1345–1347.

Randall, J. M., and J. W. Day, Jr. 1987. Effects of river discharge and vertical circulation on aquatic primary production in a turbid Louisiana (USA) estuary. Neth. J. Sea Res. **21:** 231–242.

Ravichandran, M. 1994. Investigations on the sediment chronology and trace metal accumulation in Sabine-Neches estuary, Beaumont, Texas. M.S. thesis, Texas A&M University.

———, M. Baskaran, P. H. Santschi, and T. S. Bianchi. 1995. Geochronology of sediments in the Sabine-Neches estuary, Texas, U.S.A. Chemical Geol. **125:** 281–306.

Robbins, L. G. 1977. Suspended sediment and bed material studies on the lower Mississippi River: U.S. Army Engineer District, Vicksburg, Potamology Investigations Report 300–1.

Roberts, H. H., O. K. Huh, S. A. Hsu, L. J. Rouse, and D. Rickman. 1988. Impact of cold-front passages on geomorphic evolution and sediment dynamics of the complex Louisiana coast, p. 1950–1963. *In:* Coastal Sediments '87, Proceedings of a Specialty Conference, American Society of Civil Engineers.

Ryan, J. J., and H. G. Goodell. 1972. Marine geology and estuarine history of Mobile Bay, Alabama, part 1. Contemporary sediments. Geological Society of America, Memoir 133, p. 517–553.

Slade, R. M. 1992. Status of the Gulf of Mexico: Preliminary report on inflows from streams. U.S. Environmental Protection Agency Report 800-R-92-005.

Stumpf, R. P., G. Gelfenbaum, and J. R. Pennock. 1993. Wind and tidal forcing of a buoyant plume, Mobile Bay, Alabama. Cont. Shelf Res. **13:** 1281–1301.

Ward, G. H., Jr. 1980. Hydrography and circulation processes of Gulf estuaries, p. 183–215. *In:* P. Hamilton and K. B. MacDonald [eds.], Estuaries and wetland processes with emphasis on modeling. Plenum Press.

Wells, F. C. 1980. Hydrology and water quality of the lower Mississippi River: Louisiana Office of Public Works Tech. Report 21.

Wells, J. T., and H. H. Roberts. 1980. Fluid mud dynamics and shoreline stabilization: Louisiana chenier plain. Proceedings of the 17th International Coastal Engineering Conference. ASCE.

Wilson, A., and K. Iseri. 1969. River discharge to the sea from shores of the conterminous United States. *In:* Hydrologic investigation atlas HA-282. U.S. Geological Survey.

Yorke, T. H., and J. R. Ward. 1986. Accuracy of sediment discharge estimates, p. 4-49–4-59. *In:* Proceedings, federal interagency sedimentation conference, volume 1: Interagency Advisory Committee on Water Data, Subcommittee on Sedimentation.

Section II

Nutrient Dynamics

Chapter 4

Suspended Particulate and Dissolved Nutrient Loadings to Gulf of Mexico Estuaries

R. Eugene Turner and Nancy N. Rabalais

INTRODUCTION

The concentration and loading of suspended particulate material (SPM), and both organic and inorganic nutrient concentrations, are significant variables because of their effect on both the quality and quantity of estuarine producers. SPM, defined as material retained on a 0.45-μm filter, is significant in the formation and maintenance of emergent and submerged macrophyte communities that often are a limiting habitat for animals, including commercially important species. The colonization of estuarine shallow-water areas by emergent macrophytes is strongly influenced by estuarine hydrologic conditions (e.g., tidal inundation), which are altered as sedimentation occurs. Salt marsh plants in Gulf of Mexico estuaries, for example, are usually found only where water depths are less than 0.5 m. Once formed, these marshes must accumulate enough organic and inorganic material to compensate for the relative change in water level, whether it be from a rising sea surface or land subsidence (e.g.,

Biogeochemistry of Gulf of Mexico Estuaries, Edited by Thomas S. Bianchi, Jonathan R. Pennock, and Robert R. Twilley.
ISBN 0-471-16174-8 © 1999 John Wiley & Sons, Inc.

from geological subsidence, compaction, or oxidation). The concentration of SPM affects light penetration through the water column and thereby influences both phytoplankton growth and production rates and the ecology of submerged grassbeds.

The idea of "limiting nutrients" has been applied usefully to agricultural fields and estuaries. There are several definitions of what is being limited, including the growth rate of individual species or the whole community, biomass accumulation, and primary production rate (Howarth 1988). Laboratory cultures and field experiments have shown that the elementary composition of individual cells and the species composition can be altered by changing the growth medium. Individual cells may be growing near maximum growth rates at high or low nutrient levels. Typically, their growth rate is maximum when their $C:N:Si:P$ ratio is near the Redfield ratio of $106:16:15:1$ (Redfield 1958), which constrains biological systems at the cellular, ecosystem, and evolutionary levels. The ratios of anthropogenically influenced rivers of the world are moving in the direction of Redfield ratios (Redfield 1958; Justic' et al. 1995). When other nutrients are added or when the nutrient ratio changes, the species composition changes as new species outcompete the other species. Officer and Ryther (1980) suggested that as the atomic ratio of dissolved inorganic $Si:N$ approaches $1:1$, phytoplankton communities would change because diatom growth would become Si-limited. The decline in diatom abundance has causal consequences for higher-level predator-prey relationships, such as those of, zooplankton and fish. It is possible to change the species composition from benign to noxious without changing estuarine nitrogen loading. If the new species no longer include nitrogen-fixing cyanobacteria, then net production may decrease dramatically. Under these conditions, the nutrient loading ratios could control the species composition, but not the primary production rate or biomass accumulation.

Net ecosystem production, which is gross primary production less respiration, may also be limited by nutrients. However, ecosystem respiration is dependent not only on the quality and quantity of primary production, but also on temperature, oxygen availability, and allochthonous carbon inputs. This definition of nutrient limitation is less useful than others and is not in common use.

Most aquatic ecologists and aquatic engineers are interested in nutrient limitation of the potential rate of net primary production and biomass accumulation. Estuarine and marine primary growth and accumulation is, in general, understood to be strongly influenced by nutrient concentration or loading, especially from dissolved inorganic nitrogen but also by inorganic phosphorus and silicate (Hecky and Kilham 1988; Howarth 1988; Nixon 1992; Cloern 1996; Malone et al. 1996; Nixon et al. 1996). The increased loading of nutrients to coastal systems is continuing and widespread for a variety of reasons (Rosenberg 1985; Goldberg 1995; Cloern 1996) and is the apparent causal agent of several undesirable consequences, including noxious algal blooms, low bottom water oxygen, and fisheries losses (Rosenberg 1985; Shumway

1990; Smayda 1990; Diaz and Rosenberg 1995). Understanding nutrient loading amounts and changes is, in this context, both a basic and an applied issue for coastal scientists.

The loading of sediments and of dissolved and particulate nutrient elements to the Gulf estuaries is a function of fresh-water discharge, nutrient concentration, and estuarine area: volume and thus is expected to display great variation among and within the Gulf estuaries. Human activity has greatly altered nutrient cycling on land and water in the western Atlantic watersheds (Howarth et al. 1996), and most estuaries in the northern Gulf are experiencing increased nutrient loading (e.g., Turner and Rabalais 1991; Rabalais 1992). Nutrient loading is decreasing in some estuaries, however, as a result of management intervention. For example, phosphorus loading in Tampa Bay was reduced by improving the primary sewage treatment plant and by controlling runoff from fertilizer industry activities. The apparent consequences are improved water clarity, less planktonic biomass, and a return of the seagrass beds (Johansson and Lewis 1992).

VARIABILITY IN ESTUARINE NUTRIENT LOADING

The variability of nutrient and suspended sediment supplies for the Gulf estuaries is described here using data from an anonymous NOAA Coastal Ocean Program data report prepared for the Gulf of Mexico Program [Nutrient Enrichment Potential Watershed Assessment and Comparison (NEPWAC) System Gulf of Mexico Component, Ver. 0.5]. These data were compiled from interviews, water quality monitoring programs of various sources, land use statistics, and empirical relationships used to extrapolate from sparse data sets. The estimates of nutrient loading are for circa 1987. A few of these data were not used because the loading rate did not include the interior U.S. land use data (Mississippi and Atchafalaya Rivers) or the riverine system debouched into a very small bay (e.g., Brazos River). Deegan et al. (1986) is the source for the physiography of some Mexican estuaries that unfortunately seem to have no comparable and systematically collected flow and concentration data available. The estuaries are numbered according to the map shown in Fig. 4-1.

The area of estuarine wetland and water surface varies by 100-fold across the Gulf estuaries, but depth varies by only one order of magnitude (Fig. 4-2). The average estuarine depth is only 2.5 m, and estuaries are shallower in the western (e.g., Texas) third of the Gulf compared to the eastern (Florida) third of the Gulf (Table 4-1). The surface area of both wetland and open water habitats appears to be highest in the middle of the Gulf (Fig. 4-2). The average wetland area for 26 estuaries is 1654 km^2, which is about 60% larger than the water surface area (Table 4-1). These average numbers illustrate the potential ability of wetlands to transform and store materials in the flooding waters. In the Gulf, a wetland surface area greater than that of the open water

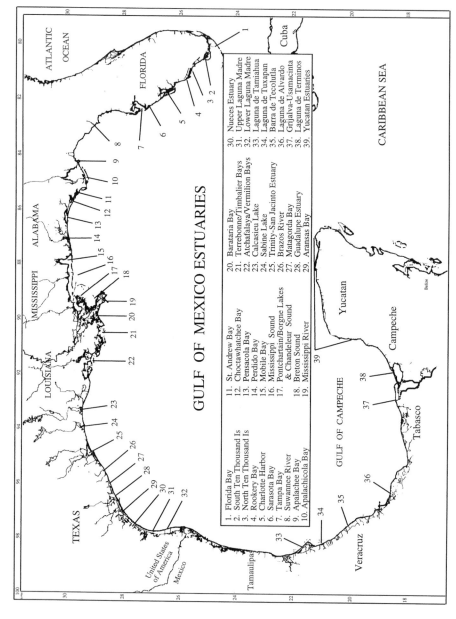

FIG. 4-1. Watersheds of the northern Gulf estuaries. The numbering system is consistent across all figures throughout this book.

GULF OF MEXICO ESTUARIES

1. Florida Bay
2. South Ten Thousand Is
3. North Ten Thousand Is
4. Rookery Bay
5. Charlotte Harbor
6. Sarasota Bay
7. Tampa Bay
8. Suwannee River
9. Apalachee Bay
10. Apalachicola Bay
11. St. Andrew Bay
12. Choctawhatchee Bay
13. Pensacola Bay
14. Perdido Bay
15. Mobile Bay
16. Mississippi Sound
17. Pontchartrain/Borgne Lakes & Chandeleur Sound
18. Breton Sound
19. Mississippi River
20. Barataria Bay
21. Terrebonne/Timbalier Bays
22. Atchafalaya/Vermilion Bays
23. Calcasieu Lake
24. Sabine Lake
25. Trinity-San Jacinto Estuary
26. Brazos River
27. Matagorda Bay
28. Guadalupe Estuary
29. Aransas Bay
30. Nueces Estuary
31. Upper Laguna Madre
32. Lower Laguna Madre
33. Laguna de Tamiahua
34. Laguna de Tuxpan
35. Barra de Tecolutla
36. Laguna de Alvardo
37. Grijalva-Usamacinta
38. Laguna de Terminos
39. Yucatan Estuaries

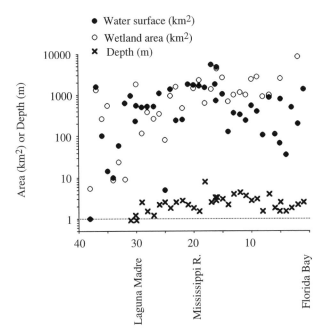

FIG. 4-2. *Average water depth (m), surface area (km²), and wetland area (km²) for northern Gulf estuaries.*

habitat is flooded aboveground, with all the possibilities for sedimentation, re-suspension, interaction with clays, and organic films in and around the plants. The ability of wetlands to affect nutrient cycling is well documented and is hardly homogeneous among wetlands (e.g., Nixon 1979; Mitsch and Gosselink 1993; Kadlec and Knight 1996). Less well known are the subtle interactions of plants on sedimentation rates and organic compounds. Salt marsh plants trap 50% of the suspended load that falls out of suspension during a tidal cycle (Stumpf 1983), for example, and organic material on clays is partially held within the clay matrix and is degradable, and the amounts are predictable across particle size (Keil et al. 1994; Mayer 1994). It is important to note that a significant tidal volume also enters the anaerobic belowground layer (about 50% in a Louisiana salt marsh; Swenson and Turner 1987). This flooding (average tidal range about 0.5 m in the Gulf) is strongly influenced by the wind regime. Seasonal changes in the physical regime therefore strongly influence the wetland-water interactions.

Land use in the watershed draining into the estuaries has a higher percentage of agricultural land (mean = 26.5%; $n = 35$) than urban land (mean = 5.6%; $n = 35$; Fig. 4-3). The eastern Gulf estuaries (e.g., in Florida) are more urbanized than the western Gulf estuaries (e.g., in Texas), but the latter estuaries have a larger proportion of the drainage basin as agricultural land.

TABLE 4-1. Statistical Description of Data for Estuaries Used in the Figures

Parameter	n	Unit	Mean	$\pm$ 1 S.E.
Wetland area	26	km^2	1654	346
Water surface	35	km^2	945.7	200
Depth	35	m	2.5	0.22
Percent urban	35	%	5.6	0.92
Percent agricultural	35	%	26.5	2.6
Suspended load to estuary	29	g m^{-2} water surface yr^{-1}	1492	381
Suspended load from watershed	29	g m^{-2} land yr^{-1}	89.5	18.2
Nitrogen fertilizer sales in watershed	29	mM N m^{-2} yr^{-1}	92.1	11.0
Nitrogen loading rom estuarine watershed as discharge	29	mM N m^{-2} yr^{-1}	17.4	3.0
Nitrogen loading to estuary surface	32	mM N m^{-2} yr^{-1}	1589	930
Phosphorus fertilizer sales in watershed	29	mM P m^{-2} yr^{-1}	16.5	2.2
Phosphorus loading from estuarine watershed as discharge	29	mM P m^{-2} yr^{-1}	1.41	0.42
Phosphorus loading to estuary surface	32	mM P m^{-2} yr^{-1}	43.0	13.6
N:P molar ratio in fertilizer	29	mol N:mol P	7.7	1.5
N:P molar ratio in river discharge	29	mol N:mol P	48.5	12.1

Note: The total number of estuaries is 32; some estuaries were subdivided into sub-basins, and some had missing or incomplete data.

The suspended load from the estuarine watershed and loading to the estuarine water surface varies 100 and 1000 times, respectively, from east (lower) to west (higher) (Fig. 4-4). In general, as the loading into the estuary increases, the loading per water surface increases. The annual mean suspended load to the estuary is 89 g SPM m^{-2} of watershed, which, when focused into the smaller area estuarine water surface, is an average annual loading of 1492 g SPM m^{-2}.

The observed variability in land use and loading rates may be useful for testing hypotheses about the interrelationships between landscape morphology and biological functions, including the effects of land use changes. For example, estuaries with higher suspended sediment loads might be expected to fill in more quickly than those with a lower suspended sediment loads. If all other factors among estuaries remain similar, then bay infilling should also lead to more emergent vegetation and less water surface as sediment loading increases. This appears to be the case because the wetland area:water surface ratio does increase with increasing loading rates (Fig. 4-5). [Note: The data from three estuaries were not used because the comparison seemed unreasonable: the Brazos estuary (a large regional river emptying into almost directly into the Gulf), the Mississippi River (draining an area 10 times larger than the next largest estuary and extending across the continental shelf), and South

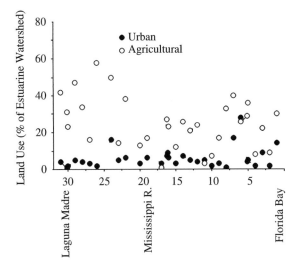

FIG. 4-3. *Land use in the estuarine watershed of northern Gulf estuaries: percentage of urban area and percentage of agricultural area.*

Ten Thousand Island, Florida (having indeterminate boundaries).] It is especially interesting that the Florida estuaries show a higher wetland : water area ratio than others throughout the Gulf, even though many of them have lost a large fraction of their historic wetlands to dredge-and-fill development. Several hypotheses could explain the differences shown in Fig. 4-5. First, the vegetation might be responding to the calcareous substrate differently, because of physiological or rooting effects, resulting in more successful wetland growth. Compared to the rest of the Gulf, there is less *Spartina alterniflora* and *S. patens* in Florida but more *Juncus roeamerianus* and mangroves (i.e., Turner 1976). Second, the dynamics of open water formation may be different because the calcareous substrate is harder and is eroding at relatively slower rates than the more western estuaries, which are mostly sedimentary deposits.

DISSOLVED INORGANIC NUTRIENT LOADING

The loading of dissolved nitrogen and phosphorus to estuaries is of significance to water quality, especially the amount and quality of phytoplankton, as has been demonstrated many times (e.g., Nixon et al. 1996). Undesirable kinds of phytoplankton may harm oysters, alter food web structure, or cause fish kills, and too much phytoplankton growth may cause low oxygen conditions when oxygen consumption during their decomposition is faster than re-aeration rates. The usual nutrients of concern are nitrogen and phosphorus.

Nitrogen sources from human activities, particularly fertilizer application and concentrated animal husbandry, wastewater treatment plants and industry,

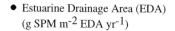

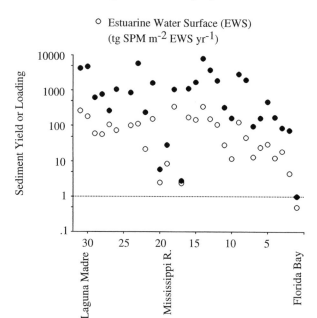

FIG. 4-4. *Sediment yield (g SPM m⁻² estuarine watershed area yr⁻¹) and loading (tg SPM m⁻² estuarine surface area yr⁻¹) in northern Gulf estuaries.*

and atmospheric deposition, have been implicated in many other systems as significant contributors to increased nutrient loading. The effects of wastewater treatment plants and industrial sources or "point sources" are easier to estimate than those of diffuse loading from land use practices or from the airshed. Among Gulf estuaries, the loading from point sources ranges from insignificant to a dominant percentage of the total watershed nutrient nitrogen and phosphorus budget (Fig. 4-6). Point source loadings of phosphorus have a greater concentration of phosphorus than of nitrogen, hence the non-linear relationship shown in Fig. 4-6.

The nitrogen fertilizer application rate (mean = 92.1 mmol N m⁻² watershed yr⁻¹) is generally higher than the discharge rate of nitrogen (mean = 17.4 mmol N m⁻² watershed yr⁻¹) for the Gulf estuaries (Fig. 4-7; Table 4-1), implying conservation within the watershed or denitrification. The nitrogen loading rate per area of estuarine surface is highest in the western Gulf estuaries (mean = 1589 mmol N m⁻² water surface yr⁻¹).

The phosphorus fertilizer application rate shows a lower value, but a similar pattern, among the estuaries compared to nitrogen loading rates (mean = 16.5 mmol P m⁻² watershed yr⁻¹). The discharge rate from the estuarine watershed is lower than the application rate (mean = 1.41 mmol P m⁻² water-

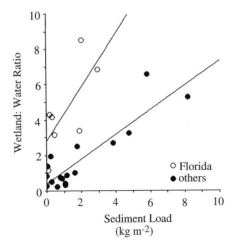

FIG. 4-5. *The relationship between the wetland : water ratio (areal basis) and the sediment loading to the estuaries in the northern Gulf of Mexico. The Mississippi, Atchafalaya, and Brazos Rivers and the south Ten Thousand Islands estuaries are excluded. The Florida estuaries are those south and east of Pensacola.*

shed yr^{-1}) (Fig. 4-7; Table 4-1). The loading rate per area of estuarine surface is highest in the western estuaries (mean $= 43.0$ mmol P m^{-2} water surface yr^{-1}) and lowest in the south Florida estuaries.

The amounts of N fertilizer applied is certainly sufficient to account for the nitrogen in the discharge waters (Fig. 4-7). Less than 20%, on average, of the fertilizer applied could account for all the nitrogen draining into the

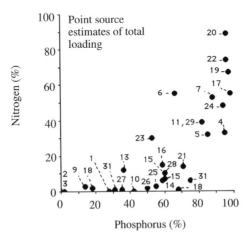

FIG. 4-6. *The percent loading from point sources (sum of wastewater treatment plants and industrial sources) of the total nitrogen and phosphorus loading for northern Gulf estuaries. The numbers correspond to the estuaries in Fig. 4-1.*

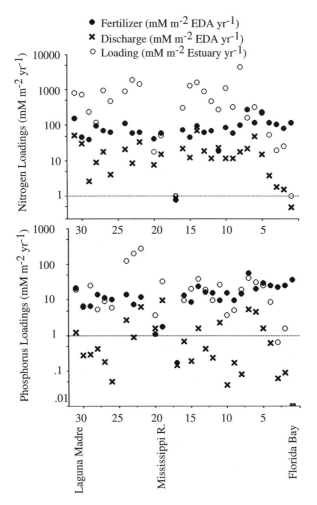

FIG. 4-7. Nitrogen and phosphorus loading for northern Gulf estuaries: fertilizer sales per area of estuarine drainage area (EDA); loading from streams per EDA; loading from streams per area of estuarine water surface.

estuaries. The same is true for phosphorus fertilizer applications, but with some exceptions. First, there are many centrally located Gulf estuaries that have phosphorus loading rates equal or nearly equal to phosphorus application rates. Second, there is more variability in P loading among estuaries than for nitrogen loading. If the leakage amounts from agriculture to water for nitrogen and phosphorus are distinctly different, as they appear to be, then perhaps the role of urban sources is important. The highly urbanized areas tend to have higher phosphorus loading rates (Fig. 4-8), consistent with the curve shown in Fig. 4-6, which suggests that urban areas (point sources) have discharges enriched in phosphorus.

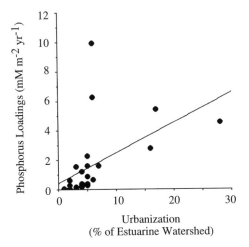

FIG. 4-8. *The relationship between the discharge of phosphorus (mM P m⁻² yr⁻¹) and the percentage of the watershed that is urban by watershed area for northern Gulf estuaries. Data for the Mississippi and Atchafalaya River deltas and the Everglades/Big Cypress areas are excluded.*

These data may be used to compare the loading rates in the Gulf estuaries with those in other systems, which is done in Fig. 4-9. The data for estuaries outside the Gulf are adapted from Boynton et al. (1995) and are for tributaries of Chesapeake Bay, south San Francisco Bay, the Baltic Sea, Tokyo Bay, Narragansett Bay, Seto Inland Sea, and Buzzards Bay. The data generally cluster around an N : P molar ratio of 16 : 1, but the spread is enormous, ranging over five orders of magnitude. Some of the Gulf estuaries appear to have a

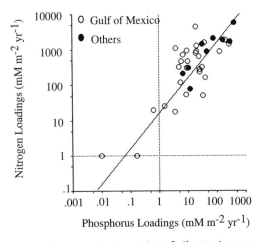

FIG. 4-9. *Nitrogen and phosphorus loading for northern Gulf estuaries compared to other estuaries in the United States. The straight line at a diagonal is the Redfield ratio of 16 : 1 :: N : P.*

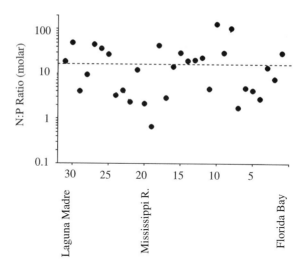

FIG. 4-10. *N:P loading ratios in discharges from the estuarine watershed area for northern Gulf estuaries. The dashed straight line is the Redfield ratio of 16:1::N:P.*

lower loading rate than the more heavily documented systems outside the Gulf. This figure does not fully illustrate the variability in the molar ratio of N:P loading rates (for total nitrogen and phosphorus). This ratio varies over two orders of magnitude across the northern Gulf estuaries (Fig. 4-10).

Fertilizer applications may be a significant source of loading into streams, as others have found elsewhere. The N:P ratio in the fertilizer applied, however, is as much as seven times higher than in the water delivered to estuaries (Table 4-1). Nitrogen is probably lost from denitrification before it reaches the estuaries. Sewer plant operations, for example, typically have a low N:P ratio (Kadlec and Knight 1996).

RELATIONSHIP BETWEEN FRESH-WATER TURNOVER TIME AND EUTROPHICATION

The NOAA Coastal Ocean Program data (Anon., no date) may also be used to examine the relationship between nitrogen loading, fresh-water turnover, and the development of hypoxia in the estuaries of the northern Gulf of Mexico. Two data groups are included in Fig. 4-11: estuaries with hypoxia and those without hypoxia [as determined by a literature review and by reports of field data (Rabalais et al. 1985)]. Two results are apparent but not proven. First, there are no reported hypoxic events in which the nitrogen loading rate is below a threshold value of 200 mmol N m^{-2} water surface yr^{-1}. Second, estuaries with a history of hypoxia tend to be those with higher loading rates

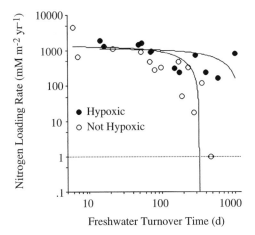

FIG. 4-11. *Nitrogen loading per estuarine surface area and turnover time of fresh water in northern Gulf estuaries. Estuaries with evidence of hypoxia (from surveys) are distinguished from those estuaries without a record of hypoxia. A regression of the untransformed data is shown.*

compared to other estuaries with the same fresh-water flushing rate. The spread between the two data groups becomes similar as fresh-water turnover time decreases (higher flushing) compared to estuaries with longer flushing times. This plot supports the hypothesis that nutrient loading increases the likelihood of eutrophication and either hypoxic or anoxic water formation. If true, this suggests that eutrophication in Gulf estuaries is at least a partially manageable phenomenon, particularly in estuaries where flushing takes longer, compared to estuaries with fast fresh-water turnover times. Hypoxia has been observed in Mobile Bay since before the 1900s, and there is no evidence of a significant increase in nitrogen loading up to the 1990s. Mobile Bay has a relatively high nutrient loading rate and a quick fresh-water turnover time. In this system, changes in nutrient loading rates seem to have had a smaller effect on the presence/absence of hypoxia than changes in fresh-water turnover time. In contrast, the estuaries with long fresh-water turnover times, like those in south Texas, are very susceptible to changes in oxygen concentration as nutrient loading fluctuates or increases.

SOURCES OF NUTRIENT ENRICHMENT TO THE GULF OF MEXICO

Nitrogen and phosphorus loading rates into Gulf estuaries from land were derived by Turner (in press), who combined literature estimates of riverine and atmospheric inputs to subsections of the continental shelf and the deeper part of the Gulf. This approach considered only sources from the land margins and the atmosphere. The nitrogen and phosphorus loading rates per shoreline source (Giga mol N or P yr^{-1}) were computed using the literature review of

Howarth et al. (1996) and were proportioned according to the watershed size. The watershed area emptying into the Gulf from Mexico could not be subdivided because suitable watershed sizes could not be found. Atmospheric sources for nitrogen and phosphorus were estimated from Cornell et al. (1995) and were calculated based on the open water area of the open Gulf, including the continental shelf. The elemental yield (Giga mol N or P yr^{-1}) is dependent on land use, precipitation, and many other factors that complicate a precise allocation of variation among basins. The general trends, however, are obvious.

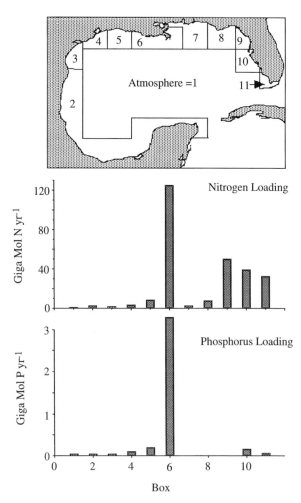

FIG. 4-12. *Nitrogen and phosphorus loading into the Gulf estuaries from land, circa 1987. Units are Giga mol N or P yr^{-1}. The dominant source is from Box 6, which received inputs from the Mississippi and Atchafalaya Rivers. The estimate for atmospheric loading is for the entire Gulf water surface.*

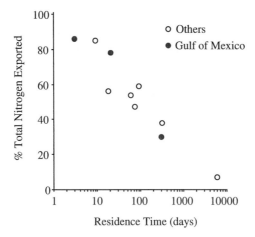

FIG. 4-13. *The relationship between estuarine residence time and the percentage of nitrogen exported from the estuary to the offshore waters. The data are from Nixon et al. (1996).*

The dominant land source of nitrogen and phosphorus to the Gulf estuaries is from the Mississippi and Atchafalaya Rivers (Fig. 4-12). The riverine load of nitrogen and phosphorus has doubled in the past few decades (Turner and Rabalais 1991), leading to enhanced diatom production (Turner and Rabalais 1994) and an increased incidence and severity of hypoxic water formation (Sen Gupta et al. 1996; a recent review is found in Rabalais et al. 1996). Compared to these sources, the atmosphere is an insignificant nitrogen and phosphorus source to the open Gulf. Of course, not all of this nutrient loading makes its way to the offshore coastal waters, and the amount from upwelling at the shelf break is not known. The relationship between the percentage of nitrogen exported to coastal waters and the estuarine residence time is shown in Fig. 4-13 for three Gulf estuaries and others (e.g., Adriatic, Baltic, Chesapeake Bay; adapted from data in Nixon et al. 1996). The water residence time is indirectly related to the nitrogen retention rate within the estuary. Denitrification and deposition are the likely causes of these losses. Note that data for the Gulf estuaries are no different from those for the other estuaries plotted. It is interesting that a 50% decline in nitrogen export is seen with a 100-fold increase in residence time. This means that a doubling in nitrogen loading will have a proportionally greater impact on the amount of nitrogen retained than a halving in residence time (resulting, for example, from weather cycles).

SUMMARY

The concentration of nutrients, the loading of suspended particulate materials, and the loading of nutrients to the Gulf estuaries vary by over three orders of magnitude among the 35 estuaries or sub-basins examined. The nitrogen and phosphorus fertilizer application rate in these estuaries is generally higher than the discharge rate of these same nutrients, and the N : P loading rate (for total nitrogen and phosphorus on a molar basis) varies by over two orders of magnitude across the northern Gulf estuaries. The N : P ratio in the fertilizer applied is much higher than in the water delivered to the estuaries, implying that nitrogen is lost or phosphorus is gained in some undetermined quantities. In general, less than 20% of the fertilizer applied could account for all the nitrogen or phosphorus appearing in water entering into the estuaries. Urban areas, which generally have a low N : P ratio, may be a more important source of phosphorus than of nitrogen to the estuary.

No hypoxic events are reported where the nitrogen loading rate is below a threshold value of 200 mmol N m^{-2} water surface yr^{-1}, and estuaries with a history of hypoxia tend to be those with the higher loading rates compared to other estuaries with the same fresh-water flushing rate. These results support the hypothesis that nutrient loading increases the likelihood of eutrophication and hypoxic or anoxic water formation, and is therefore a partially manageable phenomenon in estuaries where flushing takes longer compared to estuaries with fast turnover times.

The great potential ability of wetlands to store and transform materials within the estuary is apparent in considering that the mean wetland area for all the estuaries examined is almost equal to the mean area of the estuarine water surface. These estuaries have more emergent vegetation and less water surface as sediment loading increases. Although Gulf estuaries readily separate into two groups based on their wetland : water ratios, sediment load, and presence or absence of calcareous substrate, a causal relationship between the three is not proven.

Almost all geomorphic and nutrient flux parameters examined here varied by over two or three orders of magnitude. The estuaries of the Gulf have a range of external and internal factors that often overlap with those of estuaries elsewhere. Some Gulf estuaries are almost pristine, and others are as altered by human activities as those anywhere else; yet population expansion is, by all accounts, going to be in southern and coastal regions of the United States. With the exception of the Mississippi and Atchafalaya River estuaries, the Gulf estuaries have relatively low fresh-water turnover rates and tidal ranges compared to other U.S. estuaries. This situation is ripe for a comparative study that includes a management component. This effort requires a geographically broad and synergistic integration of social and biogeochemical studies that is somewhat rare in the estuarine sciences. The relatively lower nutrient loading of these estuaries, however, offers some hope that there is still time to apply acquired and yet-to-be-gained new knowledge before the experience of the more polluted northeastern estuaries is shared with all Gulf estuaries.

ACKNOWLEDGMENTS

We thank T. Bianchi, J. Pennock, and L. R. Pomeroy for comments on an earlier draft of the manuscript. The writing of this chapter was supported by the Louisiana Sea Grant College Program.

REFERENCES

Anonymous. No date. Nutrient Enrichment Potential Watershed Assessment and Comparison (NEPWAC) System Gulf of Mexico Component, Ver. 0.5. NOAA/ EPA Team on Near Coastal Waters. Strategic Assessment Branch, Ocean Assessments Division, National Ocean Service.

Boynton, W. R., J. H. Garber, K. Summers, and W. M. Kemp. 1995. Inputs, transformations, and transport of nitrogen and phosphorus in Chesapeake Bay and selected tributaries. Estuaries **18:** 285–314.

Cloern, J. E. 1996. Phytoplankton bloom dynamics in coastal ecosystems: A review with some general lessons from sustained investigation of San Francisco Bay, California. Rev. Geophysics **34:** 127–168.

Cornell, S., A. Rendell, and T. Jickells. 1995. Atmospheric inputs of dissolved organic nitrogen to the oceans. Nature **376:** 243–246.

Deegan, L. A., J. W. Day, Jr., J. G. Gosselink, A. Yañez-Arancibia, G. Soberón Chávez, and P. Sánchez-Gil. 1986. Relationships among physical characteristics, vegetation distribution and fisheries yield in Gulf of Mexico estuaries, p. 83–100. *In:* D. A. Wolfe [ed.], Estuarine variability. Academic Press.

Diaz, R. J., and R. Rosenberg. 1995. Marine benthic hypoxia: A review of its ecological effects and the behavioral responses of benthic macrofauna. Oceanogr. Mar. Biol. Ann. Rev. **33:** 245–303.

Elser, J. J., D. R. Dobberfuhl, N. A. Mackay, and J. H. Schampel. 1996. Organism size, life history, and N:P stoichiometry. BioScience **46:** 674–684.

Goldberg, E. D. 1995. Emerging problems in the coastal zone for the twenty-first century. Mar. Pollution Bull. **31:** 152–158.

Hecky, R. E., and P. Kilham. 1988. Nutrient limitation of phytoplankton in freshwater and marine environments: A review of recent evidence on the effects of enrichment. Limnol. Oceanogr. **33:** 796–822.

Howarth, R. W. 1988. Nutrient limitation of net primary productivity in marine ecosystems. Ann. Rev. Ecol. Syst. **19:** 89–110.

———, G. Billen, D. Swaney, A. Townsend, N. Jaworski, K. Lajtha, J. A. Downing, R. Elmgren, N. Caraco, T. Jordan, F. Berendse, J. Freney, V. Kudeyarov, P. Murdoch, and Z. Zhao-Liang. 1996. Regional nitrogen budgets and riverine N & P fluxes for the drainages to the North Atlantic Ocean: Natural and human influences. Biogeochemistry **35:** 75–139.

Johansson, J. O. R., and R. R. Lewis III. 1992. Recent improvements of water quality and biological indicators in Hillsborough Bay, a highly impacted subdivision of Tampa Bay, Florida, USA, p. 1191–1215. *In:* R. W. Vollenweider, R. Marchetti, and R. Viviani [eds.], Marine coastal eutrophication: The response of marine transitional systems to human impact: Problems and perspectives for restoration. Elsevier.

Justic', D., N. N. Rabalais, and R. E. Turner. 1995. Stoichiometric nutrient balance and origin of coastal eutrophication. Mar. Poll. Bull. **30:** 41–46.

Kadlec, R. H., and R. L. Knight. 1996. Treatment wetlands. Lewis Publishers.

Keil, R. G., D. B. Montlucon, R. G. Prahl, and J. I. Hedges. 1994. Sorptive preservation of labile organic mater in marine sediments. Nature **370:** 549–552.

Malone, T. C., D. J. Conley, T. R. Fisher, P. M. Gilbert, and L. W. Harding. 1996. Scales of nutrient-limited phytoplankton productivity in Chesapeake Bay. Estuaries **19:** 371–385.

Mayer, L. M. 1994. Surface area control of organic carbon accumulation in continental shelf sediments. Geochem. Cosmochim. Acta **58:** 1271–1284.

Mitsch, W. J., and J. G. Gosselink. 1993. Wetlands, 2nd. ed. Van Nostrand Reinhold.

Nixon, S. W. 1979. Between coastal marshes and coastal waters—a review of twenty years of speculation and research on the role of salt marshes in estuarine productivity and water chemistry, p. 437–525, *In:* P. Hamilton and K. B. Macdonald [eds.], Estuarine and wetland processes. Plenum Press.

———. 1992. Coastal marine eutrophication: A definition, social causes, and future concerns. Ophelia **41:** 199–219.

———. 1997. Prehistoric nutrient inputs and productivity in Narragansett Bay. Estuaries **20:** 253–261.

———, J. W. Ammerman, L. P. Atkinson, V. M. Berounsky, G. Billen, W. C. Boicourt, W. R. Boynton, T. M. Church, D. M. DiToro, R. Elmgren, J. H. Garber, A. E. Giblin, R. A. Jahnke, N. J. P. Owens, M. E. Q. Pilson, and S. P. Seitzinger. 1996. The fate of nitrogen and phosphorus at the land-sea margin of the North Atlantic Ocean. Biogeochemistry **35:** 141–180.

Officer, C. B., and J. H. Ryther. 1980. The possible importance of silicon in marine eutrophication. Mar. Ecol. Progr. Ser. **3:** 83–91.

Rabalais, N. N. 1992. An updated summary of status and trends in indicators of nutrient enrichment in the Gulf of Mexico. Report to Gulf of Mexico Program, Nutrient Enrichment Subcommittee. Pub. No. EPA/800-R-92-004, U.S. Environmental Protection Agency, Office of Water, Gulf of Mexico Program.

———, M. J. Dagg, and D. F. Boesch. 1985. A Gulf of Mexico review of oxygen depletion and eutrophication in estuarine and coastal waters. Final report to the U.S. Department of Commerce, NOAA National Ocean Service, Office of Oceanography and Marine Assessment, Ocean Assessments Division.

———, R. E. Turner, D. Justic', Q. Dortch, W. J. Wiseman, Jr., and B. K. Sen Gupta. 1996. Nutrient changes in the Mississippi River and system responses on the adjacent continental shelf. Estuaries **19:** 386–407.

Redfield, A. C. 1958. The biological control of chemical factors in the environment. Am. Sci. **46:** 205–221.

Rosenberg, R. 1985. Eutrophication—the future marine coastal nuisance? Mar. Poll. Bull. **16:** 227–231.

Sen Gupta, B. K., R. E. Turner, and N. N. Rabalais. 1996. Seasonal oxygen depletion in continental-shelf waters of Louisiana: Historical record of benthic foraminifers. Geology **24:** 227–230.

Shumway, S. E. 1990. A review of the effects of algal blooms on shellfish and aquaculture. J. World Aquaculture Soc. **21:** 65–104.

Smayda, T. J. 1990. Novel and nuisance phytoplankton blooms in the sea: Evidence for global epidemic, p. 29–40, *In:* E. Granneli, B. Sundstrom, R. Edler, and D. M. Anderson [eds.], Toxic marine phytoplankton. Elsevier.

Stumpf, R. P. 1983. The process of sedimentation on the surface of a salt marsh. Estuar. Coast. Shelf Sci. **17:** 495–508.

Swenson, E. M., and R. E. Turner. 1987. Spoil banks: Effects on a coastal marsh water level regime. Estuar. Coast. Shelf Sci. **24:** 599–609.

Turner, R. E. 1976. Geographic variations in salt marsh macrophyte production: A review. Contr. Mar. Sci. **20:** 47–68.

———. 1998. Inputs and outputs of the Gulf of Mexico. (In press). *In:* K. Sherman [ed.], The Gulf of Mexico, a large marine ecosystem. Blackwell Science.

———, and N. N. Rabalais. 1991. Changes in Mississippi river water quality this century: Implications for coastal food webs. BioScience **41:** 140–147.

———. 1994. Coastal eutrophication near the Mississippi River delta. Nature **368:** 619–621.

Chapter 5

Nutrient Behavior and Phytoplankton Production in Gulf of Mexico Estuaries

Jonathan R. Pennock, Joseph N. Boyer, Jorge A. Herrera-Silveira, Richard L. Iverson, Terry E. Whitledge, Behzad Mortazavi, and Francisco A. Comin

INTRODUCTION

Estuaries serve as important sites for the biogeochemical processing of terrestrially derived nutrients as they are carried from the land to the sea. In general, the rate of nutrient loading to an estuary has been found to be positively related to the level of primary, secondary, and fisheries production that is observed in the system (Nixon 1980, 1982, 1992; Nixon et al. 1986). However, significant variability in the nutrient-production relationship often occurs as a result of specific physical and hydrodynamic characteristics, residence time, and the source and form of nutrient inputs in individual estuarine ecosystems (Pennock et al. 1994b).

Nutrient input to estuaries occurs in both particulate and dissolved forms and may be both inorganic and organic. Particulate-bound terrestrial nutrients—both organic and inorganic—are relatively refractory (Benner et al. 1992) and may be readily sequestered within estuaries through direct sedimen-

Biogeochemistry of Gulf of Mexico Estuaries, Edited by Thomas S. Bianchi, Jonathan R. Pennock, and Robert R. Twilley.
ISBN 0-471-16174-8 © 1999 John Wiley & Sons, Inc.

tation (Chapter 3 this volume) or indirectly by sedimentation after flocculation. Once in the sediments, these nutrients are generally slow to react as they undergo sediment diagenesis and regeneration (Cowan et al. 1996; Chapter 6 this volume) and burial (Chapter 3 this volume). In contrast, dissolved inorganic nutrients such as nitrate (NO_3), nitrite (NO_2), ammonium (NH_4), soluble reactive phosphorus (PO_4), and soluble reactive silicate (SiO_2) are very reactive, often undergoing rapid biological processing within estuarine ecosystems.

Nutrient behavior is an important indicator of biogeochemical processing of nutrients within estuarine ecosystems. Previous studies have often found inorganic nutrient concentrations to display a non-conservative behavior with respect to salinity or other conservative property (Sharp et al. 1982; Pennock 1987; Fisher et al. 1988) indicating removal or production within an estuary. In large estuaries with long residence times, non-conservative behavior may be observed year round, indicating retention of nutrients within the system. In many cases, strong non-conservative nutrient behavior is associated with phytoplankton bloom events, as has been seen for Chesapeake Bay (Fisher et al. 1988; Conley and Malone 1992), Delaware Bay (Pennock 1987), San Francisco Bay (Peterson et al. 1985; Cloern 1996), and the Loire River estuary (Meybeck et al. 1988). Overall, phytoplankton production often serves as the major biogeochemical pathway for sequestration of inorganic nutrients and regulation of estuarine nutrient behavior.

Estuarine phytoplankton processes have been found to be regulated primarily by nutrient availability, temperature, and light in temperature aquatic ecosystems (Boynton et al. 1982; Valiela 1995). In cool-temperate estuaries and near-coastal waters, nitrogen has historically been found to be the nutrient limiting to phytoplankton production and biomass accumulation (Ryther and Dunstan 1971; Boynton et al. 1982; Paasche and Kristiansen 1982; Dortch and Postel 1989). However, recent studies have frequently found different nutrients [dissolved inorganic nitrogen (DIN), PO_4, and SiO_2] to be limiting to phytoplankton production and biomass in the same estuary at different periods of the year or in different regions of the estuary (D'Elia et al. 1986; Fisher et al. 1992; Pennock and Sharp 1994).

Temperature is also important in controlling seasonal productivity patterns in some estuaries (Williams and Murdoch 1966), and, as a result, cool-temperate estuaries generally show low production (<0.1 gC m^{-2} d^{-1}) during the winter and a production maximum during the summer (Boynton et al. 1982). These same systems, however, generally have a biomass maximum that occurs during the late winter and spring, when water temperature is less than 12°C (Boynton et al. 1982; Pennock and Sharp 1986; Fisher et al. 1992). These phytoplankton bloom events result in rapid processing of water-column nutrients and can have both positive and negative effects on the ecology of estuarine ecosystems, depending on whether the material moves into the food web or is removed to the sediments where microbial processes dominate (Pennock et al. 1994b; Cloern 1996).

Finally, numerous investigators have found light to be an important regulator of estuarine phytoplankton dynamics. The presence of an estuarine turbidity maximum is often an important regulator of the spatial distribution of phytoplankton growth (Cloern et al. 1985; Pennock 1985; Fisher et al. 1988; Pennock and Sharp 1994). Similarly, increase in light availability as a result of flow-induced stratification was shown to be important to the initiation of the winter-spring bloom in Delaware Bay (Pennock 1985).

Estuaries of the Gulf of Mexico are characterized by a diverse range of hydrodynamic characteristics (Chapter 1 this volume), freshwater discharge and residence times (Chapter 2 this volume), and nutrient loading rates (Chapter 4 this volume). Nutrient enrichment of these systems is at least partially responsible for the rich fisheries found in this region (Deegan et al. 1986), but at the same time can result in the formation of extensive areas of hypoxia and anoxia (May 1973; Schroeder and Wiseman 1988; Rabalais et al. 1994). As such, Gulf estuaries provide an excellent opportunity to examine estuarine nutrient biogeochemistry and primary production processes across a range of physical, chemical, and biological conditions that is as great as that found in any region of the world.

Research on nutrient biogeochemistry and phytoplankton processes in warm-temperate and sub-tropical estuaries is currently under-represented in the scientific literature compared with better-studied cool-temperate estuaries, on which much of our current understanding is based. Numerous recent and ongoing studies of Gulf of Mexico estuaries provide important new insights into how these ecosystems function and to how they may respond to human perturbations. In this chapter, we combine a review of historical nutrient data with "case studies" that incorporate previously unpublished data from several systems, including Florida Bay, Florida (a carbonate system with low freshwater input), Apalachicola Bay, Florida (a river-dominated and micro-tidal estuary), Mobile Bay, Alabama (a river-dominated estuary with a seasonally dominant coastal plume), the Nueces River estuary, Texas (an event-driven river system with low average discharge), and the Celestun, Chelem, Dzilam, and Rio Lagartos lagoons of the northern Yucatan, Mexico (a series of freshwater spring-fed lagoons with strong seasonal and regional variations in freshwater input). These systems span the range of geomorphological and climatological conditions found in Gulf of Mexico estuaries (Table 5-1) and provide a background from which to compare nutrient behavior and phytoplankton processes with better-studied cool-temperate systems.

COASTAL EMBAYMENTS AND ESTUARIES OF FLORIDA

Estuaries along the west coast of Florida—Florida Bay, Charlotte Harbor, Sarasota Bay, and Tampa Bay (Fig. 5-1)—are generally characterized by low fresh-water discharge and significant seagrass cover (Zieman et al. 1989). The differences between Florida Bay and Tampa Bay illustrate the impact of

TABLE 5-1. Physical Characteristics of Gulf Estuaries Examined in Case Studies

Site	Area (km^2)	Volume (10^6 m^3)	Mean Depth (m)	Maximum Length (km)	Maximum Width (km)
Florida Bay	1800	3200	2.0	70.0	40.0
Tampa Bay	896	3490	3.9	61.1	16.1
Apalachicola Bay	593	1600	2.7	12.9	33.8
Mobile Bay	1060	3200	3.0	48.5	40.0
Fourleague Bay	56	72.8	1.3	16.0	3.5
Nueces River Estuary	538	1290	2.4	28.2	12.9
Celestun Lagoon	28	33	1.2	20.7	2.1
Chelem Lagoon	13.6	16.3	1.0	14.7	1.8
Dzilam Lagoon	9.4	11.2	0.8	12.9	1.6
Rio Lagartos Lagoon	96	76.8	0.8	80.0	1.5

watershed geomorphology and land use on downstream ecosystem function and development. Both are lagoons with carbonate sediments, but because of the long-term loading of phosphorus from mining activities, Tampa Bay has become a nitrogen-limited system, whereas Florida Bay, with very low phosphorus inputs from the Everglades, remains phosphorus limited. Both estuaries have a history of seagrass losses but while the effects are similar, the causes are very different.

Florida Bay

Florida Bay (Fig. 5-2) is a wedge-shaped, shallow estuary located off the southern tip of the Florida peninsula, bounded by the Everglades to the north, the Pleistocene reef of the Florida Keys to the south, and open to the Gulf of Mexico along its western margin. The sediments of Florida Bay are composed mostly of biogenic carbonate muds that have formed shallow mud banks subdividing the Bay, thereby restricting water movement between basins and attenuating both tidal range and current. Florida Bay has a sub-tropical savanna climate characterized by a dry season from November to April and a wet season from May to October in which ~80% of the precipitation occurs.

Recent ecological changes in this region have brought attention to the sensitivity of the ecosystem to disturbances, including the invasion of exotic species in the fresh-water wetlands and uplands (Bodle et al. 1994), periods of prolonged hypersalinity of coastal embayments (Fourqurean et al. 1993), a poorly understood seagrass die-off (Robblee et al. 1991), sponge mortality events (Butler et al. 1995), and elevated phytoplankton abundance (Phlips and Badylak 1996). The seagrass die-off did not follow the typical cultural eutrophication model wherein elevated nutrients fuel increased phytoplankton abundance followed by increased shading, epiphytization, and senescence. Other factors that have been explored are sulfide toxicity (Carlson et al. 1994), over-development of seagrass beds, chronic hypersalinity of Florida Bay, and pathogens (Durako and Kuss 1994). The overall concern has been the potential

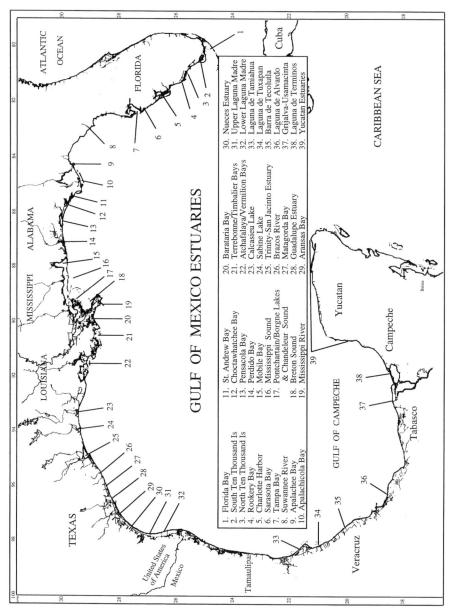

GULF OF MEXICO ESTUARIES

1. Florida Bay
2. South Ten Thousand Is
3. North Ten Thousand Is
4. Rookery Bay
5. Charlotte Harbor
6. Sarasota Bay
7. Tampa Bay
8. Suwannee River
9. Apalachee Bay
10. Apalachicola Bay

11. St. Andrew Bay
12. Choctawhatchee Bay
13. Pensacola Bay
14. Perdido Bay
15. Mobile Bay
16. Mississippi Sound
17. Pontchartrain/Borgne Lakes
    & Chandeleur Sound
18. Breton Sound
19. Mississippi River

20. Barataria Bay
21. Terrebonne/Timbalier Bays
22. Atchafalaya/Vermilion Bays
23. Calcasieu Lake
24. Sabine Lake
25. Trinity-San Jacinto Estuary
26. Brazos River
27. Matagorda Bay
28. Guadalupe Estuary
29. Aransas Bay

30. Nueces Estuary
31. Upper Laguna Madre
32. Lower Laguna Madre
33. Laguna de Tamiahua
34. Laguna de Tuxapan
35. Barra de Tecolutla
36. Laguna de Alvardo
37. Grijalva-Usamacinta
38. Laguna de Terminos
39. Yucatan Estuaries

FIG. 5-1. Major estuaries of the Gulf of Mexico.

113

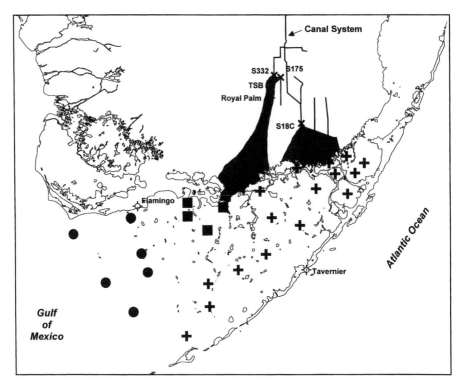

FIG. 5-2. *Sampling locations in Florida Bay, Florida. Samples were collected at 25 stations in the bay over a 6-year period between July 1989 and July 1995. The Eastern region (✚), Central Region (■), and Western Region (●) were defined using principal component analysis to represent environmental variables and subsequently delineating regions using cluster analysis. All data are for surface samples grouped by month and displayed using a box and whisker plot in which the middle horizontal line represents the median, the upper and lower lines of the boxes represent the 25th and 75th percentiles, and the ends of the whiskers are the 5th and 95th percentiles. The notch in the box is the 95% confidence interval of the median.*

for a shift in primary production from submergent macrophytes to phytoplankton in these ecosystems.

The ecosystem is generally phosphorus limited (Fourqurean et al. 1992; Phlips and Badylak 1996; Boyer et al. 1997) and has significant seagrass communities. Research carried out over the past decade has helped define three distinct zones in Florida Bay (Boyer et al. 1997): the eastern, central, and western regions (Fig. 5-2). The eastern region receives fresh water from the Everglades and is characterized by a wide range in salinity (0.2–45 g liter^{-1}; median = 28.6 g liter^{-1}), while the central region may be hypersaline for extended periods (9–63 g liter^{-1}; median = 32.8 g liter^{-1}), especially during the summer months, as a result of evaporation and restricted circulation (Fig. 5-3). The western region has more typical marine salinities (Table 5-2) and is dominated by interactions with the Gulf.

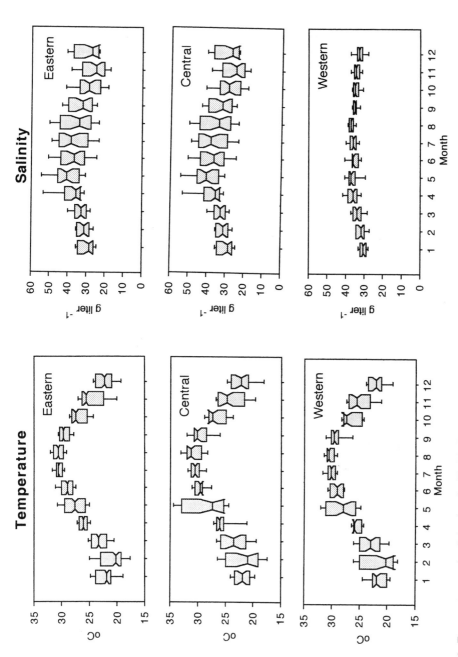

FIG. 5-3. Temperature and salinity data for Florida Bay presented by month for the Eastern, Central, and Western Regions. See Fig. 5-2 for sample locations and discussion of statistical presentation.

TABLE 5-2. Annual Mean and Range of Salinity and Nutrient Concentrations for Gulf Estuaries Examined in the Case Studies

Site	Salinity (g liter⁻¹)	NO₃ (μM)	NH₄ (μM)	PO₄ (μM)	SiO₂ (μM)
Florida Bay					
Eastern Region	29 (0.2–45)	0.62 (0.01–10.0)	3.19 (0.03–82.1)	0.03 (0.01–0.51)	16.0 (0.18–122)
Central Region	33 (9–63)	0.25 (0.01–5.70)	5.31 (0.01–120)	0.04 (0.01–0.84)	65.6 (0.06–109)
Western Region	35 (25–51)	0.12 (0.01–7.25)	0.12 (0.01–7.25)	0.03 (0.01–0.39)	18.6 (0.13–57.1)
Tampa Bay	— (20–35)	—	—	—	—
Apalachicola Bay	16.6 (0–37.3)	7.6 (1–35.7)	1.4 (0.1–7.1)	0.16 (0.1–0.62)	51.8 (3.0–136)
Mobile Bay					
Upper Bay	5.2 (0–19)	6.6 (0–40)	3.2 (0–17)	0.58 (0.05–1.4)	02.1 (13–131)
Mid Bay	9.1 (0–23)	4.1 (0–19)	2.4 (0–13)	0.46 (0.05–1.7)	59.7 (8–110)
Lower Bay	15.3 (0–32)	3.3 (0–18)	1.9 (0–12)	0.38 (0.05–1.6)	40.5 (2–104)
Fourleague Bay	—	42	2.5	0.8	—
Nueces River Estuary					
Nueces Bay	18.8 (3–30)	4.6 (0.5–23)	4.63 (0.5–19)	1.99 (0.6–4.6)	38.2 (10.5–60.8)
Corpus Christi Bay	32.1 (28.6–37.9)	3.2 (1.5–11.3)	7.82 (1–90)	0.82 (0.02–7.2)	54.4 (5–220)
Celestun	25 (5–37)	4.82 (0.9–1.5)	7.31 (1–38)	0.41 (0.1–6)	36.8 (4–50)
Chelem	36 (27–43)	1.89 (1–6)	2.5 (1–15)	1.45 (0.2–8.1)	163 (12–210)
Dzilam	31 (30–37)	7.07 (1–10)	8.5 (2–21)	1.55 (0.3–11)	24.7 (5–75)
Rio Lagartos	57 (20–100)	0.7 (0.2–5)			

Inorganic nutrient concentrations are generally low throughout Florida Bay, with median NO_3 and PO_4 concentrations being 0.40 and 0.03 μM, respectively (Fig. 5-4). Nitrate is highest in the eastern region (Table 5-2) as a result of loading from the Everglades watershed (Boyer and Jones in press). Ammonium concentrations in the estuary are much higher than NO_3 concentrations (Table 5-2), particularly in the central region (median = 5.3 μM) where concentrations during the winter have reached 120 μM. Nitrate and NH_4 display a distinct seasonal pattern, with the lowest concentrations during the summer period; however, PO_4 shows little seasonal periodicity, being kept at or below the kinetic threshold of most phytoplankton.

Organic nutrient concentrations are generally an order of magnitude greater than those of inorganic forms, rising to as high as 300 μM total organic nitrogen (TON) and 4 μM total phosphorus (TP) in the central region (Fig. 5-5). Median TP concentration in eastern Florida Bay (0.3 μM) is at or below the concentration of fresh-water inputs from the Everglades (Boyer and Jones in press). As PO_4 is so low, organic phosphorus becomes the important source of phosphorus in the system, as evidenced by the observation of high alkaline phosphatase activity (Boyer, unpublished data). The TON : TP ratio is consistently high and shows an increasing trend toward phosphorus limitation from west to east (medians of 58, 138, and 193 for the western, central, and eastern regions, respectively; Boyer, unpublished data; Fourqurean et al. 1993).

The median phytoplankton chlorophyll-a concentration in eastern Florida Bay is 0.76 μg liter^{-1} and shows no obvious seasonality (Fig. 5-5). Chlorophyll-a concentrations are higher in both central (median = 2.0 μg liter^{-1}) and western (median = 1.8 μg liter^{-1}) regions, with peaks of up to 11 μg liter^{-1}. Phytoplankton biomass displays significant variance, with highest concentrations during late summer-fall (Fig. 5-5). The chlorophyll-a concentration is significantly related to the TP concentration for all regions of the Bay ($p <$ 0.0001); NO_3 and NH_4 add little predictive power to the regression.

Phytoplankton primary production measurements (Tomas 1996) have shown a positive relationship between chlorophyll-a and production rates, with highest rates of 400 gC m^{-2} y^{-1} being observed in the central region and 200–300 gC m^{-2} yr^{-1} in the western region. The eastern region is the least productive, with an annual production of 75 gC m^{-2} as a result of chronic phosphorus limitation.

Boyer et al. (1997) have found significant trends in nutrient and phytoplankton dynamics over the past 7 years. Salinity, PO_4, and TP all have declined significantly throughout Florida Bay, while turbidity in all areas has increased. In the central region, both NO_3 and NH_4 increased dramatically over the period of record. The central region was also characterized by a large increase in chlorophyll-a during 1991–1994, which has since subsided to previous levels. Temporally, these increases correspond with the timing of the seagrass die-off, although no definitive cause and effect can be shown. In the eastern region chlorophyll-a has declined by 63%, but it has increased by 50% in the western region. As a result of these changes, phosphorus is even more likely to be

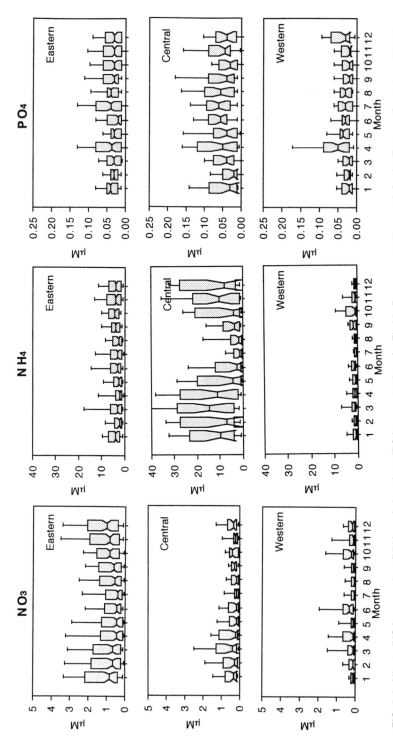

FIG. 5-4. Nitrate (NO_3), ammonium (NH_4), and phosphate (PO_4) concentrations for Florida Bay presented by month for the Eastern, Central, and Western Regions. See Fig. 5-2 for sample locations and discussion of statistical presentation.

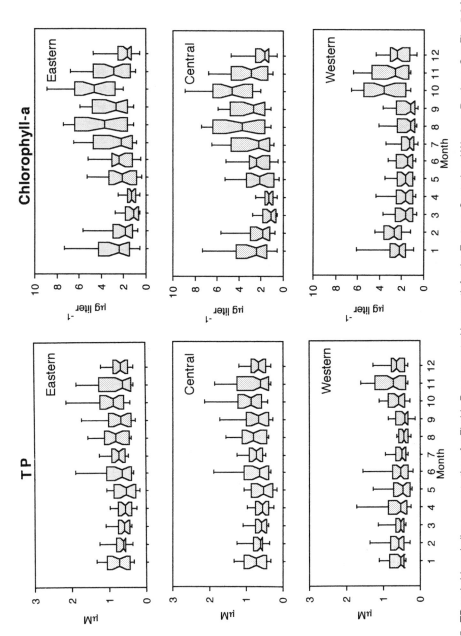

FIG. 5-5. TP and chlorophyll-a concentrations for Florida Bay presented by month for the Eastern, Central and Western Regions. See Fig. 5-2 for sample locations and discussion of statistical presentation.

the limiting nutrient for primary productivity in Florida Bay than it was histori-cally.

Tampa Bay

Tampa Bay is a marine-dominated estuary with a salinity range of 29–35 g liter^{-1} in the lower bay and 20–30 g liter^{-1} in the upper embayments of Hillsborough Bay and Old Tampa Bay (Boler 1995). Point-source inputs from the city of Tampa contribute to slightly elevated nutrient concentrations in Hillsborough Bay (TN = 35–130 μM and TP = 13 μM for 1992–1994) com-pared with the lower bay (TN = 20–80 μM and TP = 3 μM for 1992–1994). Of these nutrients, only 2–10% of the TN is in the form of NO_3 and NH_4, while 50–80% of the TP is found as PO_4. As a result of this partitioning, $DIN:PO_4$ ratios are usually below 16, and nitrogen is thought to be the limiting nutrient for phytoplankton production. Chlorophyll-a concentrations in Tampa Bay during 1992–1994 averaged 10–18 μg liter^{-1} in Hillsborough Bay (maximum = 45 μg liter^{-1}) and 4–5 μg liter^{-1} in the lower bay (Table 5-2). Seasonally, chlorophyll-a displays a minimum during the winter-spring period and late summer-fall maxima, with inter-annual variations associated with precipitation inputs (Boler 1995).

As a result of management of point-source nutrient inputs from domestic sewage and fertilizer manufacturing (Boler 1995; Dunn 1996), Tampa Bay showed significant long-term decreases in the concentrations of TP and chloro-phyll-a between 1974 and 1994. Over this same period, TN inputs showed no long-term trend, but nitrogen remains the potential limiting nutrient for phytoplankton growth in the bay (Boler 1995).

RIVER-DOMINATED ESTUARIES OF THE NORTHERN GULF

In the northern Gulf, Apalachicola Bay and Mobile Bay are river-dominated estuaries that display seasonal discharge patterns similar to those of better-studied temperate estuaries. Both systems are shallow and receive only minor tidal energy input as a result of weak diural tides. As a result of their small volume, these systems have low residence times, particularly under high dis-charge conditions. Compared with other Gulf estuaries, these systems receive greater seasonal forcing in the form of river discharge and temperature than do estuaries to the south and to the west.

Apalachicola Bay

Apalachicola Bay (Fig. 5-6), a river-dominated, bar-built estuary, is one of the most productive estuaries along the northern Gulf of Mexico. In addition to its importance for commercial fisheries of oysters, shrimp, and blue crabs, the bay serves as a nursery for many species of invertebrates and fin-fishes.

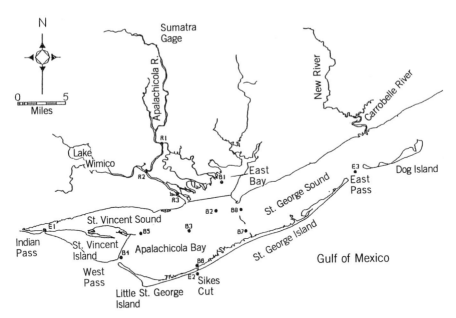

FIG. 5-6. *Sampling locations in Apalachicola Bay, Florida. Data are for surface samples collected at eight bay stations between 1993 and 1996 and are represented as the mean ± standard deviation for samples averaged by cruise.*

All of these higher trophic level species depend on the transport of particulate organic matter and nutrients into the bay from the Apalachicola River (Livingston 1984). Despite high discharge per unit area of estuary, river inputs to Apalachicola Bay are currently in dispute as a result of the opposing needs for fresh water in the Atlanta metropolitan area and for the general health of the estuary. Intensive studies carried out from 1993 to 1996 provide important insights into the nutrient and production processes in the ecosystem.

Discharge from the Apalachicola River is generally greater during winter than during summer, resulting in increasing salinity and marine influence during summer periods (Fig. 5-7). Under high discharge conditions, NO_3 concentrations range from >37 μM in the upper estuary to <8 μM at the mouth of the bay as a result of a combination of mixing between Apalachicola River and Gulf of Mexico end members and biological uptake processes (Fig. 5-8). While NH_4 exhibited a concentration increase near the end of the 1993–1996 sample period, neither NH_4 nor NO_2 is typically a significant fraction of the total DIN (Fig. 5-8). During low-discharge summer conditions, DIN concentrations were reduced to ~2 μM throughout the system.

As with NO_3, the SiO_2 concentration varies directly with salinity (Fig. 5-8). For 1995, a trend toward decreasing concentration with time was interrupted by a pronounced mid-summer SiO_2 maximum in response to increased river input during June relative to May and July (Fig. 5-8). Overall, SiO_2

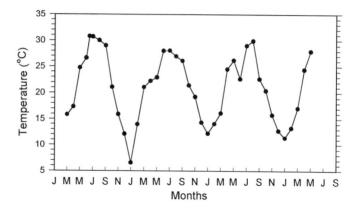

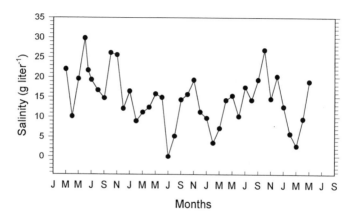

FIG. 5-7. Temperature and salinity for Apalachicola Bay, Florida, averaged for all stations for the period March 1992 through May 1996. See Fig. 5-6 for discussion of sample locations.

concentrations were >102 μM in the upper estuary and under high-discharge conditions and seldom decreased below 15 μM over the annual cycle. Reactive phosphate concentrations were variable over time (0–0.65 μM), with the highest concentration observed during summer (Fig. 5-8). The summer maximum is consistent with experimental results from Apalachicola Bay showing that wind-mixing events caused increased water-column PO_4 concentrations during summer as sediment-bound PO_4 desorbed from reduced sediments upon re-suspension into the water column (Myers 1977).

Phytoplankton production in Apalachicola Bay ranged from 96 to 1812 mg C m^{-2} d^{-1} (mean = 771 mg C m^{-2} d^{-1}; Iverson et al. 1997), with maxima during the late spring and summer months (Fig. 5-9). This pattern is consistent with a positive relation between productivity and water temperature observed for the system (Eastbrook 1973). Chlorophyll-a concentrations generally display winter-spring maxima (6–11 μg $liter^{-1}$) coincident with low salinity values

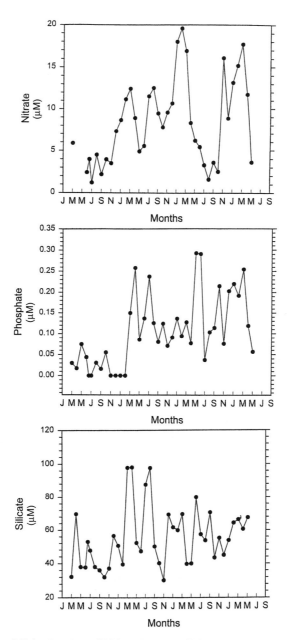

FIG. 5-8. Nitrate (NO₃), phosphate (PO₄) and silicate (SiO₂) concentrations for Apalachicola Bay, Florida, averaged for all stations for the period March 1992 through May 1996. See Fig. 5-6 for discussion of sample locations.

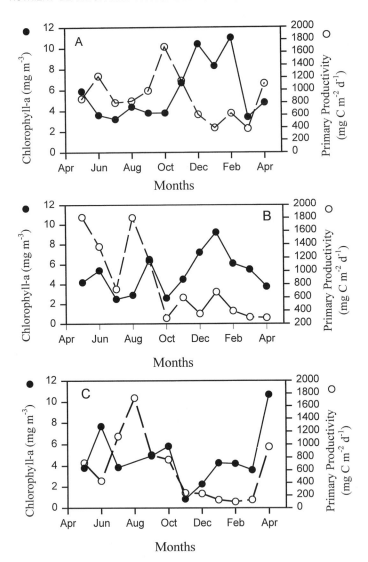

FIG. 5-9. *Chlorophyll-a and primary production data for Apalachicola Bay, Florida, averaged for all stations for three periods: May 1993 to April 1994 (top), May 1994 to April 1995 (middle), and May 1995 to April 1996 (bottom). See Fig. 5-6 for discussion of sample locations.*

and high NO_3 values, although summer maxima coincident with salinity minima have been observed (Fig. 5-9). In general, the relation of chlorophyll-a to salinity in Apalachicola Bay is similar to that reported for the Chesapeake Bay, where the concentration of chlorophyll-a is a function of Susquehanna River flow (Malone et al. 1996). These results are also consistent with those of Myers and Iverson (1981), who reported that nitrogen limitation of phytoplankton productivity occurred more frequently than phosphorus limitation in Apalachicola Bay. These data are similar to those obtained in studies of 19 estuaries and fjords (Boynton et al. 1982), which found that variations in nitrogen input were more important in controlling annual phytoplankton production than were variations in phosphorus input.

Mobile Bay

Mobile Bay receives fresh water from the Alabama and Tombigbee Rivers, resulting in an average discharge of 2245 m^3 s^{-1} (range = ~500–13,000 m^3 s^{-1}), making it the fourth highest discharge in North America behind the Mississippi, Yukon, and Columbia Rivers. Nutrient and phytoplankton production processes in Mobile Bay (Fig. 5-10) were examined between 1989 and 1995 (Pennock et al. 1994a). Climatological forcing is provided by a typical warm-temperate annual temperature cycle (10°–30°C; Fig. 5-11) and a winter-spring maximum in fresh-water discharge that freshens the upper bay and typically results in salinities <10 g $liter^{-1}$ even in the lower bay, 60 km to the south (Fig. 5-11). Nutrient input to Mobile Bay is regulated primarily by upstream inputs delivered to the bay by the rivers (Pennock et al. 1994a). Inorganic nitrogen inputs are dominated by NO_3, which displays a seasonal variation of ~2–25 μM in the upper bay and 0–10 μM in the lower bay (Fig. 5-12), with maximum concentrations occurring during periods of high river discharge. Ammonium concentrations are generally <5 μM, although regenerated NH_4 from sediments (Cowan et al. 1996) and micro-zooplankton (Lehrter et al. in press) is often the predominant DIN species during summer, when NO_3 concentrations are low. Although TP inputs to Mobile Bay are high (Pennock et al. 1994a) and N:P loading ratios are phosphorus-enriched (~12:1), PO_4 concentrations are low, averaging <0.3 μM throughout the estuary and seldom rising above 1 μM (Fig. 5-12). As a result of the high variability in both DIN and PO_4 in Mobile Bay, DIN:PO_4 ratios are highly variable over the seasonal cycle. In the upper estuary and during periods of high fresh-water inflow, DIN:PO_4 ratios are generally above 30, while in the lower estuary and during low-inflow summer periods, DIN:PO_4 ratios fall to ~10. Silicate concentrations in Mobile Bay are generally high and strongly related to fresh-water inputs, ranging from >120 μM in low-salinity waters to a minimum that seldom decreases to below 20 μM in the lower estuary.

Despite strong river forcing, chlorophyll-a distributions in Mobile Bay show little of the clear seasonality typical of cool-temperate systems but rather are dominated by high variance. Chlorophyll-a concentrations generally range from 2 to 12 μg $liter^{-1}$, with aperiodic blooms ranging as high 50 μg $liter^{-1}$, particularly in the shoal regions of the mid-day (Fig. 5-13, Table 5-3). Diagnostic pigment analysis has shown aperiodic dominance by Bacillariophyceae, Cyanophyceae, and Pyrrophyceae species (McManus, unpublished data).

Phytoplankton production rates in Mobile Bay range from 200 to 2000 mg C m^{-2} d^{-1}, resulting in annual production rates of 194–325 gC m^{-2} y^{-1} between

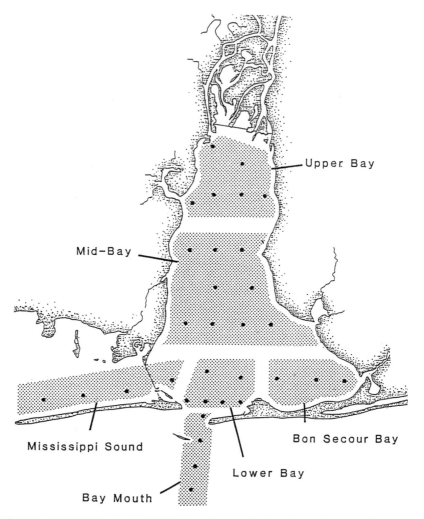

FIG. 5-10. *Sampling locations in Mobile Bay, Alabama. Data are for surface samples collected at 8 to 25 bay stations (●) between 1989 and 1995 and are represented as the mean ± standard deviation for samples averaged by region for each cruise. Regions of the bay are defined geographically as shown.*

1991 and 1995. While phytoplankton production displays more seasonality—with a summer-fall maximum—than chlorophyll, production patterns in the bay are also characterized by high variance (Fig. 5-13). Studies of micro-phytobenthos production has not been carried out in Mobile Bay. However, experiments have been conducted in Weeks Bay—a shallow tributary estuary on the eastern side of Mobile Bay—and have shown that the micro-phytobenthos contribute 20% of the micro-algal production to the system (Schreiber and Pennock 1995). Similar contributions should be expected in shallow regions of Mobile Bay and other Gulf estuaries.

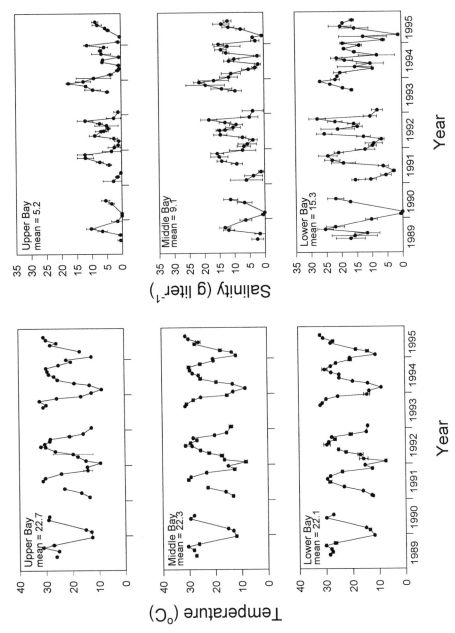

FIG. 5-11. Temperature and salinity data for Mobile Bay, Alabama, averaged by region and by cruise for the period June 1989 to November 1995. See Fig. 5-10 for sample locations and a discussion of statistical presentation.

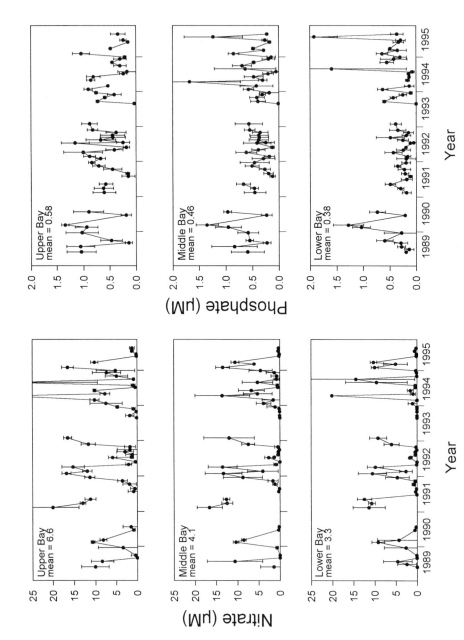

FIG. 5-12. NO_3 and PO_4 concentrations for Mobile Bay, Alabama, averaged by region and by cruise for the period June 1989 to November 1995. See Fig. 5-10 for sample locations and a discussion of statistical presentation.

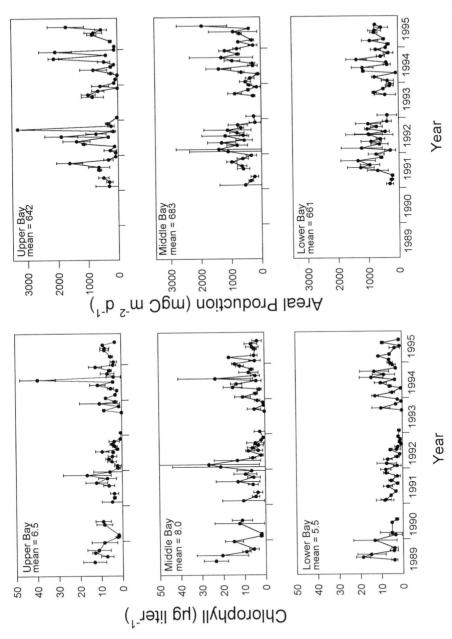

FIG. 5-13. Chlorophyll-a and primary production data for Mobile Bay, Alabama, averaged by region and by cruise for the period June 1989 to November 1995. See Fig. 5-10 for sample locations and a discussion of statistical presentation.

TABLE 5-3. Annual Mean and Range of Chlorophyll-a and Primary Production for Gulf Estuaries Examined in the Case Studies

Site	Chlorophyll (μg liter^{-1})	Volume (mg C liter^{-1} d^{-1})	Primary Production Area (mg C m^{-2} d^{-1})	Annual (gC m^{-2} yr^{-1})
Florida Bay				
Eastern Region	0.76 (0.04–11.3)	—	—	75
Central Region	2.04 (0.21–11.6)	—	—	400
Western Region	1.83 (0.23–11.6)	—	—	250
Tampa Bay	7.5 (0–45)	—	—	—
Apalachicola Bay	5.6 (0–37.6)	21.4 (0.2–137.1)	771 (96–1812)	240 (188–300)
Mobile Bay	—			242 (194–325)
Upper Bay	6.5 (0.2–41)		642	
Mid Bay	8.0 (0.2–55)		683	
Lower Bay	5.5 (0.2–20)		661	
Fourleague Bay	—			
Nueces River Estuary	—			370 (270–400)
Nueces Bay	21.92	2.2 (0.2–7)		
Corpus Christi Bay	8.4 (5.3–11.2)	1.2 (0.1–6)		
Celestun	5.8 (0.5–28.5)	1.2 (0.21–2.1)	1440 (300–2520)	525 (109–919)
Chelem	2.8 (1.4–9)	0.42 (0.3–0.55)	420 (300–500)	153 (109–182)
Dzilam	2.7 (2–4)	0.53 (0.057–0.922)	420 (40–730)	153 (14–266)
Rio Lagartos	4.9 (2–10)	0.74 (0.054–0.875)	540 (40–700)	215 (14–255)

While the high variance in phytoplankton production estimates limits our ability to analyze statistically the factors that limit phytoplankton production, light-limitation and rapid flushing rates appear to limit production during periods of high river discharge. In contrast, nutrient limitation and grazing in combination are more important regulators of phytoplankton dynamics during periods of low discharge. Preliminary nutrient bioassay experiments (Pennock, unpublished data) carried out during the summer show nitrogen limitation in the lower bay, when $DIN:PO_4$ ratios are ~ 10.

Inorganic nutrient concentrations and loading to Mobile Bay have shown little change over the past 25 years (Paulson et al. 1993). This lack of trend is most likely the result of land-use practices that remain focused on managed timberland rather than intensive agriculture and urban growth. Over the same period, TN loading has shown a minimal increase, while TP loading has decreased (Dunn 1996). For systems such as Mobile Bay, it will be important to understand more clearly the roles that dissolved organic nitrogen (DON), particulate organic nitrogen (PN), and particulate phosphorus (PP) play in the process of eutrophication. For Mobile Bay, DON and PN each make up $\sim 40\%$ of the TN in the estuary, while PP makes up $\sim 75\%$ of the TP (Pennock, unpublished data).

MISSISSIPPI RIVER PLUME AND COASTAL DISTRIBUTARIES

The Mississippi River ranks among the top 10 rivers in the world in fresh-water and sediment discharge (Milliman and Meade 1983). Concentrations of inorganic nutrients are high in the lower river, averaging ~ 114 μM NO_3, 7.7 μM TP, and 108 μM SiO_2 during the 1981–1987 period (Rabalais et al. 1996). As a result, the Mississippi is the dominant source of nutrient input to the Gulf of Mexico (Chapter 4 this volume). The Mississippi discharges into both coastal wetland systems—through the Atchafalaya River—and the coastal ocean—through Southwest Pass—where it forms an "extended estuary" in its coastal plume (Fig. 5-1). This region supports one of the most productive fisheries in the United States; as a result, nutrient enrichment and nutrient-enhanced production have attracted much attention (Lohrenz et al. 1994; Smith and Hitchcock 1994; Turner and Rabalais 1994; Rabalais et al. 1996).

Nutrients delivered by the Mississippi River are removed by biogeochemical processes—in particular, phytoplankton uptake—within the plume and show non-conservative behavior with respect to salinity (Rabalais et al. 1996; Shiller 1993). This pattern is caused by a slowing of the mixing process at high salinity (Shiller 1993) and increased light transparency that stimulates production at the interface between plume and oceanic waters (Lohrenz et al. 1990). Season-ally, nutrient concentrations—and the position of the transition between high-nutrient river waters and low-nutrient oceanic waters—in the near-shore wa-ters of the Louisiana shelf are regulated by the discharge of the river, extending between 10 and 100 km from the river mouth (Nelson and Dortch 1996).

Smith and Hitchcock (1994) found nutrient concentrations in the Mississippi plume during March 1991 to range from 0 to 75 μM for NO_3, 0 to 3 μM for PO_4, and 0 to 50 μM for SiO_2 under high-discharge conditions, resulting in N:P and SiO_2:DIN ratios that were well above the Redfield ratio (Redfield et al. 1963). During low-discharge conditions in September, nutrient concentrations throughout the region were <2 μM for NO_3, 0 to 1 μM for PO_4, and 0 to 5 μM for SiO_2, resulting in much lower N:P ratios.

Chlorophyll-a concentrations in the Mississippi River plume are highly variable in space and time, ranging between 1.1 and 14.4 μg liter^{-1} in the outer plume during four cruises in 1990–1992 (Redalje et al. 1994), but ranging as high as 40–80 μg liter^{-1} in near-coastal waters during the late spring of 1993 (Nelson and Dortch 1996; Table 5-3). As a result of high nutrient loading and phytoplankton biomass, phytoplankton production in the Mississippi River plume is high, with a range of 0.5–5 gC m^{-2} d^{-1} at salinities between 20 and 35 g liter^{-1} (Lohrenz et al. 1990). Although high primary production rates are characteristic, nutrient limitation is observed as a result of high nutrient demand by phytoplankton populations. Silicon limitation of diatom growth has been observed during spring (Smith and Hitchcock 1994; Nelson and Dortch 1996), while phosphorus and nitrogen have been shown to limit potential phytoplankton production during spring and summer, respectively (Smith and Hitchcock 1994).

As a result of the nutrient-enhanced production reported above and particulate nutrient input (Trefry et al. 1994), the plume of the Mississippi River is a region of both high secondary production (Deegan et al. 1986) and ecosystem degradation, as evidenced by extensive bottom water hypoxia (Rabalais et al. 1994, 1996). The relative importance of these contrasting processes ultimately determines the biogeochemical pathways that dominate in the system and the fate of nutrients and phytoplankton organic matter in the ecosystem. Of primary concern is the observed increase in NO_3 loading, decreased silicate flux, and decreasing SiO_2:DIN ratios observed in the river between 1960 and 1985 (Rabalais et al. 1996). These changes may be related to the increased area of the hypoxic region reported since 1993 and potential changes in phytoplankton species composition (Rabalais et al. 1996).

Fourleague Bay

Over one-third of the discharge of the Mississippi River is discharged through the Atchafalaya River, where it flows through shallow coastal ecosystems inshore of the extended plume discussed previously. Fourleague Bay, a shallow, turbid, wetland-dominated estuary just east of the mouth of the Atchafalaya, receives high nutrient inputs (Stern et al. 1991). Nitrate distributions in the estuary are non-conservative, averaging 42 μM (range = 0.4–87 μM) in the upper bay and 6 μM (range = 0–45 μM) in the lower bay (Table 5-2). At the same stations, NH_4 averaged 2.7 μM and 2.4 μM, while PO_4 averaged

1.2 μM and 0.6 μM at the upper and lower bay stations, respectively (Madden 1986; Randall and Day 1987).

Chlorophyll-a concentrations in Fourleague Bay range from 3.2 to 19.8 μg liter^{-1} (mean = 13.4 μg liter^{-1}) in the upper bay and from 4.0 to 26.7 μg liter^{-1} (mean = 14.5 μg liter^{-1}) in the lower bay (Randall and Day 1987). Phytoplankton production rates in Fourleague Bay are related to both turbidity and nutrient concentration, averaging 120 gC m^{-2} y^{-1} in the light-limited upper bay and 317 gC m^{-2} y^{-1} in the less turbid lower bay (Randall and Day 1987).

TEXAS ESTUARIES

Fresh-water inflows have been described as the defining factor for biological productivity in the estuaries of Texas (Copeland 1966; Armstrong 1987). Starting in 1987, monthly samples were collected throughout the Corpus Christi-Nueces Estuary in order to quantify biological responses to fresh-water inflow from the principal river (Whitledge 1989). Somewhat later, in 1990, the Texas Water Commission ordered monthly fresh-water releases in order to increase biological productivity. The releases of fresh water were based on analyses of (1) the firm yield of the reservoir system; (2) the relationship to fisheries yield and ecological need; and (3) the probability of failure using historical rainfall and inflow data. The following section addresses some of the relationships that have been observed between fresh-water inflows and biological productivity at the base of the food chain, particularly with reference to nutrients and plankton growth.

In general, nutrient and primary productivity studies have identified nitrogen as the most important nutrient governing the growth of phytoplankton. The overall DIN:PO$_4$ ratio collected throughout Corpus Christi Bay was observed to be 6.1 (by atoms), which clearly indicates an excess of PO$_4$. Nutrient amendments of nitrogen, phosphorus, and silicate to primary productivity samples also showed no enhancement by PO$_4$ or SiO$_2$ additions, but DIN always showed a response.

The Nueces River Estuary and Corpus Christi Bay

The Nueces River discharge into Nueces Bay and Rincon Delta is dominated by extreme inflow events (Fig. 5-14). The normal spring precipitation in April and May produces the largest median inflows, while the sporadic fall tropical storms contribute somewhat less. Periods of adequate rainfall are interspersed with droughts that last for 1–2 years. In the early 1980s, the construction of a new dam and water reservoir greatly reduced the fresh-water discharge to the estuary for over a decade. Water sampling for the entire Corpus Christi and Nueces Bay at 35 stations was carried out during 1987–1988 to study the effects of fresh-water inflows on nutrient and primary productivity processes.

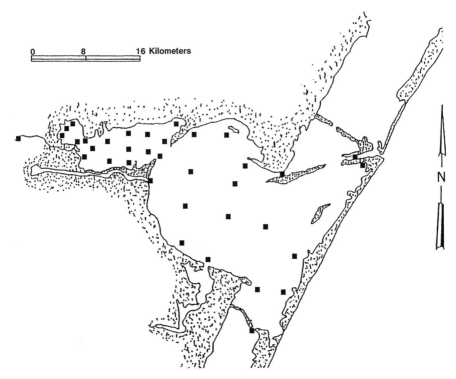

FIG. 5-14. *Sampling locations in the Nueces River Estuary/Corpus Christi Bay, Texas. Data are the mean values averaged by cruise between May 1990 and December 1994. The samples were collected at 25 stations (■) for hydrography, nutrients, and plankton pigments, and 5 of the sites included primary production rate measurements. At each site, discrete water bottle samples were taken at the surface and near bottom except in water depths of less than 30 cm, where only surface samples were collected.*

Subsequently, water releases were started in 1990 to return some of the low-discharge inflows that had been eliminated by the operation of the dam. Between May 1990 and December 1994, monthly sampling was carried out in Nueces Bay at 17 stations in order to examine the effects of these releases and the potential effect of wastewater discharge to the delta and open bay waters.

Mean salinity in the Corpus Christi Bay complex was 32.1 g liter^{-1} (range = 28.6 to 37.9 g liter^{-1}), while Nueces Bay showed a variation from 3 to 30 g liter^{-1} (mean = 18.8 g liter^{-1}), with the highest values occurring in the winter, when precipitation and releases are minimal (Fig. 5-15). The lowest salinity values occurred in late spring and summer during periods of direct precipitation and fresh-water releases. The high salinity in the winters of 1990 and 1991 varied by ~4 g liter^{-1}, while the summer low values went from 3 to 16 g liter^{-1} for 1990–1992. These summer variations of 13 g liter^{-1} reflect the

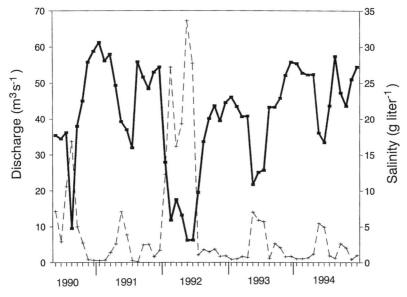

FIG. 5-15. *Mean monthly salinity and river discharge for the Nueces River Estuary/Corpus Christi Bay for the period May 1990 to December 1994. See Fig. 5-14 for sample locations and a discussion of sampling protocols.*

overflow of the upstream dams when the reservoirs were full. When overflow from the dams did not occur in 1991, a mean salinity of about 16 g liter^{-1} was the lowest for the year.

Mean monthly nitrate concentrations have several maxima related to either fresh-water inflow or regeneration/nitrification processes (Fig. 5-15). Nitrate concentrations in the entire Corpus Christi Bay complex averaged 3.2 μM (range = 1.5 to 11.3 μM), while Nueces Bay averaged 4.6 μM (range = 0.5 to 23 μM). Nitrite in Corpus Christi Bay averaged 0.43 μM (range = 0.4 to 1.5 μM), and Nueces Bay had a mean of 0.76 μM (range = 0.2 to 2.5 μM). The largest NO_2 concentrations occurred during December 1991 and January 1992, when the river waters had elevated concentrations. The other nitrogen nutrient, NH_4, ranged from 0.5 to 19 μM (mean = 4.63 μM). The high ammonium concentrations in late summer almost certainly result from regeneration of organic matter in situ while, in later winter months the ammonium maximum corresponds to a probable river source (e.g., fertilizer). For comparison purposes, mean PO_4 and SiO_2 concentrations for Corpus Christi Bay were 1.99 and 38.2 μM, respectively (ranges = 0.6 to 4.6 μM PO_4; 10.5 to 60.8 μM SiO_2: Fig. 5-16, Table 5-2).

The mean DIN concentration for the Corpus Christi Bay complex was 12.1 μM (range = 5.1 to 25.1 μM), and in Nueces Bay the range of mean DIN concentrations was 2 to 28 μM (mean = 10.0 μM). The lowest DIN concentrations may be small enough to retard phytoplankton growth, but

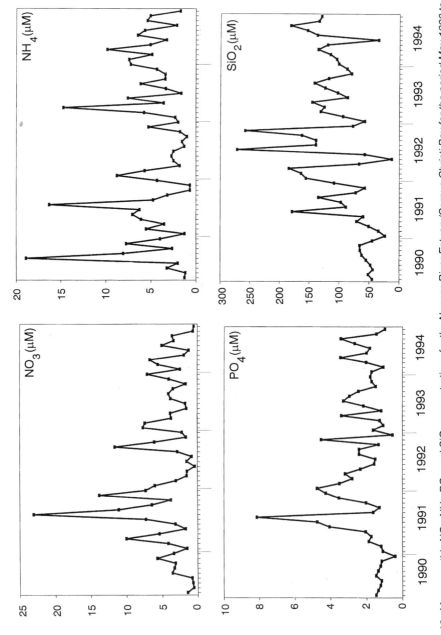

FIG. 5-16. Mean monthly NO_3, NH_4, PO_4, and SiO_2 concentrations for the Nueces River Estuary/Corpus Christi Bay for the period May 1990 to December 1994. See Fig. 5-14 for sample locations and a discussion of sampling protocols.

values >5 to 10 μM are certainly large enough to promote maximal growth rates if sufficient light is also available. There are few obvious trends in mean DIN concentrations compared to the mean salinity values, except that the abrupt increases in DIN in August 1990 and 1991 occurred concurrently with abrupt increases in salinity, a possible result of reduced denitrification rates. The unusually long time period with low DIN concentrations from April through October 1992 resulted from large phytoplankton biomass and high primary productivity rates.

The mean chlorophyll-a concentration for the Corpus Christi Bay ecosystem was 8.4 μg liter^{-1} (range = 5.3 to 11.2 μg liter^{-1}), while the mean monthly chlorophyll-a concentration in Nueces Bay displayed periods of high values in the fall but at other times ranged between 10 and 20 μg liter^{-1}, with an overall mean of 21.92 μg liter^{-1}. Size fractionation experiments have indicated that 75–90% of the phytoplankton are present in the size class <20 μm, which consists of small diatoms with an occasional bloom of coccoid blue-green algae (D. Stockwell, personal communication). Both of the enhanced chlorophyll concentrations occurred after large fresh-water inflow events. The mean monthly primary productivity rate measurements in the Corpus Christi Bay complex ranged from 0.1 to 6.0 gC m^{-3} day^{-1} (mean = 1.2 gC m^{-3} day^{-1}), and those in Nueces Bay varied from 0.2 to 7 gC m^{-3} day^{-1}, with a mean of 2.2 gC m^{-3} day^{-1} (Fig. 5-17). The lowest primary production rates were observed during the winter, when the day is short and overcast skies are com-

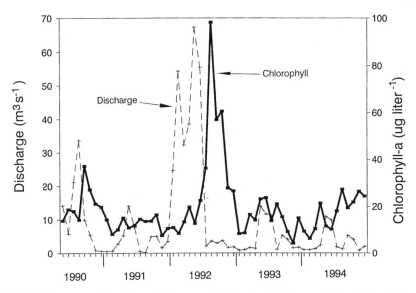

FIG. 5-17. *Mean monthly chlorophyll-a concentration for the Nueces River Estuary/Corpus Christi Bay for the period May 1990 to December 1994. See Fig. 5-14 for sample locations and a discussion of sampling protocols.*

mon. The highest primary production rates were measured during each of the three summers shortly after fresh-water inflows from releases and natural spills and are approximately a factor of 2 larger than previous measurements (Odum and Wilson, 1966; Flint 1984). Overall, the rates of primary production measured throughout Nueces Bay and Rincon Delta are not influenced strongly by salinity. The rates of primary production are lowest at very low salinity (<3 g liter^{-1}) or very high salinity (>35 g liter^{-1}), while maximal rates occur at 5 to 35 g liter^{-1}. Salinity does not directly influence phytoplankton growth rates but co-occurs with nutrient concentration that has direct effects.

The primary production rates measured at the six locations in Nueces Bay and Rincon Delta are subject to variations with respect to the environmental conditions conducive to plant growth. The mean primary production rates show the highest values in the lower portion of the Nueces River and in the interior of Rincon Delta in upper Nueces Bay (Fig. 5-18, Table 5-4). Another significant aspect of the primary productivity measurements is the maxima in rates at all stations during summer periods. Station 4A, located in the lower Nueces River downstream from the Allison Wastewater Treatment Plant, enjoyed obvious nutrient enrichment, while Station 51, located about 5 km up in the Rincon Delta, responded to fresh-water inflows, spills, and overbanking events.

Marsh Ponds and Delta Channels

The primary production rates of phytoplankton populations in marsh ponds and connecting channels in the interior of Rincon Delta (Stations 51 and 54) were monitored by monthly sampling of temperature, salinity, nutrients

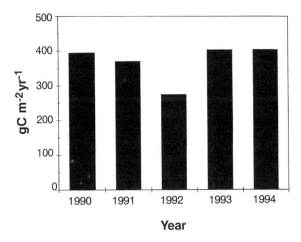

FIG. 5-18. Annual primary production estimates for the Nueces River Estuary/Corpus Christi Bay for 1990 through 1994. Data are averaged from monthly data collected at five stations in the estuary.

TABLE 5-4. Mean Primary Production Rates and Number of Samples at Stations in Nueces Bay and Rincon Delta

	Number of Samples	Mean (mg C liter^{-1} h^{-1})
Station 2	24	244.1
Station 4A	24	410.1
Station 7	29	196.1
Station 13	29	178.6
Station 43	18	125.9
Station 51	28	389.5

(nitrate, nitrite, ammonium, phosphate, and silicate), water transparency, plant pigments, and primary productivity. In 1992, the large amounts of direct precipitation and the overbanking events of the Nueces River delivered unusually large quantities of fresh water to the Rincon Delta. The mean salinity for all samplings at these sites in 1992 was 7.8 g liter^{-1} compared to 23.9 g liter^{-1} in 1991. The primary production rates in the delta varied from 0.49 to 8.22 (mean $=$ 3.40 gC m^{-3} day^{-1}) compared to 0.10 to 4.78 gC m^{-3} day^{-1} (mean $=$ 1.82 gC m^{-3} day^{-1}) for the reference station (Station 7) in open Nueces Bay near Whites Point (Table 5-5). The maximum primary production rate in the delta in 1991 was 7.22 gC m^{-3} day^{-1}, but it increased to 8.22 gC m^{-3} day^{-1} in 1992 as a result of the increased fresh-water discharge.

The primary production of phytoplankton in the surface sediments (micro-phytobenthos) was measured at four sites in marsh ponds in order to include this previously omitted productivity component. The pigment content of sediment ranged from 22.32 to 35.95 mg chlorophyll-a m^{-2}, which is approximately equivalent to the water quantity in the water column (Table 5-2). The primary productivity of the surface sediments ranged from 4.0 to 27.7 mg C m^{-2} day^{-1} (mean $=$ 19.1 mg C m^{-2} day^{-1}). In comparison, the water column had productivity values of 529 to 1473 mg C m^{-2} day^{-1}, with a mean of 806 mg C m^{-2} day^{-1}. The primary productivity of the micro-phytobenthos was therefore about 3.8% of the water column phytoplankton production rates, which is unusually small compared to

TABLE 5-5. Primary Productivity Rates and Pigment Biomass in the Water and Sediments in Rincon Delta

Station Sediment	Water Chlorophyll	Sediment Production	Percent Production	Production in Sediments
51	22.32	1473	4.02	0.27
53	24.33	601	17.94	2.99
54	26.28	529	26.55	5.02
55	35.95	623	27.72	4.45
Mean	27.22	806	19.06	3.18

Note: Units for chlorophyll are mg m^{-2}, and units for productivity are mg C m^{-2} d.

published reports (DeJonge and Colijn 1994; Schreiber and Pennock 1995). The absolute rates of production in the sediments are not low; rather, the water column rates are large. Monitoring of the sediment productivity rates is continuing in order to include higher salinity conditions when the water column rates decrease. It is expected that 25–50% of primary productivity occurs in shallow waters, where light can penetrate to the bottom.

This study showed that increased primary production of phytoplankton and micro-phytobenthos could be expected to occur if Nueces River or Allison wastewater treatment plant waters were diverted into the Rincon Delta rather than flowing down the river and bypassing the brackish marsh area, as currently exists. Additional measurements are still being collected in order to obtain measurements under a wide range of salinity, nutrient, and incident radiation conditions.

The quantity of primary production of the water column in the channels and marsh ponds in Rincon Delta and open Nueces Bay waters was compared at the same time during the years 1991 and 1992 to produce an estimate of productivity enhancement in the delta. Overall productivity enhancement was 8.5 times higher in Rincon Delta than in upper Nueces Bay, although the value declines to 3.2 times higher when two data values are omitted during an unusual "brown tide" bloom of *Aeroumbra lagunensis* Stockwell (De Yoe et al. 1997) that occurred in August and September 1992.

The quantity of primary production of phytoplankton in the Nueces River downstream of the Allison Wastewater Treatment Plant and upper Nueces Bay was compared to estimate the effect of wastewater. The overall mean productivity factor for all data collected in 1991 and 1992 was 2.6 times larger in the Nueces River than in Nueces Bay. However, several data points collected at low salinities, when the flow rates of the river were large, are probably not representative because the transit time was so short. When the data below a salinity of 4 are omitted, the productivity factor ranges from 1.7 to 10.6, with a mean value of 5.0.

The most direct comparison of fresh-water and waste-water effects is found in the Nueces River stations, where primary productivity rates are nearly always elevated compared to those in Nueces Bay. The productivity factor of ~5 was estimated for the Nueces River during 1991–1992. The introduction of additional quantities of fresh water into Rincon Delta by direct precipitation and overbanking of the Nueces River in 1992 increased the estimated 1991 productivity factor from ~2 to 3.2. This confirms that ambient river water diversions to Rincon Delta can be expected to increase primary production by a factor 2 or 3, while wastewater diversions can be expected to increase it by a factor of 5.

COASTAL LAGOONS OF THE YUCATAN

Four major lagoons (Celestun, Chelem, Dzilam, and Rio Lagartos) are located along the coast of the northern Yucatan, Mexico (Fig. 5-19). These lagoons

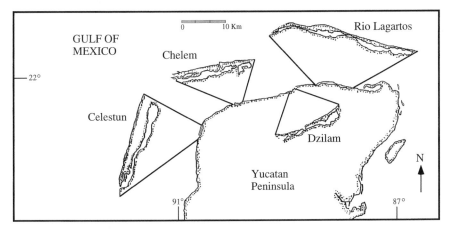

FIG. 5-19. *Map of the northern Yucatan, Mexico, showing the location of Celestun, Chelem, Dzilam, and Rio Lagartos lagoons. Samples were collected during three seasons: dry (March–June), wet (July–October), and nortes (November–February). Data are for surface samples by station for each sampling date. For Celestun, 10 stations were occupied monthly between February 1988 and January 1992. Data for Chelem (10 stations), Dzilam (11 stations), and Rio Lagartos (21 stations) were collected between 1988 and 1992.*

are coastal depressions connected with the sea through inlets in sand barriers and bordered by the land surface on their inner margins (Lankford 1977). They are classified as "choked" lagoons and/or "restricted" lagoons (Kjerfve 1994) and are formed through deposition of biogenic materials transported by the coastal currents. The region is hot and semi-arid, with an annual mean temperature of 25°–28°C and a mean annual rainfall of 550–800 mm. Climatologically, there are two main seasons, the dry season (March–May) and the rainy season (June–October), which are separated by a period known locally as "nortes," which lasts from November to February. Nortes consists of frequent 3-d events characterized by strong winds (>80 km h^{-1}), little rainfall (20–60 mm), and low temperatures (<22°C) during frontal passages from the north.

The dominant karst geomorphology of the soil and rocks in the Yucatan results in the absence of surface/riverine inputs of fresh water to the coastal lagoons. Instead, ground-water discharges from springs predominate. Ground-water discharges differ from river inputs in their characteristically low suspended sediment and dissolved oxygen concentrations (<1 mg liter^{-1}). With respect to nutrient input, the ground-water discharges are characterized by high concentrations of NO_3 (20 to 160 μM) and SiO_2 (25 to 400 μM) and low concentrations of NH_4 and PO_4 (0.1 to 4 μM and 0.02 to 2 μM, respectively). A seasonal pattern of low concentrations during the dry season and high concentrations during the late rainy and nortes seasons has been observed (Herrera-Silveira 1994).

In spite of a common climatic and geological pattern for all the coastal lagoons in the northern Yucatan, slight differences in the ground-water circulation between land and sea play an important role in establishing differences between the salinity, nutrient pattern, and phytoplankton production characteristics of close coastal lagoons.

Celestun Lagoon

Celestun lagoon displays a strong horizontal salinity gradient during all seasons (Fig. 5-20), with salinity varying from <20 g liter^{-1} in the inner zone to >30 g liter^{-1} in the seaward zone as a result of the fresh-water input from many springs. Nitrate and SiO_2 concentrations display variations that are related spatially to the source of fresh-water input and temporally to periods of rainfall and wind/re-suspension during the late wet and early nortes seasons (Figs. 5-21 to 5-24). In general, maximum concentrations of NO_3 (13 μM), PO_4 (1.5 μM), and SiO_2 (225 μM) are indicative of modest nutrient enrichment in the inner lagoon. Ammonium concentrations are significantly higher and more variable (5 to 100 μM) as a result of re-mineralization of organic matter and subsequent re-suspension by high winds during the wet and nortes seasons.

The annual range of net phytoplankton productivity in Celestun was 0.2 to 2.1 mg C l^{-1} d^{-1}, with a chlorophyll-a range from 0.9 to 30 μg liter^{-1}. Phytoplankton productivity and chlorophyll-a showed similar seasonal patterns characterized by two peaks. The first one occurs early in the rainy season (June = 30 μg chl-a l^{-1}; 2.15 mg C l^{-1} d^{-1}), and the second one occurs during the early nortes season (November = 10 μg chl-a l^{-1}; 0.45 mg C l^{-1} d^{-1}). Spatially, the higher rates of phytoplankton production occur in the middle zone of the lagoon, primarily associated with the classes Bacillariophyceae (58%) and Pyrrophyceae (31%). This distribution in the phytoplankton responsible for primary production is markedly different from that in the inner zone—where there is co-dominance of Crypthtophyceae (45%), Cyanophyceae (33%), and Pyrrophyceae (10%)—and the outer zone, where the phytoplankton is clearly dominated by Bacillariophyceae (83%).

The relationship between nutrient inputs and primary production in Celestun has also been shown to be affected by the levels of phenolic materials (tannins) brought into the inner zone of the lagoon during the nortes season. Field and laboratory bioassays (Herrera-Silveira and Ramirez 1996) demonstrated that tannin concentrations >18 mg liter^{-1} depressed phytoplankton production. As a result, the rates of phytoplankton production are significantly lower (three to four times less) during the nortes period than during the rainy period, in spite of high PO_4 and ammonium concentrations during this period.

Chelem Lagoon

Chelem lagoon displays low rates of ground-water discharge and is strongly influenced by sea water. As a result, Chelem has little horizontal salinity

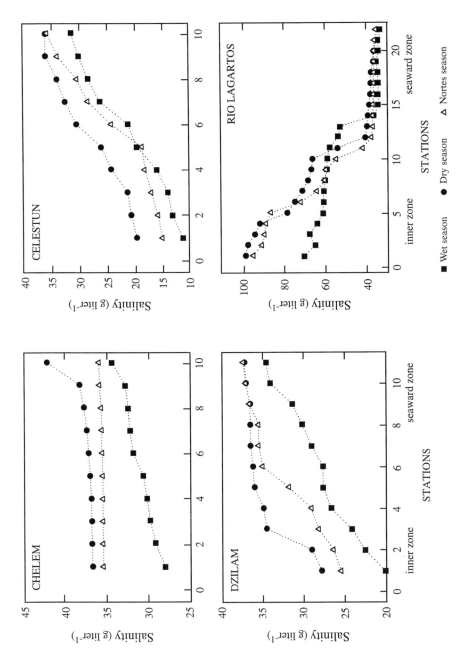

FIG. 5-20. Salinity distributions in Celestun, Chelem, Dzilam, and Rio Lagartos lagoons in the northern Yucatan, Mexico, during 1992. Data are presented by season for stations along the longitudinal axis of the lagoons from the inner zone to the seaward zone at the mouth of the lagoon. See Fig. 5-19 for the location of the lagoons and a discussion of seasonal sampling protocols.

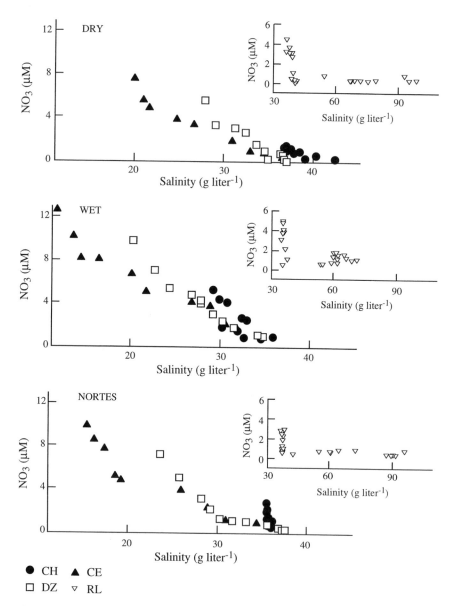

FIG. 5-21. Nitrate NO_3 concentration in Celestun, Chelem, Dzilam, and Rio Lagartos (see inset) lagoons in the northern Yucatan, Mexico, during 1992. Data are presented versus salinity and by season. See Fig. 5-19 for the location of the lagoons and a discussion of seasonal sampling protocols.

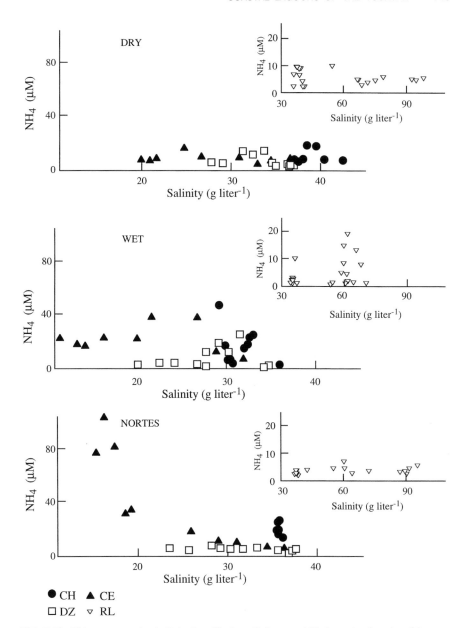

FIG. 5-22. NH$_4$ concentration in Celestun, Chelem, Dzilam, and Rio Lagartos (see inset) lagoons in the northern Yucatan, Mexico, during 1992. Data are presented versus salinity and by season. See Fig. 5-19 for the location of the lagoons and a discussion of seasonal sampling protocols.

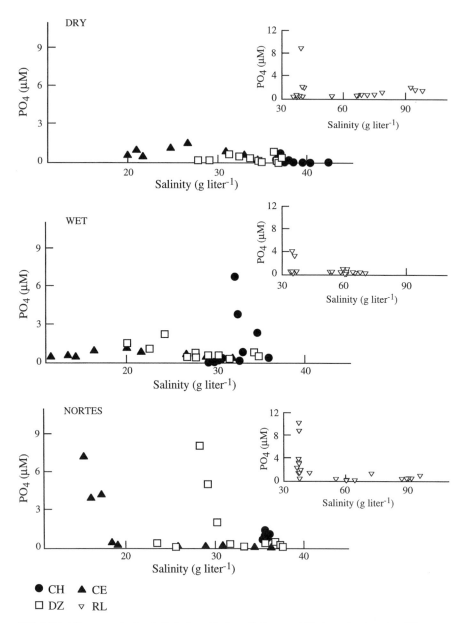

FIG. 5-23. *PO₄ concentration in Celestun, Chelem, Dzilam, and Rio Lagartos (see inset) lagoons in the northern Yucatan, Mexico, during 1992. Data are presented versus salinity and by season. See Fig. 5-19 for the location of the lagoons and a discussion of seasonal sampling protocols.*

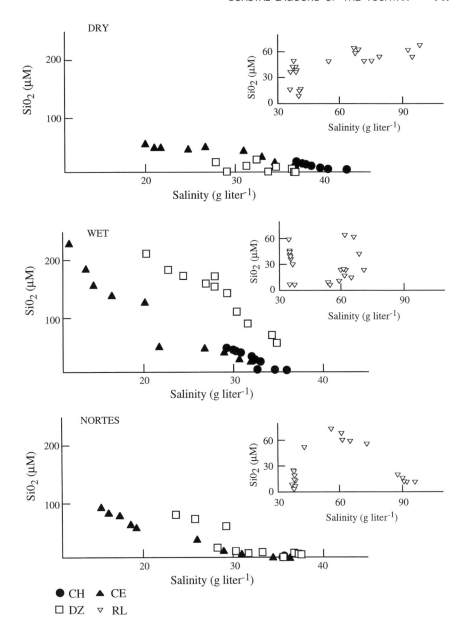

FIG. 5-24. SiO_2 concentration in Celestun, Chelem, Dzilam, and Rio Lagartos (see inset) lagoons in the northern Yucatan, Mexico, during 1992. Data are presented versus salinity and by season. See Fig. 5-19 for the location of the lagoons and a discussion of seasonal sampling protocols.

gradient except during the wet season, when ground-water discharges increase (Fig. 5-20). The mean salinity is highest during the dry season, when salinity >40 g liter^{-1} can be observed. This pattern of low ground-water discharge results in lower concentrations of NO_3 (1 to 6 μM), PO_4 (0.02 to 6 μM), and SiO_2 (4 to 50 μM) than three observed in Celestun (Figs. 5-21 to 5-24). As with Celestun, NH_4 (1 to 38 μM) dominates the DIN pool as a result of re-mineralization of organic matter within the lagoon. However, these concentrations are less than those found in Celestun as a result of low external inputs from ground-water discharges and only modest re-mineralization of detritus from submerged aquatic vegetation.

Chelem also displays low concentrations of chlorophyll-a and low rates of phytoplankton production, with annual ranges of 1.4 to 9 μg chl-a l^{-1} and 0.03–0.5 gC m^{-3} d^{-1}, respectively. For Chelem, phytoplankton production and chlorophyll-a patterns were not spatially congruent, with high rates of production observed in the shallow inner zone of the lagoon near the mangrove forest and high chlorophyll-a concentrations observed at the lagoon inlet. These patterns appear to be regulated by the interactions of light and nutrients, as has been seen in temperate estuaries (Pennock and Sharp 1994; Cloern 1996). In Chelem, as a result of minimal fresh-water input, the inner zone has high light, which penetrates to the bottom of the lagoon, while the inlet zone displays high light extinction coefficients. As a result of the marine influence in this lagoon, the phytoplankton is dominated throughout by Bacillariophyceae (80%). Overall, phytoplankton production in Chelem is the lowest found in the lagoons of the coastal Yucatan as a result of low nutrient inputs.

Dzilam Lagoon

Dzilam lagoon displays an intermediate pattern between estuarine and marine conditions, with a salinity ranging from 20 to 27 g liter^{-1} in the inner zone and >30 g liter^{-1} in the seaward zone throughout the year (Fig. 5-20). While Dzilam is characterized by many ground-water discharges, water-column nutrient concentrations are lower than in Celestun. Nitrate and SiO_2 follow an inverse pattern with salinity, with mean concentrations of 4.1 and 163 μM, respectively, in the inner zone (Figs. 5-21 and 5-24). Ammonium and PO_4 show generally low concentrations throughout the year; however, the seasonal pattern of high concentrations during rainy and early nortes seasons is maintained for most of the nutrients (Fig. 5-23 and 5-24).

The annual range of phytoplankton production is 0.05 to 0.9 gC m^{-3} d^{-1}, with a chlorophyll-a range from 1.5 to 20 μg liter^{-1}. The average phytoplankton production and chlorophyll-a show the same seasonal and space pattern, which is characterized by high rates and high concentrations in the late rainy season (October = 20 μg liter^{-1}; 0.9 gC m^{-3} d^{-1}) in the middle zones of the lagoon and low rates and low concentrations during the early dry season (March = <1 μg liter^{-1}; <1 gC m^{-3} d^{-1}) in the inner and seaward zones. Of the four coastal lagoons, Dzilam has the highest coverage and biomass of submerged aquatic vegetation (SAV), which is the principal primary producer. The phytoplankton community is dominated by Bacillariophyceae (48%) and Pyrrophyceae (36%).

Rio Lagartos

Rio Lagartos lagoon is a hypersaline system and is unique in the Yucatan Peninsula. Salinity displays a horizontal gradient throughout the year, ranging from >70 g liter^{-1} in the inner zone to typical marine values of <40 g liter^{-1} in the seaward zone (Fig. 5-20). The nutrient concentrations in this lagoon are generally low and are not clearly related to the salinity gradient (Figs. 5-21 to 5-24). Nitrate and NH_4 show mean concentrations of 0.7 and 8.5 μM, respectively. Concentrations of SiO_2 vary from 5 to 75 μM, and PO_4 concentrations range from 0 to 11 μM. Overall, nutrient concentrations in Rio Lagartos generally show maxima in the inner and middle zones of the lagoon during both the dry and rainy seasons as a result of inputs from a large population of flamingos (>10,000) that use the lagoon for reproduction.

For Rio Lagartos, the annual range of net phytoplankton production is low, 0.05 to 0.8 gC m^{-3} d^{-1}, with a chlorophyll-a range from 2 to 10 μg liter^{-1}. Phytoplankton production and chlorophyll-a show the same seasonal pattern, characterized by one peak during the rainy season. Spatially, the higher rates of phytoplankton production are observed in the middle zone of the lagoon. Thus, phytoplankton processes do not appear to be related to fresh-water-derived nutrients—as in the other systems—but rather display a strong relationship with nutrient concentrations resulting principally from bioturbation and fecal pellets from the flamingos (Comin et al. 1997).

The phytoplankton in the Rio Lagartos lagoon is dominated by Pyrrophyceae in the middle zone and Bacillariophyceae in the seaward zone. The inner zone of Rio Lagartos is unique in that the salinity in the bottom waters can rise to >100 g liter^{-1} during the wet season and >200 g liter^{-1} during the dry season. In this region, nutrient levels are low and phytoplankton is dominated by a few species of cyanophytes. However, an association of microphytobenthic red bacteria and blue-green algae—known as "estromatholite"—covers almost all of the bottom of the lagoon in this area and is responsible for high rates of primary production.

The monitoring of nutrients in Celestun lagoon between 1987 and 1992 (Fig. 5-25) allows examination of the relative influence of external and internal factors on nutrient behavior. During rainy and early nortes periods, precipitation and winds play the most important role in regulating nutrient distributions. High concentrations of NO_3 and SiO_2 can be found as a result of groundwater discharge inputs and re-suspension.

However, during the late nortes and dry seasons, the re-mineralization of organic matter in the inner zone, and competition between phytoplankton and macroalgae, are the most important factors controlling nutrient dynamics in the lagoon (Herrera-Silveira, 1994). Interestingly, the highest concentrations of nutrients during the period were observed in September 1988 as a result

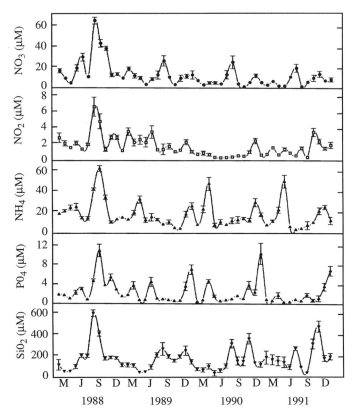

FIG. 5-25. Time series of nutrient concentrations in Celestun lagoon, Yucatan, Mexico, between April 1988 and January 1992. Nutrient concentrations are generally related to rainfall, as shown by the peaks in NO_3 in the wet period of each year and the effect of Hurricane Gilbert in September 1988.

of the passage of Hurricane Gilbert. Subsequent to this event, normal seasonal nutrient patterns were restored in the system.

The comparative results presented here show strong differences in coastal lagoonal ecosystems of the same geomorphological and climatic region that can be attributed to differences in hydrologic and biogeochemical processes. Where ground–fresh-water input of nutrients occurs (Celestun and Dzilam lagoons), phytoplankton production is relatively high and increases as a result of climatic events (late wet and early nortes periods) and the intensity of the fresh-water inputs. Phytoplankton production in these systems can be limited by light, PO_4, or high concentrations of tannins in the inner zones of the lagoons and high water turnover rates in the seaward zones. As a result, phytoplankton production is generally at a maximum in the central zone of the lagoons, where higher water residence time, transparency, and nutrient inputs from the sediment are observed. Competition with macrophytes for light and nutrients can be an additional factor limiting phytoplankton growth in these lagoons, particularly in the central zones, where environmental factors permit the development of very dense, extensive meadows of macro-algae and seagrass.

Where no significant fresh-water inputs occur (Chelem and Rio Lagartos lagoons), no spatial patterns of nutrients are observed and lower phytoplankton biomass and production are observed. While the highest phytoplankton production still occurs during the rainy season, in the absence of external nutrient inputs other biogeochemical processes become important forcing factors for primary production. For example, sediment/water-column exchange and dense populations of aquatic birds are directly related to elevated concentrations of ammonium in lagoons with limited external inputs.

These data show that, in the absence of strong riverine or anthropogenic forcing, the nutrient-phytoplankton relationship is tightly coupled with biogeochemical processes such as re-mineralization, re-suspension, bioturbation, and competition with other primary producers. Similarly, the reduced seasonality in temperature and hydrologic processes compared with cooler temperate ecosystems results in a variable annual cycle in phytoplankton production and biomass accumulation.

The long-term observations in the Celestun lagoon suggest that this system is regulated primarily by natural forcing functions and very little by external forcing from factors (e.g., nutrient-rich fresh-water discharges from agricultural or urban areas) that could favor toxic algal blooms and other eutrophic responses (Caddy 1993). As a result, the coastal lagoons of the north Yucatan are one of the few areas in the Gulf of Mexico estuaries where natural nutrient behavior and phytoplankton responses can be found. These regions are thus very important for understanding natural nutrient and phytoplankton dynamics in similar estuarine ecosystems.

DISCUSSION

The estuarine systems described in this chapter provide examples of the diversity of estuaries that exist in the Gulf of Mexico. They can be divided into three categories based on their geomorphologic and physical characteristics: river-dominated estuaries (Apalachicola Bay, Mobile Bay, the Mississippi River plume and Fourleague Bay, Nueces Bay, and Corpus Christi Bay), marine-dominated estuaries (Florida Bay and Tampa Bay), and coastal lagoon systems (Celestun, Chelem, Dzilam, and Rio Lagartos).

River-Dominated Estuaries

Nutrient input to the river-dominated estuaries of the Gulf of Mexico occurs mainly from river systems that are relatively pristine compared to more heavily populated and urbanized estuaries on the east and west coasts of the United

States (Bowman 1977; Sharp et al. 1982; Fisher et al. 1988; Magnien et al. 1992; Cloern 1996), with the Mississippi River and its plume being the exceptions. As a result, while nutrient concentrations in systems such as Apalachicola Bay and Mobile Bay are highest during winter high river discharge periods, maximum concentrations (20–30 μM DIN, 0.2–0.5 μM PO$_4$) are often an order of magnitude lower than those found in these other systems. In addition, the generally shallow nature of Gulf estuaries (Chapter 1 this volume) and the low residence time of these systems during periods of high fresh-water discharge (Chapter 2 this volume) result in the flushing of nutrients through the estuaries and out into the near-coastal waters of the Gulf during high discharge winter periods. The net result of these processes is that the river-dominated estuaries of the Gulf do not accumulate high standing stocks of dissolved inorganic nutrients during the winter, as occurs in estuaries that have more concentrated inputs, larger estuarine volumes, and longer fresh-water residence times (e.g., the Chesapeake Bay and Delaware Bay estuaries: Sharp et al. 1982; Fisher et al. 1988).

A majority of the river-dominated estuaries of the Gulf of Mexico lie above 26°N, where seasonal temperature cycles range from a minimum of ~10°C during the winter to a maximum of ~30°C during summer. This temperature range is capable of supporting high rates of phytoplankton production year round (Eppley 1972) and—importantly—does not provide a period during which low rates of production and nutrient uptake occur. As a result, phytoplankton are important regulators of nutrient dynamics in these systems throughout the year. Phytoplankton production rates in the river-dominated estuaries of the Gulf are average (200–350 gC m^{-2} yr^{-1}) compared with those of other estuaries in the United States (Boynton et al. 1982). While the highest production rates are observed during summer in most Gulf estuaries, there is a notable lack of recurring high chlorophyll concentrations (>50 μg liter^{-1}) such as those that characterize winter-spring blooms in more nutrient-enriched estuarine systems (Boynton et al. 1982; Pennock 1985). As a result, the general pattern in phytoplankton biomass and production over the annual cycle in most Gulf estuarine systems is dominated by variability.

Phytoplankton production and biomass in the river-dominated estuaries of the Gulf appear to be regulated by several factors, including short residence time, light availability, and nutrient availability, as seen in other river-dominated estuaries (D'Elia et al. 1986; Pennock and Sharp 1994). During periods of high fresh-water discharge, residence times are short and turbidity is high, resulting in the displacement of phytoplankton processes to the coastal plume (e.g., Apalachicola Bay and Mobile Bay). Since nutrient loading to river-dominated Gulf systems is primarily a result of high discharge rate rather than high nutrient concentration, it suggests that the Gulf estuaries will fall below other systems in the established relationship between phytoplankton production/chlorophyll concentration and nutrient loading (Nixon 1982, 1992; Nixon et al. 1986). At the same time, the seasonal pulsing of nutrients through these shallow systems is important to sustaining the productivity of a region that is among the most productive coastal ecosystems in the world (Deegan et al. 1986; Nixon et al. 1986).

During periods of low river discharge, nutrient limitation of phytoplankton production occurs, varying both seasonally and spatially among river-dominated Gulf estuaries. For example, phytoplankton production in Apalachicola Bay, Mobile Bay, Corpus Christi Bay, and the Mississippi River plume exhibited nitrogen limitation during summer, while phosphorus limitation has been observed in Apalachicola Bay and the Mississippi River plume during winter and spring, respectively. Silicon limitation of phytoplankton has also been reported during spring in the Mississippi River plume. In shallow Gulf estuaries, temporal alternation between nitrogen and phosphorus limitation is affected by patterns of sediment nutrient flux and wind mixing (Fernandez 1995; Chapter 6 this volume). While these processes undoubtedly contribute to the high degree of variability in nutrient and phytoplankton distributions observed in Gulf estuaries, additional research is required to establish the linkage, if any, between observed nutrient limitation and secondary production in these systems.

Marine-Dominated Estuaries

Florida Bay and Tampa Bay represent a class of estuaries that—by definition—has limited fresh-water input. As a result, residence times in these systems are often extremely long, and retention of nutrients within these systems is great. Florida Bay is the largest marine-dominated system in the North American subtropical and temperate zones. Nutrient input to the system occurs mainly through fresh-water discharge from the Everglades; however, internal nutrient cycling dominates nutrient distributions in the system. As a result, DIN distributions are generally dominated by NH_4 rather than NO_3, such as occurs in the river-dominated estuaries of the northern Gulf. The lack of a dominant fresh-water source and the carbonate geology of Florida Bay are important regulators of phosphorus dynamics in the system. As a result, PO_4 concentrations are extremely low throughout the year, and phosphorus is the limiting nutrient for both phytoplankton and seagrass production in the system.

As in many estuarine ecosystems, the experimental assessment of nutrient limitation in an estuary does not necessarily imply low rates of primary production. In fact, the production rate of $400 \, gC \, m^{-2} \, y^{-1}$ observed under phosphorus-limiting conditions in the central region of Florida Bay is among the highest in the Gulf. In this case, it appears that mechanisms that foster the retention of organic matter, particularly TP, result in rapid cycling of phosphorus and high rates of production despite low PO_4 concentrations. In contrast, the eastern region is fivefold less productive ($75 \, gC \, m^{-2} \, yr^{-1}$) as a result of chronic phosphorus limitation. The role that nutrient limitation plays in regulating phytoplankton production and biomass can be seen in the response of Tampa Bay to PO_4 loading from phosphate mining operations. As a result of these inputs, water-column chlorophyll concentrations increased one to two orders of magnitude during the 1970s and 1980s (Boler 1995). These data provide evidence that marine-dominated systems such as Florida Bay and Tampa Bay are particularly susceptible to nutrient over-enrichment.

Coastal Lagoons

Coastal lagoons are extremely important in the processing of nutrients and in primary and secondary production in regions with minor fresh-water input (Nixon 1982; Espino and Medina 1993; Herrera-Silveira 1996). The coastal lagoons of the Yucatan Peninsula discussed herein have seasonal nutrient cycles similar to those of many of the river-dominated estuaries of the northern Gulf as a result of seasonal precipitation cycles. Although considerable inter-annual variability occurs in maximum values, nutrient concentration maxima in these lagoons occur during the fall wet period. Comparative data from Celestun, Chelem, Dzilam, and Rio Lagartos lagoons show generally low water-column chlorophyll concentrations and phytoplankton production values that are correlated with fresh-water—ground-water—input. As with Florida Bay, PO_4 is the nutrient most likely to limit phytoplankton production, although light availability and high tannin concentrations have been shown to regulate production at different times.

Of the estuarine ecosystems in the Gulf of Mexico, the Yucatan lagoons provide the best example of natural nutrient and production patterns. These patterns show a natural seasonal pulsing of nutrients from precipitation cycles in the watershed, the influence of natural biological communities (waterfowl rookeries), and aperiodic natural events such as hurricanes. While these "pulsing events" can result in nutrient concentrations in excess of 20 μM DIN and 4 μM PO_4, they appear to be able to be assimilated into these estuaries without causing the dramatic effects that can result from cultural eutrophication.

Comparison of Gulf and Cool-Temperate Estuaries

In cool-temperate estuaries, phytoplankton productivity maxima generally occur during spring (e.g., Hudson River Estuary: Malone 1984; Delaware Bay: Pennock and Sharp 1986) as a result of nutrient-enhanced accumulation of high biomass and during mid-summer periods of maximal water temperature (e.g., Chesapeake Bay: Boynton et al. 1982). Data presented herein show different patterns of nutrient distribution and phytoplankton production for Gulf estuaries compared with higher-latitude North American estuaries. These differences are primarily attributable to the fact that, except for the Mississippi River, nutrient concentrations in Gulf estuaries are generally lower—particularly at the end of the winter—than those in these other systems. This difference results in lower chlorophyll concentrations during bloom events and less accumulation of unassimilated phytoplankton material in stratified bottom waters than observed in cool-temperate estuaries in the United States (Cowan et al. 1996).

In addition, the fact that winter temperatures do not approach freezing in Gulf estuarine waters results in year-round patterns of production that are better denoted by high rates of intra-annual variability than by predictable seasonal maxima and minima. Overall, annual phytoplankton production rates exhibit significant variability, ranging from 75 to 400 g C m^{-2} yr^{-1} among all Gulf estuaries (Table 5-3), with the river-dominated estuaries ranging from 188 ± 53 g C m^{-2} yr^{-1} (1 S.D.) to 336 ± 44 g C m^{-2} yr^{-1}.

Finally, these comparisons show that Gulf estuaries do not cleanly fit our predicted relationship between DIN loading and phytoplankton production (Nixon 1992). For example, Mobile Bay receives higher DIN and PO_4 loading per unit area and volume than Chesapeake Bay (Pennock et al. 1994b; Chapter 4 this volume), although phytoplankton production rates in mid-Mobile Bay (249 g C m^{-2} yr^{-1}: Table 5-3) are significantly less than those in mid-Chesapeake Bay (459 g C m^{-2} yr^{-1}: Boynton et al. 1982). We hypothesize that this difference is due to flushing of significant quantities of nutrients offshore during periods of high river discharge and low residence time. Similar flushing patterns occur in Apalachicola Bay and other Gulf estuaries.

Ecosystem Responses to Nutrient Enrichment

Nutrient enrichment of estuaries and near-coastal waters is responsible for the high productivity of these ecosystems, but it may also initiate a cascade of detrimental food web changes that are often collected under the general term of "eutrophication" (Caddy 1993). The Mississippi River provides an example of a river-dominated Gulf estuary in which anthropogenic nutrient enrichment is blamed for causing phytoplankton over-production, hypoxic bottom waters, and associated fisheries declines (Rabalais et al. 1994, 1996; Turner and Rabalais 1994). There is also concern over the increase in the ratio of NO_3 to SiO_2 in the Mississippi, which has risen to ~1:1 over the past 50 years. Further increases in DIN and/or reductions in SiO_2 could alter species composition from the typically diatom-dominated community that serves as the base of the coastal food web (Turner and Rabalais 1991, 1994; Chapter 4 this volume). In the case of the Mississippi, the strong correlation between increased dissolved nutrient loading and increased area affected by bottom water hypoxia between 1950 and 1990 resulted in major research and management initiatives in the 1990s.

In contrast to the Mississippi, many Gulf estuaries—such as Mobile Bay— have not displayed similar increases in nutrient concentration and loading over the past 25 years (Paulson et al. 1993). However, many of these systems have long-term records of hypoxia going back to the 1800s. This suggests that while the river-dominated estuaries of the Gulf of Mexico are not as nutrient enriched as the higher-latitude estuaries of the United States, they may be more susceptible to the negative effects of nutrient over-enrichment. Several factors contribute to this susceptibility, including (1) strong stratification caused by high rates of fresh-water discharge, (2) generally low tidal mixing energy throughout the Gulf, (3) high rates of nutrient loading from watersheds that drain a majority of the continental United States, and (4) rapid population growth in the watersheds that is projected for the next several decades.

Similarly, Florida Bay, the Laguna Madre in Texas, and the Yucatan lagoons provide examples of estuarine systems that are clearly susceptible to anthropogenic influences. This susceptibility is due, at least in part, to the relative roles that macrophytes and phytoplankton play in the production of these systems. In the Gulf, a switch from seagrass- to plankton-dominated habitats has been observed as a result of eutrophication (Tampa Bay), fresh-water diversion (Florida Bay), and the onset of a "brown tide" (Laguna Madre).

It is apparent from this review that ecosystems with wide heterogeneity and seasonal variability in nutrient loading and production processes, but with very low levels of impacts—for example, Celestun lagoon—show the ability to recover after major events of high intensity and low frequency. At the same time, many impacted systems in the Gulf show less resilience, suggesting the need for working management plans. Unfortunately, many Gulf estuaries do not clearly fit the established eutrophication paradigm currently applied to management decisions. For example, existing paradigms would not predict that a shallow (3-m-deep) estuary such as Mobile Bay would be strongly stratified and hypoxic or that estuaries such as Florida Bay would suffer from nutrient-enhanced species shifts despite the lack of obvious external nutrient loading. These differences make it essential to gain a better understanding of the diversity of ecosystem responses to nutrient over-enrichment in Gulf estuaries and the role that biogeochemical cycling plays in these responses. Such efforts will also be important in evaluating the responses to longer-term forcing such as the El Nino/Southern Oscillation (ENSO) and other global changes processes. Gulf estuaries receive fresh-water input from watersheds that receive significantly greater amounts of precipitation during ENSO events. Therefore, Gulf estuaries might be expected to exhibit greater ENSO-related nutrient input from fresh-water sources than North American estuaries located at higher latitudes. This hypothesis is testable but requires longer-term data collection than occurs during conventional short-term estuarine research funding.

ACKNOWLEDGMENTS

Funding for the research presented in this chapter was provided through the South Florida Water Management District (C-7919) and Everglades National Park (CA5280-2-9017) for Florida Bay, the Northwest Florida Water Management District and the Florida Department of Environmental Protection for Apalachicola Bay, the U.S. Geological Survey (14-08-0001-A0775UA) and

National Science Foundation (OSR-9108761) for Mobile Bay, and CONACYT and CONABIO for the coastal lagoons of Yucatan.

This is contribution 61 of the Southeast Environmental Research Program at Florida International University and contribution 299 from the Dauphin Island Sea Lab.

REFERENCES

Armstrong, N. E. 1987. The ecology of open-bay bottoms of Texas: A community profile. Biological Report No. 85(7.12). National Wetlands Research Center, Fish and Wildlife Service, U.S. Department of the Interior.

Benner, R., J. D. Pakulski, M. McCarthy, J. I. Hedges, and P. G. Hatcher. 1992. Bulk chemical characteristics of dissolved organic matter in the ocean. Science **255:** 1561–1564.

Bodle, M. J., A. P. Ferriter, and D. D. Thayer. 1994. The biology, distribution, and ecological consequences of *Melaleuca quinqueneria* in the Everglades, p. 341. *In:* S. M. Davis and J. C. Ogden [eds.], Everglades, the ecosystem and its restoration. St. Lucie Press.

Boler, R. 1995. Surface water quality 1992–1994: Hillsborough County, Florida. Environmental Protection Commission of Hillsborough County.

Bowman, M. J. 1977. Nutrient distributions and transport in Long Island Sound. Estuar. Coast. Marine Sci. **5:** 531–548.

Boyer, J. N., J. W. Fourqurean, and R. D. Jones. 1997. Spatial characterization of water quality in Florida Bay and Whitewater Bay by multivariate analysis: Zones of similar influence (ZSI). Estuaries **20:** 743–758.

———, and R. D. Jones. (In press). Effects of freshwater inputs and loading of phosphorus and nitrogen on the water quality of Eastern Florida Bay. *In:* K. R. Reddy (ed.), Phosphorus biogeochemistry in sub-tropical ecosystems. CRC Press.

Boynton, W. R., W. M. Kemp, and C. W. Keefe. 1982. A comparative analysis of nutrients and other factors influencing estuarine phytoplankton production, p. 69–91. *In:* V. S. Kennedy [ed.], Estuarine comparisons. Academic Press.

Butler, M. J. IV, J. V. Hunt, W. F. Herrnkind, M. J. Childress, R. Bertleson, W. Sharp, T. Matthews, J. M. Field, and H. G. Marshall. 1995. Cascading disturbances in Florida Bay, USA: cyanobacteria blooms, sponge mortality and implications for juvenile spiny lobsters *Panulirus argus*. Mar. Ecol. Proc. Ser **129:** 119–125.

Caddy, J. F. 1993. Toward a comparative evaluation of human impacts on fishery ecosystems of enclosed and semi-enclosed seas. Rev. Fisheries Sci. **1:** 57–95.

Carlson, P. R., L. A. Yarbro, and T. R. Barber. 1994. Relationship of sediment sulfide to mortality of *Thalasia testudinum* in Florida Bay. Bull. Mar. Sci. **54:** 733–746.

Cloern, J. E. 1996. Phytoplankton bloom dynamics in coastal ecosystems: A review with some general lessons from sustained investigation of San Francisco Bay, California. Rev. Geophys. **34:** 127–168.

———, B. E. Cole, R. L. J. Wong, and A. E. Alpine. 1985. Temporal dynamics of estuarine phytoplankton: A case study of San Francisco Bay. Hydrobiologia **129:** 153–176.

Comin, F. A., J. A. Herrera-Silveira, and M. Martin. 1997. Flamingo footsteps enhance nutrient release from the sediment to the water column. Wetland International Publication **43:** 211–227.

Conley, D. J., and T. C. Malone. 1992. Annual cycle of dissolved silicate in Chesapeake Bay: Implications for the production and fate of phytoplankton biomass. Mar. Ecol. Prog. Series **81:** 121–128.

Copeland, B. J. 1966. Effects of decreased river flow on estuarine ecology. Estuar. Ecol. **38:** 1831–1839.

Cowan, J. L. W., J. R. Pennock, and W. R. Boynton. 1996. Seasonal and inter-annual patterns of sediment-water nutrient and oxygen fluxes in Mobile Bay, Alabama (USA): Regulating factors and ecological significance. Mar. Ecol. Prog. Series **141:** 229–245.

Deegan, L. A., J. W. Day, J. G. Gosselink, A. Yanez-Arancibia, G. Soberon-Chavez, and P. Snachez-Gil. 1986. Relationships among physical characteristics, vegetation distributions and fisheries yield in Gulf of Mexico Estuaries, p. 83–100. *In:* D. A. Wolfe [ed.], Estuarine variability. Academic Press.

DeJonge, V. N., and F. Colijn. 1994. Dynamics of microphytobenthos biomass in the Ems estuary. Mar. Ecol. Prog. Series **104:** 186–196.

D'Elia, C. F., J. G. Sanders, and W. R. Boynton. 1986. Nutrient enrichment studies in a coastal plain estuary: Phytoplankton growth in large-scale, continuous cultures. Can. J. Fish. Aquat. Sci. **43:** 1945–1955.

DeYoe, H. R., D. A. Stockwell, R. R. Bidigare, M. Latasa, P. W. Johnson, P. E. Hargraves, and C. A. Suttle. 1997. Description and characterization of the algal species *Aeroumbra lagunensis* Gen. ET. sp. Nov. and referral of *Aureoumbra* and *Aureococcus* to the Pelagophyceae. J. Phycol. **33:** 1042–1048.

Dortch, Q., and J. R. Postel. 1989. Phytoplankton-nitrogen interactions, p. 139–173. *In:* M. R. Landry and B. M. Hickey [eds.], Coastal oceanography of Washington and Oregon. Elsevier.

Dunn, D. D. 1996. Trends in nutrient inflows to the Gulf of Mexico from streams draining the conterminous United States, 1972–93. Water-Resources Investigations Report No. 96-4113. U.S. Geological Survey.

Durako, M. J., and K. M. Kuss. 1994. Effects of *Labyrinthula* infection on the photosynthetic capacity of *Thalassia testudinum* in Florida Bay. Bull. Mar. Sci. **54:** 727–732.

Eastbrook, R. H. 1973. Phytoplankton ecology and hydrography of Apalachicola Bay. MS thesis, Florida State University.

Eppley, R. W. 1972. Temperature and phytoplankton growth in the sea. Fishery Bull. **70:** 1063–1085.

Espino, G. d. l. L., and M. A. R. Medina. 1993. Nutrient exchange between subtropical lagoons and the marine environment. Estuaries **16:** 273–279.

Fernandez, F. 1995. Nitrogen and phosphorus fluxes across the sediment-water interface during summer oxic and hypoxic/anoxic periods in Mobile Bay, Alabama. MS thesis, University of South Alabama.

Fisher, T. R., J. L. W. Harding, D. W. Stanley, and L. G. Ward. 1988. Phytoplankton, nutrients, and turbidity in the Chesapeake, Delaware, and Hudson estuaries. Estuar. Coast. Shelf Sci. **27:** 61–93.

————, E. R. Peele, J. W. Ammerman, and L. W. Harding. 1992. Nutrient limitation of phytoplankton in Chesapeake Bay. Mar. Ecol. Progr. Series **82:** 51–63.

Flint, R. W. 1984. Phytoplankton production in the Corpus Christi Bay estuary. Contrib. Mar. Sci. **27:** 65–83.

Fourqurean, J. W., R. D. Jones, and J. C. Zieman. 1993. Processes influencing water column nutrient characteristics and phosphorus limitation of phytoplankton biomass in Florida Bay, Florida, USA: Inferences from spatial distributions. Estuar. Coast. Shelf Sci. **36:** 295–314.

———, V. N. Powell, and J. C. Zieman. 1992. Phosphorus limitation of primary production in Florida Bay: Evidence from the C:N:P ratios of the dominant seagrass, *Thalassia testudinum.* Limnol. Oceanogr. **37:** 162–171.

Herrera-Silveira, J. A. 1994. Nutrients from underground discharges in a coastal lagoon Celestun, Yucatan, Mexico. Verh. Internat. Verein. Limnol. **25:** 1398–1401.

———. 1996. Salinity and nutrients in a tropical coastal lagoon with groundwater discharges to the Gulf of Mexico. Hydrobiologia **321:** 165–176.

———, and J. R. Ramirez. 1996. Effects of natural phenolic material (tannin) on phytoplankton growth. Limnol. Oceanogr. **41:** 1018–1023.

Iverson, R., W. Landing, B. Mortazavi, J. Fulmer, and F. G. Lewis. 1997. Apalachicola River and Bay Freshwater Needs Assessment—nutrient transport and primary productivity in the Apalachicola River and Bay. No. Florida State University.

Kjerfve, B. 1994. Coastal lagoon processes. Elsevier.

Lankford, R. R. 1977. Coastal lagoons of Mexico: Their origin and classification, p. 182–215. *In:* Estuarine processes, volume 2. Academic Press.

Lehrter, J. C., J. R. Pennock, and G. B. McManus. (In press). Microzooplankton grazing and nitrogen excretion across an estuarine/coastal interface. Estuaries.

Livingston, R. J. 1984. The ecology of the Apalachicola Bay system: An estuarine profile. Technical Report No. FWS/OBS-82-05. U.S. Department of the Interior, Fish and Wildlife Service.

Lohrenz, S. E., M. J. Dagg, and T. E. Whitledge. 1990. Enhanced primary production at the plume/oceanic interface of the Mississippi River. Cont. Shelf Res. **10:** 639–664.

———, G. L. Fahnenstiel, and D. G. Redalje. 1994. Spatial and temporal variation of photosynthetic parameters in relation to environmental conditions in coastal waters of the northern Gulf of Mexico. Estuaries **17:** 779–795.

Madden, C. J. 1986. Distribution and loading of nutrient in Fourleague Bay, a shallow Louisiana estuary. MS thesis, Louisiana State University.

Magnien, R. E., R. M. Summers, and K. G. Sellner. 1992. External nutrient sources, internal nutrient pools, and phytoplankton production in Chesapeake Bay. Estuaries **15:** 497–516.

Malone, T. C. 1984. Anthropogenic nitrogen loading and assimilation capacity of the Hudson River estuarine system, USA, p. 291–311. *In:* V. S. Kennedy [ed.], The estuary as a filter. Academic Press.

———, D. J. Conley, T. R. Fisher, P. M. Glibert, and L. W. Harding. 1996. Scales of nutrient-limited phytoplankton productivity in Chesapeake Bay. Estuaries **28:** 371–385.

May, E. B. 1973. Extensive oxygen depletion in Mobile Bay, Alabama. Limnol. Oceanogr. **18:** 353–366.

Meybeck, M., G. Cauwet, S. Dessery, M. Somville, D. Gouleau, and G. Billen. 1988. Nutrients (organic C, P, N, Si) in the eutrophic river Loire (France) and its estuary. Estuar. Coast. Shelf Sci. **27:** 595–624.

Milliman, J. D., and R. H. Meade. 1983. World-wide delivery of river sediment of the oceans. J. Geol. **91:** 1–21.

Myers, V. B. 1977. Nutrient limitation of phytoplankton productivity in north Florida coastal ecosystems, technical considerations, spatial patterns and wind mixing effects. Ph.D. thesis, Florida State University.

Myers, V. B., and R. L. Iverson. 1981. Phosphorus and nitrogen limited phytoplankton production in the Gulf of Mexico, p 569–582. *In:* B. J. Neilson and L. E. Cronin [eds.], Estuaries and nutrients. Humana Press.

Nelson, D. M., and Q. Dortch. 1996. Silicic acid depletion and silicon limitation in the plume of the Mississippi River: Evidence from kinetic studies in spring and summer. Mar. Ecol. Prog. Series **136:** 163–178.

Nixon, S. W. 1981. Fresh water inputs and estuarine productivity, pp. 31–58. *In:* R. D. Cross and D. L. Williams [eds.]. Proceedings of the National Symposium on Freshwater Inflow to Estuaries, Vol. 1. FWS/OBS-81/04.

———. 1982. Nutrient dynamics, primary production and fisheries yelds of lagoons. Oceanologica Acta. **Spec. Ed.:** 357–371.

———. 1992. Quantifying the relationship between nitrogen input and the productivity of marine ecosystems, pp. 107–137. *In:* H. Kawanabe, Y. Date, T. Kikuchi, E. Harada, I. Miyata, F. Koike, and H. Kunii [eds.], For richer ecological systems of brackish water zones. Kokuchosha Publishers.

———, C. A. Oviatt, J. Frithsen, and B. Sullivan. 1986. Nutrients and productivity of estuarine and coastal marine ecosystems. J. Limnol. Soc. South Africa **12:** 43–71.

Odum, H. T., and R. F. Wilson. 1961. Further studies on reaeration and metabolism of Texas bays, 1958–1960. Contrib. Mar. Sci. **8:** 23–55.

Paasche, E., and S. Kristiansen. 1982. Nitrogen nutrition of the phytoplankton in the Oslofjord. Estuar. Coast. Shelf Sci. **14:** 237–249.

Paulson, R. W., E. B. Chase, J. S. Williams, and D. W. Moody. 1993. National Water Summary 1990–91. Water-Supply Paper No. 2400. U.S. Geological Survey.

Pennock, J. P. 1985. Chlorophyll distributions in the Delaware estuary: Regulation by light-limitation. Estuar. Coast. Shelf Sci. **21:** 711–725.

———. 1987. Temporal and spatial variability in phytoplankton ammonium and nitrate uptake in Delaware Bay. Estuar. Coast. Shelf Sci. **24:** 841–857.

———, F. Fernandez, and W. W. Schroeder. 1994a. Mobile Bay Data Report—MB-01 to MB-34 cruises (May 1989–January 1993). Dauphin Island Sea Lab Technical Report No. 94-001. Dauphin Island Sea Lab.

———, and J. H. Sharp. 1986. Phytoplankton production in the Delaware Estuary: Temporal and spatial variability. Mar. Ecol. Prog. Series **34:** 143–155.

———, and J. H. Sharp. 1994. Temporal alternation between light- and nutrient-limitation of phytoplankton production in a coastal plain estuary. Mar. Ecol. Prog. Series **111:** 275–288.

———, J. H. Sharp, and W. W. Schroeder. 1994b. What controls the expression of estuarine eutrophication? Case studies of nutrient enrichment in the Delaware Bay and Mobile Bay estuaries, USA, p. 139–146. *In:* K. R. Dyer and R. J. Orth [eds.], Changes in fluxes in estuaries. International Symposium Series. Olsen & Olsen.

Peterson, D. H., R. E. Smith, S. W. Hager, D. D. Harmon, R. E. Herndon, and L. E. Schemel. 1985. Inter-annual variability in dissolved inorganic nutrients in northern San Francisco Bay estuary. Hydrobiologia **129:** 37–58.

Phlips, E. J., and S. Badylak. 1996. Spatial variability in phytoplankton standing crop and composition in a shallow inner-shelf lagoon, Florida Bay, Florida. Bull. Mar. Sci. **58:** 203–216.

Rabalais, N. N., R. E. Turner, Q. Dortch, W. J. Wiseman, and B. K. Sen Gupta. 1996. Nutrient changes in the Mississippi River and system responses on the adjacent continental shelf. Estuaries **19:** 386–407.

———, W. J. Wiseman, and R. E. Turner. 1994. Comparison of continuous records of near-bottom dissolved oxygen from the hypoxia zone along the Louisiana coast. Estuaries **17:** 850–861.

Randall, J. M., and J. W. Day, 1987. Effects of river discharge and vertical circulation on aquatic primary production in a turbid Louisiana (USA) estuary. Neth. J. Sea Res. **21:** 231–242.

Redalje, D. G., S. E. Lohrenz, and G. L. Fahnenstiel. 1994. The relationship between primary production and the vertical export of particulate organic matter in a river-impacted coastal ecosystem. Estuaries **17:** 829–838.

Redfield, A. C., B. H. Ketchum, and F. A. Richards. 1963. The influence of organisms on the composition of sea-water, p. 26–77. *In:* M. N. Hill [ed.], The sea. Wiley.

Robblee, M. B., T. R. Barber, P. R. Carlson, Jr., M. J. Durako, J. W. Fourqurean, L. M. Muehlstein, D. Porter, L. A. Yabro, R. T. Zieman and J. C. Zieman. 1991. Mass mortality of the tropical seagrass *Thalassia testudinum* in Florida Bay (USA). Mar. Ecol. Progr. Ser. **71:** 297–299.

Ryther, J. H., and W. M. Dunstan. 1971. Nitrogen, phosphorus, and eutrophication in the coastal marine environment. Science **3975:** 1008–1013.

Schreiber, R. A., and J. R. Pennock. 1995. The relative contribution of benthic microalgae to total microalgal production in a shallow sub-tidal estuarine environment. Ophelia **42:** 335–352.

Schroeder, W. W., and W. J. Wiseman. 1988. The Mobile Bay Estuary: Stratification, oxygen depletion, and jubilees, p. 41–52. *In:* B. Kjerfve [ed.], Hydrodynamics of estuaries, Volume 2. CRC Press.

Sharp, J. H., C. H. Culberson, and T. M. Church. 1982. The chemistry of the Delaware Estuary. General considerations. Limnol. Oceanogr. **27:** 1015–1028.

Shiller, A. M. 1993. A mixing rate approach to understanding nutrient distributions in the plume of Mississippi River. Mar. Chem. **43:** 211–216.

Smith, S. M., and G. L. Hitchcock. 1994. Nutrient enrichments and phytoplankton growth in the surface waters of the Louisiana Bight. Estuaries **17:** 740–753.

Stern, M. K., J. W. Day, and K. G. Teague. 1991. Nutrient transport in a riverine-influenced tidal freshwater bayou in Louisiana. Estuaries **14:** 382–394.

Tomas, C. 1996. The role of nutrient in initiating and supporting Florida Bay microalgal blooms and primary production, p. 89–92. *In:* 1996 Florida Bay conference. University of Florida Sea Grant.

Trefry, J. H., S. Metz, T. A. Nelsen, R. P. Trocine, and B. J. Eadie. 1994. Transport of particulate organic carbon by the Mississippi River and its fate in the Gulf of Mexico. Estuaries **17:** 839–849.

Turner, R. E., and N. N. Rabalais. 1991. Changes in Mississippi River water quality this century. BioScience **41:** 140–147.

———. 1994. Changes in Mississippi River nutrient supply and offshore silicate-based phytoplankton responses, p. 147–150. *In:* K. R. Dyer and R. J. Orth [eds.], Changes in fluxes to estuaries: Implications from science to management. Olsen and Olsen.

Valiela, I. 1995. Marine ecological processes, 2nd ed. Springer-Verlag.

Whitledge, T. E. 1989. Data synthesis and analysis—nitrogen processes study (NIPS): Nutrient distributions and dynamics in Lavaca, San Antonio and Nueces/Corpus Christi Bays in relation tofreshwater inflow; Part I—results and discussion. Technical Report No. 89-007. University of Texas, Marine Science Institute.

Williams, R. B., and M. B. Murdoch. 1966. Phytoplankton production and chlorophyll concentration in the Beaufort Channel, North Carolina. Limnol. Oceanogr. **11:** 73–82.

Zieman, J. C., J. W. Fourqurean, and R. L. Iverson. 1989. Distribution, abundance, and productivity of seagrasses and macroalgae in Florida Bay. Bull. Mar. Biol. **44:** 292–311.

Chapter **6**

Benthic Nutrient Fluxes in Selected Estuaries in the Gulf of Mexico

Robert R. Twilley, Jean Cowan, Tina Miller-Way, Paul A. Montagna, and Behzad Mortazavi

INTRODUCTION

Material cycling associated with benthic communities plays an important role in the ultimate fate of inorganic nutrients and organic matter in estuarine ecosystems (Nixon et al. 1976, 1980; Nixon 1981; Glibert 1982; Harrison et al. 1983; Kemp and Boynton 1984; Aller et al. 1985; Jensen et al. 1990). Benthic nutrient regeneration contributes to elevated rates of primary productivity in shallow coastal margin ecosystems by sequestering nutrients during periods of high river discharge and releasing these nutrients during seasons of increased productivity in coastal waters. Rates of benthic nutrient re-mineralization can be equivalent to allochthonous inputs of inorganic and organic nutrients (Nixon 1981; Fisher et al. 1982; Nixon and Pilson 1983; Pennock 1987; Cowan et al. 1996), thus influencing nutrient stoichiometry in estuarine ecosystems. Over short time scales, nutrient re-mineralization from the sediments and water column can provide a significant proportion of the nutrients required to support primary production (Zeitzschel 1980; Boynton et al. 1982; Fisher

Biogeochemistry of Gulf of Mexico Estuaries, Edited by Thomas S. Bianchi, Jonathan R. Pennock, and Robert R. Twilley.
ISBN 0-471-16174-8   © 1999 John Wiley & Sons, Inc.

et al. 1982; Klump and Martens 1983; Kemp and Boynton 1984; Boynton and Kemp 1985). Over longer time periods, the balance between organic matter deposition and nutrient re-mineralization determines the availability and fate of nutrients in estuarine ecosystems (McKee et al. 1983; DeMaster et al. 1985).

Conceptual models of nutrient dynamics in shallow temperate estuaries predict that seasonal deposition of organic matter during spring, followed by elevated rates of water column and benthic re-mineralization as water temperatures increase, sustain peak rates of primary productivity during the summer (Boynton et al. 1982; Kemp and Boynton 1984). The majority of studies, which have led to this conceptual model of benthic nutrient flux, have been conducted in temperate estuaries with strong seasonal fluctuations of river discharge and water temperature. The 39 estuaries of the Gulf of Mexico region (see Fig. 1-1 in Chapter 1) are different in many respects, such as climate and geomorphology, to these temperate estuaries. This partially landlocked Mediterranean-type sea indents the southeastern periphery of North America, encompassing 4000 km of shore-line that traverses a diversity of shallow coastal bays and estuaries. Regional climate varies from warm-temperate to subtropical and experiences warmer and less seasonably variable water temperatures. The Gulf of Mexico region has one of the largest ranges in fresh-water discharge from the seventh largest river in the world, the Mississippi River, to hypersaline lagoons with minor discharge in lower Texas (Deegen et al. 1986; Chapter 2 this volume). Geomorphological types range from deltaic systems of the central Gulf of Mexico coast characterized by extremely high rates of terrigenous input (Chapter 1 this volume). Most coastal bays in the Gulf of Mexico are shallow, and residence times among estuaries can vary from 2 to 200 d (Chapter 2 this volume). Thus the Gulf of Mexico estuaries represent significant variations in climate and geophysical forcings to test conceptual models of benthic nutrient regeneration.

The goal of this chapter is to describe the spatial and seasonal variation in benthic nutrient fluxes among nine selected estuaries in the Gulf of Mexico region (Fig. 6-1). These estuaries encompass a wide variety of estuarine types including typical coastal plain estuaries and lagoons of the Alabama, Mississippi, and north Florida coasts, deltaic systems of the Louisiana and north Texas coasts (including the extended estuary of the Mississippi River delta), and hypersaline lagoons of south Texas. Variations in temperature, precipitation, and evapotranspiration result in a strong gradient of local climate among the Gulf of Mexico estuaries (Table 6-1), as has been discussed in detail in Chapter 1. Superimposed on these regional differences are site-specific differences in size, mean depth, tidal range, fresh-water input, nutrient loading, and turnover time that may affect patterns of benthic flux (Table 6-1).

ECOLOGY OF BENTHIC NUTRIENT FLUXES

Factors Controlling Benthic Fluxes

The depositional flux of organic matter to bottom sediments is one of the dominant factors controlling benthic nutrient regeneration (Hargrave 1973; Nixon et al. 1976; Nixon 1981; Klump and Martens 1983; Kelly and Nixon 1984; Jensen et al. 1990; Billen et al. 1991; Blackburn 1991; Cowan and Boynton 1996), suggesting that there may be distinct differences in regeneration rates associated with specific geomorphological types of estuaries. In deltaic environments, the reactivity of organic matter deposited and the ratio of organic to inorganic sediment input influence rates of benthic nutrient regeneration (Westrich and Berner 1984; Canfield 1989). In river-dominated coastal environments, sedimentation of terrigenous materials and associated water-column turbidity dilute the labile or reactive organic matter reaching bottom sediments and inhibit in situ production of particulate material in the overlying water column by decreasing light penetration (DeMaster and Nittrouer 1983; Aller et al. 1985). The reactivity (relative to decomposition) of bottom sediments in these environments is inversely proportional to the sedimentation rate because terrigenous materials dominate sediment flux (Shokes 1976; Berner 1978; Aller et al. 1985).

In river-dominated estuaries with two-layer circulation caused by tides, there are distinct patterns of benthic nutrient fluxes along the salinity gradient associated with rates and quality of organic matter deposition (Boynton and Kemp 1985; Cowan and Boynton 1996; Middelburg et al. 1996). Peak ammonium recycling rates in the top few centimeters of sediment have been shown to occur where and when organic substrates are fresh and of higher quality, particularly following deposition of phytoplankton blooms (Blackburn 1991; Conley and Johnstone 1995). There is a lag in nutrient regeneration rates following depositional events that seems to be related to the quality of organic material (Rudnick and Oviatt 1986), and more labile nutrients are returned to the water column within 30 d (Gardner et al. 1991). In lagoons with minor river discharge and more wind and tidal energies, benthic regeneration rates are generally lower due to the lack of organic matter deposition and the utilization by epipelic algae of nutrients regenerated at the sediment-water interface (Corredor and Morell 1989). However, bioturbation by both macrophytes and macrofauna in these lower-energy environments can enhance benthic fluxes to rates similar to those at sites with high organic matter input.

As temperatures increase during the summer, regeneration processes in temperate estuarine sediments liberate large quantities of re-mineralized nutrients to the water column (Hargrave 1969; Kemp and Boynton 1984). Yet in deeper estuaries, summer is the period of maximum vertical stability, and because of minimum mixing, little of these regenerated nutrients is transported back to the surface water column (Malone et al. 1986). Often summer stratifi-

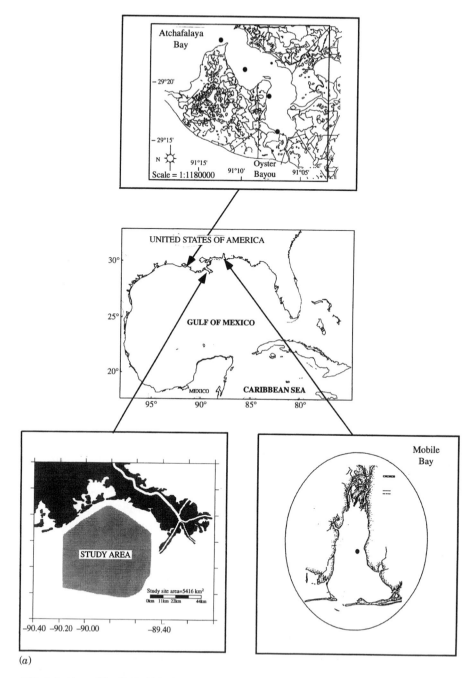

FIG. 6-1. Map of the Gulf of Mexico region with inserts of the selected estuaries where benthic nutrient regeneration rates have been studied. (A) Estuaries of the eastern and western sectors of the Gulf of Mexico. (B) Estuaries of the central region of the Gulf of Mexico.

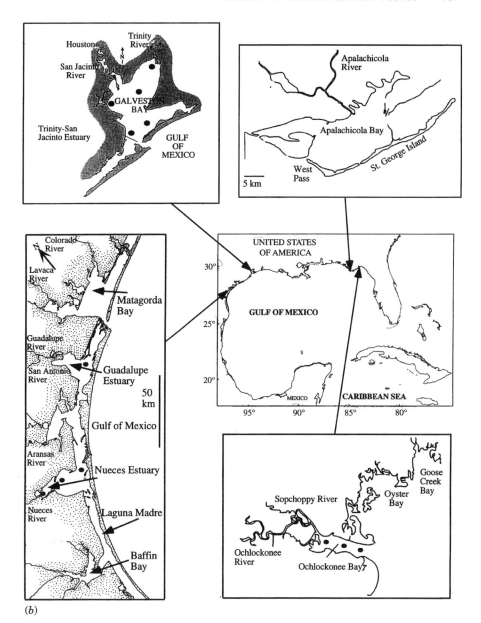

(b)

FIG. 6-1. (Continued).

TABLE 6-1. Climate, Geomorphology, and Geophysical Characteristics of the Eight Bounded Estuaries in this Survey of Benthic Nutrient Regeneration (excluding Mississippi River Bight)

Source	Parameter	Unit	Apalachee Bay (9)	Apalachicola Bay (10)	Mobile Bay (15)	Atchafalaya-Vermilion Bay (22)	Trinity-San Jacinto Estuary (25)	Guadalupe Estuary (28)	Nueces Estuary (30)	Laguna Upper (31)	Madre Lower (32)
1	Freeze-free period	days	270	300	270	300	300	300	300	330	330
1	Precipitation	mm yr^{-1}	1465	1432	1614	1533	1137	888	677	686	761
1	Potential Evapotranspiration	mm yr^{-1}	1025	1080	1040	1098	1125	1173	1118	1243	1283
1	Avg daily water deficit	mm yr^{-1}	3	9	1	2	85	285	441	557	522
2	Open water area	km^2	411.8	554.3	1059.3	1820.8	1359.8	531.0	497.3	1500.6	2009.8
1	Submersed aquatic vegetation	km^2	128.06	37.96	20.24	0	73.27	66.19	51.61	773.27	ND
2	Mean annual river discharge	m^3 s^{-1}	150	824	2246	6638	430	116	34	1.3	25
1	Mean tide range	m	1	0.72	0.36	0.36	0.3	0.18	0.39	0.42	0.42
1	Mean depth	m	1.3	2.9	3.0	2	2.3	1.1	1.2	1	ND
2	Estuary volume	km^3	1.28	1.50	3.20	3.88	2.57	0.70	1.19	1.50	7.22
2	Fresh-water fraction flushing	days	18	10	10	6	40	36	223	25	90
	Salinity zones										
2	Tidal fresh zone	km^2	2.59	20.72	75.11	492.10	33.67	0	0	0	0
2	Mixing zone	km^2	88.06	416.99	846.93	1585.08	1232.84	414.40	435.12	0	0
2	Seawater zone	km^2	321.16	116.55	137.27	0	132.09	116.55	62.16	558.92	948.46

Note: Numbers in parentheses refer to the location of the estuary in the Gulf of Mexico (see Fig. 1-1 in Chapter 1).

Source: Modified from Deegan et al. (1986)[1] and NOAA/EPA (1989).[2]

cation results in trapping of nutrients below the pycnocline, which mixes to the surface only infrequently (Kemp and Boynton 1984; Malone et al. 1986). In more shallow bays, stronger geophysical forcings reduce stratification in the water column, and resuspend surface sediments and pore water materials to the overlying water (Søndergaard et al. 1992; Vidal 1994). The lack of stratification in the water column allows materials to be mixed into surface waters, where they may enhance rates of primary production.

Stratification of estuaries also controls the redox conditions of bottom waters and sediment, largely by influencing the distribution of dissolved oxygen (DO). Reducing conditions at the sediment-water interface can alter the nature of nutrient flux by changing specific nutrient processes. Microbial processes such as nitrification (the oxidation of NH_4 to NO_3) are particularly sensitive to the absence of DO (Henriksen et al. 1981; Kemp et al. 1982; Henriksen and Kemp 1986; Kemp et al. 1990), and low-redox environments have higher regeneration of reduced N compared to either NO_3 or N gas (Seitzinger 1988; Kemp et al. 1990). Reduced sediments can also control the chemical sorption/desorption processes of sediments; and while most of these are associated with the cycling of P (Patrick and Khalid 1974; Sundby et al. 1992), there is some suggestion that N exchange may also be linked to chemical processes (Gardner et al. 1991). In most cases, the presence of hypoxia in bottom waters will result in higher fluxes of NH_4 and PO_4 from sediments to overlying water, promoting the input of nutrients that can contribute to poor water quality conditions.

There are also spatial patterns that are associated with benthic communities, particularly in the littoral zone of estuaries. The effect of both macroinfauna and macrophyte communities on the physical, chemical, microbial, and redox conditions of sediments is thus a very complex interaction with nutrient transformations that control benthic nutrient fluxes. Bioturbation associated with macroinfaunal communities can have a strong spatial effect by elevating rates of nutrient release from sediments in shallow estuaries (Henriksen et al. 1980, 1983; Banta et al. 1995). These increased fluxes are linked to both the metabolic effects of animals on water exchange and nutrient concentration between the water column and sediment, as well as the effect of benthos on sediment properties and microbial processes that can enhance nutrient exchange (Miller-Way 1994; Pelegri et al. 1994; Pelegri 1995). Benthic macrophyte communities can also modify the redox and chemical nature of estuarine sediments and alter patterns of nutrient exchange compared to unvegetated areas (McRoy et al. 1972; Caffrey and Kemp 1990; Rizzo 1990). Submersed vegetation enhances sediment deposition in the littoral zone and controls the physical matrix of sediments (Ward et al. 1984).

Benthic Fluxes and Ecosystem Stoichiometry

Shifts in the N:P ratio of regenerated nutrients from the expected ratio of 106C:16N:1P (Redfield 1958) have been postulated as controlling the metabolism of shallow estuarine ecosystems (Smith 1991). Stoichiometric signals of elements buried in sediments that vary from this theoretical ratio provide insights into the processes that have dominated an estuarine system over decadal time scales. Stoichiometric models of Narragansett Bay demonstrated that N : P ratios <16 occur because benthic re-mineralization of organic matter yields lower inorganic N relative to P fluxes, and lower amounts of N relative to depositing organic matter (Nixon and Pilson 1983). The amount of P recycled in Narragansett Bay was 125% of the input to the sediment, whereas N recycling rates were only 70% of the inputs (Nixon 1981). Material deposited to sediments had an N : P ratio of 13.3, yet the ratio of sediment re-mineralization ranged from 3.8 to 7.5. This pattern was also observed in the Patuxent River estuary, where N flux back to the water column was only 19% of that expected based on the N : P ratio of organic matter deposited from the water column to the sediments (Boynton et al. 1980).

The low ratios of N flux relative to P regeneration from sediments to the water column are primarily the result of N losses (as N_2 or N_2O) via denitrification (Seitzinger et al. 1980; Nixon 1981; Kemp et al. 1990). In sediments, the O : N ratio of fluxes across the sediment-water interface should be 13 if NH_4 is the end product (ammonification) and 17 if oxidation goes to NO_3 (ammonification coupled to nitrification). If ratios are high, then NH_4 flux out of sediments is less than expected based on rates of metabolism, indicating that N loss may be occurring via denitrification. O : N ratios were >100 in spring compared to ratios near the theoretical model of 13 in summer in tributaries of Chesapeake Bay (Boynton and Kemp 1985). Coupled nitrification and denitrification could account for the increase in O relative to N in fluxes during spring due to the transformation of regenerated NH_4 to N_2. The relative contribution of direct versus coupled denitrification rates to total N_2 loss from estuaries has seldom been determined (Seitzinger 1988). More important, the influence of these processes has not been interpreted relative to patterns of N regeneration from sediments (Rizzo et al. 1992).

REGIONAL DESCRIPTION OF SELECTED ESTUARIES

This survey of benthic nutrient fluxes in nine estuaries of the Gulf of Mexico includes Ochlockonee Bay, Apalachicola Bay, Mobile Bay, Mississippi River Bight, Fourleague Bay, Trinity-San Jacinto estuary, Guadalupe estuary, Nueces estuary, and Laguna Madre (Fig. 6-1). The geomorphology of these estuaries ranges from deltas with extremely high inputs of terrigenous sediments, as in coastal Louisiana, to coastal plain estuaries with limited runoff. Subtropical climates are indicated in Terminos lagoon and Florida Bay that have 365 freeze-free days, compared to Mobile Bay, which has the lowest number at 270 days (Table 6-1). Most of the Louisiana and Texas bays have around 300 freeze-free days, whereas south Texas systems are intermediate, with 330 days. Regional precipitation ranges from 1600 mm yr^{-1} in Mobile Bay to 686 mm

yr^{-1} in south Texas, compared to a range of only 1040 to 1586 mm yr^{-1} for evapotranspiration (Table 6-1). This establishes a strong moisture gradient, as indicated by water deficits ranging from only 1–2 mm yr^{-1} in the central Gulf of Mexico region to 557 mm yr^{-1} in south Texas (Table 6-1). River discharge varies among the selected estuaries from a high of 1879 $m^3 s^{-1}$ in Fourleague Bay to <5 $m^3 s^{-1}$ in the Upper Laguna Madre.

The regional geomorphology and circulation of the estuaries in this survey of benthic nutrient regeneration have been generally described in Chapter 1. Ochlockonee Bay is a sub-estuary of the Apalachee Bay system located in the panhandle region of west Florida. Together with the Apalachicola Bay, benthic fluxes from these two systems represent coastal plain lagoons and bar-built estuaries with significant runoff from mainly forested watersheds. The multiple connections of these estuaries with the Gulf of Mexico and the shallow depths (2–3 m) relative to upland runoff result in water residence times <1 mo owing to strong baroclinic and wind forcings (Chapters 1 and 2). Just to the west in Alabama is a more defined drowned river valley, Mobile Bay, with an area of 1070 km^2 and a mean depth just under 3.0 m. Approximately 85% of flow into the bay originates from the Mobile River system, consisting of the Mobile, Tombigbee-Black Warrior, and Alabama-Coosa-Tallapoosa river systems. This distributary system is the fourth largest in the United States (Morisawa 1968). The watershed is dominated by agricultural and silviculture activities. Periods of strong vertical stratification are common in this estuarine system (Chapter 1 this volume).

The coastal margin of Louisiana is a delta characterized by high rates of terrigenous input and an extensive shallow bay/wetland complex adjacent to a low-energy shallow continental shelf (Dagg et al. 1991; Day et al. 1994). Fourleague Bay is located on the eastern boundary of the Atchafalaya/Vermilion Bay complex that is part of the westernmost extent of the Mississippi River Deltaic Plain. The Atchafalaya River is the third largest river in the United States, consisting of flow at the confluence of the Red River and about 30% of the Mississippi River flow. The bay covers about 93 km^2, with an average depth of 1.5 m, and is surrounded by fresh, brackish, and saline marshes. This bay has little vertical stratification, and a large part of the estuary has salinities <5 g $liter^{-1}$ much of the year (Madden et al. 1988; Day et al. 1994).

The estuaries of Texas represent lagoons with a gradient in the amount of fresh-water input from Trinity-San Jacinto estuary (Galveston Bay) to Laguna Madre (Chapter 2, this volume). Guadalupe estuary (San Antonio Bay) occupies 530 km^2, and there is minimal exchange with the Gulf of Mexico through Cedar Bayou (Table 6-1). Pass Cavallo is shared with Matagorda Bay to the north, and the estuary also connects with Aransas Pass through Aransas Bay to the south. The average depth is approximately 1 m, and there are numerous oyster reefs in the shallow areas of the middle to lower regions. Guadalupe estuary receives most of its water from the Guadalupe River (70% of gaged flow) and the San Antonio River (26% of gaged flow) (Orlando et al. 1993).

Nueces estuary is composed of Nueces and Corpus Christi Bays (Fig. 6-1). The estuary is approximately 500 km^2, and its boundaries include the Calallen Dam and the Aransas Pass (Table 6-1). The estuary connects with Aransas Bay to the north via Redfish Bay and with Laguna Madre to the south via the Intracoastal Waterway. The average depth is about 1.3 m in Nueces Bay and 3 m in Corpus Christi Bay. The estuary receives 99% of its fresh water from the Nueces River (Orlando et al. 1993). Evaporation exceeds precipitation due to the semi-arid climate, and hypersalinity occurs during dry periods.

Hypersalinity also dominates the Laguna Madre estuary farther south along the Gulf of Mexico coast. This is the largest estuary in Texas, with an area of about 1500 km^2, a length of 170 km, and an average depth of 1 m. It is two systems, the Upper and Lower Laguna Madre, which are separated by a large mud-sand flat land bridge called the Land Cut. This land bridge is inundated intermittently by wind-driven tides. The two parts of the estuary are somewhat autonomous, and current studies were confined to the Upper Laguna Madre. The Upper Laguna Madre estuary consists of the Laguna Madre and Baffin Bay. Intermittent creeks drain into the secondary bays of Baffin Bay, which include Alazan Bay, Cayo del Grullo, and Laguna Salada. Direct precipitation contributes 65% of the total fresh-water discharge to the estuary, which averages only 1 m^3 s^{-1}, and gaged inflow represents only 17% of the discharge (Orlando et al. 1993). Exchange with the Gulf of Mexico occurs through Corpus Christi Bay to the north.

DESCRIPTION OF BENTHIC STUDIES

Experimental Designs

The studies included in this survey vary considerably in spatial and temporal coverage, as well as in methodological differences (Table 6-2). The specific details of each experimental design, such as core dimensions and sampling methods, can be found in references to each estuary. Continuous flow methodology used in Fourleague Bay and Mississippi Bight is described in Miller-Way et al. (1994) and Miller-Way and Twilley (1996). Closed system methods, used in most of the other estuaries, are the more traditional batch incubations described in Cowan and Boynton (1996) and Cowan et al. (1996). While most flux determinations were made shipboard or in the laboratory, the survey includes two studies using in situ deployments (Table 6-2), which are described in more detail below. Data from all studies were selected such that fluxes are representative of unvegetated substrates, static hydrodynamic conditions (no flow), and the absence of ambient light. Sampling sites in most studies were chosen as representative of the range of water salinities and organic carbon contents of sediments in the respective estuaries (Fig. 6-1). The most intensive

TABLE 6-2. Summary of Sampling Stations, Frequency, and Spatial Coverage of Sampling Stations in Each Estuary, with Relevant Information on the Design to Measure Benthic Fluxes, Including Pertinent References.

Station	Latitude	Longitude	Depth (m)	Dates	Methodology	Reference
Ochlockonee Bay (9)						
A	NA		1–2	4, 6, 11 '84; 3 '85	Batch incubations of cores in laboratory, 10-d incubations	Seitzinger et al. (1980); Seitzinger (1987)
B			1.5	4, 6, 11 '84; 3 '85		
C			1–2	4, 6, 11 '84; 3 '85		
D			2	4, 6, 11 '84; 3 '85		
E			1–3	4, 6, 11 '84; 3 '85		
Apalachicola Bay (10)						
A	29 40.43	84 57.64	2	9, 11, 12 '94; 1, 4, 8, 11 '95; 2, 6 '96	Batch incubations in the laboratory, 6-h dark incubations	Mortazavi et al. (submitted)
Mobile Bay (15)						
DR7	30 34.42	87 56.00	3.1	1–12 '93; 1–12 '94; 1, 4, 8 '95	Batch incubations in the laboratory, 4- to 6-h dark incubations	Cowan et al. (1996); Miller-Way et al. (unpublished data)
PC1	30 27.51	88 04.00	3.2	7, 10 '94; 1, 4, 8 '95		
WB7	30 22.27	87 55.21	2.8	7, 10 '94; 1, 4, 8 '95		
FM7	30 16.21	87 59.15	3.4	7, 10 '94; 1, 4, 8 '95		
Mississippi River Bight (19)						
B20	28 55.60	89 28.18	20	4 '88; 4 '89	Continuous flow with shipboard incubations, 4- to 6-h dark incubations	Bourgeois (1994); Miller-Way et al. (1994); Twilley and McKee (1996)
B50	28 50.62	89 29.06	50	4, 9 '89; 4, 10 '90; 11, '93; 6, '94		
C20	28 58.00	89 23.50	20	8 '87; 4, '88; 4 '89		

TABLE 6-2. (*Continued*).

Station	Latitude	Longitude	Depth (m)	Dates	Methodology	Reference
C50	28 52.50	89 31.50	50	8 '87; 4 '88; 4 '89; 4 '90; 10 '90; 11 '93; 6 '94		
C80	28 49.70	89 32.50	80	8 '87; 4 '88; 4 '89;		
D20	29 01.80	89 36.60	20	8 '87; 4 '88		
D50	28 56.00	89 35.60	50	8 '87; 4 '88; 4 '89; 4 '90; 10 '90; 11 '93; 6 '94		
E20	29 07.24	89 44.63	20	8 '87; 4 '88; 4 '89; 10 '90; 4 '92; 11 '93; 6 '94		
E50	28 56.75	89 43.55	50	8 '87; 4 '88; 4 '89; 10 '90; 4 '92; 11 '93; 6 '94		
E80	28 46.50	89 43.50	80	8 '87; 4 '88; 4 '89		
F20	29 08.25	89 50.00	20	11 '93		
F50	28 52.50	89 50.00	50	11 '93; 6 '94		
G50	28 43.05	90 01.97	50	11 '93; 6 '94		
H50	28 43.00	90 10.00	50	11 '93; 6 '94		
Fourleague Bay (22)						
1	29 15.70	91 08.30	1.5	4, 9, '86; 3, 8, '87; 4, 8, '88; 5, 8, '89	Continuous flow with shipboard incubations, 2- to 4-h dark incubations	Miller-Way (1994); Miller-Way and Twilley (1996); Twilley, unpublished data
2	29 17.50	91 07.00	1.5	4, 9, '86; 3, 8, '87 4, 8, '88; 5, 8, '89		
3	29 20.50	91 10.00	1.5	9, '86; 8, '87; 4, 8, '88; 8, '89		
4	29 23.50	91 14.05	1.5	8, '87; 4, 8, '88		

Trinity-San Jacinto estuary (25)

	Latitude	Longitude		Months	Method	Reference
1	29 42.06	94 44.38	2.1	3, 5, 7, '93	Batch incubations in the laboratory 2- to 3-d incubations	Yoon and Benner (1992); Zimmerman and Benner (1994)
2	29 37.45	94 49.43	2.9	3, 5, 7, '93		
3	29 33.22	94 59.44	2.8	3, 5, 7, '93		
4	29 22.59	94 50.34	2.4	3, 5, 7, '93		
5	29 26.27	94 42.54	1.8	3, 5, 7, '93		

Guadalupe estuary (28)

	Latitude	Longitude		Months	Method	Reference
A	28 23.61	96 46.34		1, 4, 7, '87; 4 '89	In situ benthic chambers with no flow, 2- to 3-h incubations	Montagna, unpublished data
B	28 20.87	96 44.74		1, 4, 7, '87; 4 '89		
C	28 14.92	96 45.62		1, 4, 7, '87; 4 '89		
D	28 18.13	96 41.06		1, 4, 7, '87; 4 '89		

Nueces River estuary (30)

	Latitude	Longitude		Months	Method	Reference
A	27 50.99	97 28.25		10, 12, '87; 2, 4, 5 '88	In situ benthic chambers with no flow, 2- to 3-h incubations	Montagna, unpublished data
B	27 50.86	97 23.97		10, 12, '87; 2, 4, 5 '88		
C	27 49.31	97 21.08		10, 12, '87; 2, 4, 5 '88		
D	27 42.60	97 10.73		10, 12, '87; 2, 4, 5 '88		
E	27 47.00	97 45.00		10, 12, '87; 2, 4, 5 '88		

Upper Laguna Madre (31)

	Latitude	Longitude		Months	Method	Reference
24	27 15.83	97 33.09	2.1	3, 5, 7, 9, 11 '89; 1 '90	Batch incubations of sediment cores in laboratory, 4- to 6-h incubations	Montagna, unpublished data
6	27 16.61	97 25.61	2.2	3, 5, 7, 9, 11 '89; 1 '90		
189	27 20.99	97 23.54	1.0	3, 5, 7, 9, 11 '89; 1 '90		
155	27 25.45	97 20.48	1.2	3, 5, 7, 9, 11 '89; 1 '90		

Note. Month are identified numerically (1 = January, 12 = December). NA = not available.

temporal coverage is for Station DR7 in Mobile Bay, while the most extensive spatial coverage is for the Mississippi River Bight (Table 6-2).

Benthic fluxes were measured using in situ chambers at four sites in Guadalupe Estuary, two in the upper bay and two in the lower bay. Benthic measurements were made twice each day, once in the morning and once in the afternoon. Diel differences were never found, so the two deployments are treated as replicates in the analyses. Only clear chambers were deployed; however, changes in DO and nutrients are probably due only to bacterial metabolic processes since turbidity is so high at the sediment surface that no photosynthesis occurs (MacIntyre and Cullen 1988).

Four stations were occupied along the axis of the Nueces Estuary to measure benthic fluxes; two stations were in Nueces Bay, and two stations were in Corpus Christi Bay. Results for dark chambers are reported only in this survey. Benthic flux measurements in Guadalupe and Nueces estuaries were made with benthic chambers designed like an annular flume (Fisher et al. 1982; Taghon et al. 1984). Only data for the dark chambers that serve as controls, with no current flow, are reported here. Chambers were deployed with four or eight synoptic replicates. Studies of benthic flux in Upper Laguna Madre included six stations, two in Baffin Bay and four in Laguna Madre (Fig. 6-1). The Laguna Madre stations were composed of two sites each, one over seagrasses and one over a bare, sandy patch. Samples were taken three times per day, in the morning, afternoon, and early evening.

Environmental Conditions of Benthic Studies

Water Column Characteristics. The lowest temperature of bottom waters was 7.8°C, recorded at Station DR7 in Mobile Bay, while most of the low temperatures were 15°C to 19°C (Fig. 6-2A). Higher temperatures ranged from 29.5°C to 32.0°C and were associated with the shallow bay in coastal Louisiana. The annual range in temperature among the eight shallow estuaries varied from 17°C in Mobile Bay to 10°C in Nueces estuary. The deeper stations in Mobile Bay (4 m) had greater variation in temperature, while the shallow bays in Texas and Louisiana (about 1 m) showed some of the least variation in water temperature in this survey. Average temperatures of benthic studies in the Gulf of Mexico estuaries ranged from 20°C to 26°C, reflecting observations dominated by spring and fall cruises. These temperatures are similar to those recorded in more spatially and temporally complete surveys of estuaries in this region (NOAA/EPA 1989). Stations in the Mississippi River Bight range from 20 to 50 m and had an average temperature of 20.5°C, which is much less variation than that found in the more shallow estuaries at 15°C to 30°C.

Average bottom water salinities among five of the estuaries from Ochlockonee and Mobile Bays to Guadalupe estuary were similar at about 8 g liter^{-1}, with a range from 0 to 20.0 g liter^{-1} (Fig. 6-2B). Of these five estuaries, salinities <3 g liter^{-1} were more frequent in the shallow coastal bays associated with deltaic geomorphology, represented by Fourleague Bay. Average salinity

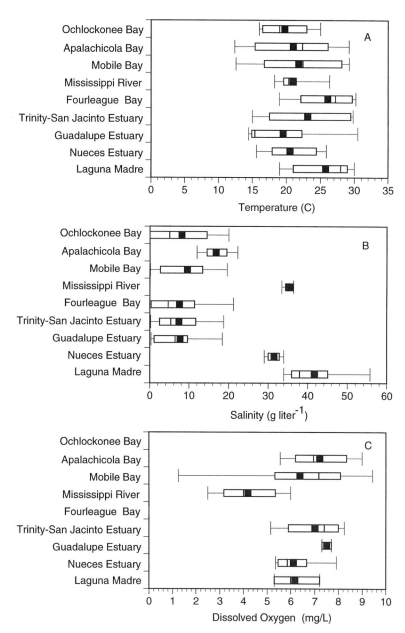

FIG. 6-2. *Average and range of bottom water-column characteristics for the different stations and sampling dates where benthic regeneration rates were determined in the respective estuaries. (A) Water column temperature. (B) Salinity. (C) Dissolved oxygen.*

was slightly higher in Apalachicola Bay at 18 g liter^{-1}, compared to much higher salinities in two southwest Texas estuaries that averaged 31.4 and 41.7 g liter^{-1} in Nueces estuary and Laguna Madre, respectively (Fig. 6-2B). The least variable salinity was observed in bottom waters of the Mississippi River Bight, at about 35 g liter^{-1}, in a region where surface salinities fluctuate with river discharge. Salinities observed in each respective estuary during benthic surveys reflect the diverse climatic and hydrographic conditions along the Gulf coast (Chapter 2 this volume). The arid coasts with limited river discharge in south Texas to the deltaic estuaries of the Louisiana coast are representative of the extreme residence times of water in the estuaries of the Gulf of Mexico.

Estuaries in the central Gulf of Mexico sector have large watersheds with seasonal fresh-water discharge, causing changes in salinity that are indicative of the short residence times of materials. In Fourleague Bay, measurements of net water fluxes through the upper bay entrance indicate a residence time of 6 d during high flow from February to April compared to 61 d during low flow from August to December (Denes 1983; Madden 1986). Approximately 70% of the annual sediment discharge to Fourleague Bay occurs during the period from February to May (Cratsley 1975; Caffrey and Day 1986). Peak fresh-water inflow to Mobile Bay occurs in April (4000 m^3 s^{-1}) compared to low flow in August to September (600 m^3 s^{-1}); thus, salinity can range from 0 to 35 g liter^{-1} in any one region of the bay over an annual cycle in response to river flow and frontal passages (Schroeder and Wiseman 1986; Cowan et al. 1996). The average daily input of fresh-water volume to Trinity-San Jacinto estuary from 1924 to 1990 was about 450 m^3 s^{-1} in May compared to 50 m^3 s^{-1} in August (NOAA/EPA 1989). The same seasonal pattern of fresh-water input occurred in Guadalupe estuary, and because the estuary is small and Gulf exchange is restricted, floods can completely replace brackish water and salinity can reach zero for extended periods (Montagna and Kalke 1992). In contrast, high inflow to the Nueces estuary occurs between September and November, when discharge flows can range between 30 and 600 m^3 s^{-1} compared to 10 to 140 m^3 s^{-1} during low inflow between June and August. The salinity in Laguna Madre is strongly influenced by isolated fresh-water pulses between September and November and high evaporation rates from June to August (Orlando et al. 1993).

Only one of eight shallow estuaries in this review recorded hypoxia in bottom waters during benthic flux measurements, and that was in the well-stratified Mobile Bay. In addition, hypoxia occurred at one of the 20-m-deep stations in the Mississippi River Bight (Fig. 6-2C). Despite its shallow depth, Mobile Bay is moderately stratified throughout the year, with the strongest gradients during the spring due to increased river discharge (NOAA/EPA 1989). Moderate stratification also occurs within central and western Trinity Bay and the Houston shipping channel, where 70% of the fresh-water inflow occurs to the bay from the Trinity River. The remainder of the estuary remains vertically homogeneous. Vertical stratification is thought to be rare in Nueces

Estuary (Orlando et al. 1993). However, it is common in the southeastern part of Corpus Christi Bay during June–August, when salinities are high, and hypoxia does occur in this region (Ritter and Montagna in review). Dissolved oxygen was >5 mg liter^{-1} for all of the other estuaries during studies of benthic fluxes. Hypoxia seldom occurs in any of these other estuaries owing to the well-mixed conditions of shallow water columns that are characteristic of the Gulf of Mexico estuaries (Chapter 1 this volume).

Average $NO_2 + NO_3$ concentrations during the different flux studies were higher in Guadalupe estuary at 28.3 μM, followed by an average of 16.1 μM in Trinity-San Jacinto estuary (Fig. 6-3A). Oxidized inorganic N concentrations in the other estuaries were <15 μM. The range in $NO_2 + NO_3$ concentrations in Guadalupe estuary was 1 to 92 μM, and the next highest range was in Trinity-San Jacinto Estuary at about 40 μM. Although there are high seasonal concentrations of $NO_2 + NO_3$ in the Mississippi River that can reach 100 μM (Smith et al. 1985; Madden et al. 1988; Teague et al. 1988), the highest concentration observed during the benthic flux studies in either Fourleague Bay or Mississippi River Bight was 32 μM. Average $NO_2 + NO_3$ concentrations were lowest in the most eastern and western estuaries of this survey, averaging <2 μM in Ochlockonee Bay and Laguna Madre, with very little seasonal or site variation during the benthic surveys.

Average NH_4 concentrations among the five coastal bays with average salinities <10 g liter^{-1} were <4 μM (Fig. 6-3B). However, two estuaries in south Texas with much higher bottom water salinities had average NH_4 concentrations of 6.0 and 5.3 μM at Corpus Christi and Laguna Madre, respectively. In Nueces estuary, NH_4 concentrations ranged as high as 15 μM during the benthic nutrient flux study. PO_4 concentrations followed the trend of $NO_2 + NO_3$, with higher average concentrations of 6.5 μM occurring at Guadalupe estuary, compared to average concentrations <1.5 μM in Laguna Madre, Fourleague, Mobile, and Apalachicola Bays. Average concentrations in Trinity-San Jacinto and Nueces estuaries ranged from 2 to 3 μM. The lowest average NH_4 and PO_4 concentrations among the eight estuaries occurred at Apalachicola estuary, Florida.

Sediment Characteristics. Average total sediment carbon (SC) concentrations were lower in Trinity-San Jacinto estuary, at 1.01% dry mass (dm), compared to the other estuaries in this survey (Fig. 6-4A). Average SC concentration was highest in the plume region of the Mississippi River Bight at 15.5%, compared to 3.1% dm for the highest concentration among the shallow estuaries, occurring in Apalachicola Bay. Variation in SC concentrations was greater within most of the estuaries than among the different systems, particularly when considering the variation in Guadalupe estuary, which ranged from 0.6 to 3.8% dm (Fig. 6-4A).

Average sediment nitrogen (SN) concentrations exhibited more variation among estuaries compared to within-system ranges (Fig. 6-4B), with the exception of the Mississippi River Bight stations. As observed for SC, this coastal

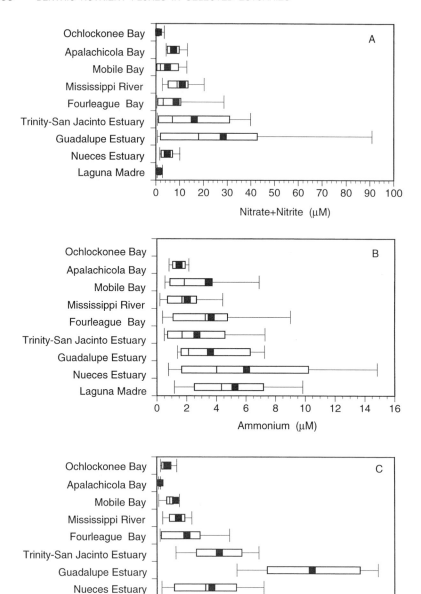

FIG. 6-3. *Average and range of bottom water-column characteristics for the different stations and sampling dates where benthic regeneration rates were determined in the respective estuaries. (A) Nitrate + nitrite. (B) Ammonium. (C) Phosphate.*

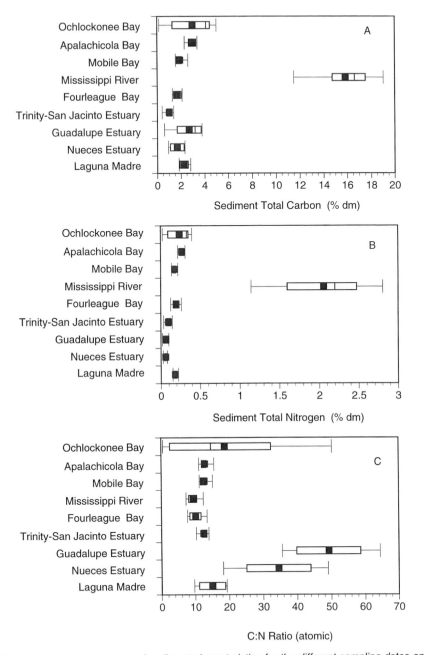

FIG. 6-4. *Average and range of sediment characteristics for the different sampling dates and stations where benthic regeneration rates were determined in the respective estuaries. (A) Sediment total carbon. (B) Sediment total nitrogen. (C) C:N atomic ratios.*

system had much higher N concentrations and more variation than was observed in the other estuaries. There were basically two groups of estuaries: those with concentrations from 0.18 to 0.25% dm, including Apalachicola Bay, Mobile Bay, Fourleague Bay, and Laguna Madre, and those with average SN concentrations <0.1% dm, occurring in Trinity-San Jacinto estuary, Guadalupe estuary, and Nueces estuary. C:N ratios were similar, at about 16, for the four estuaries that had higher SN concentrations (>0.15% dm), but Trinity-San Jacinto estuary is included in this range of C:N ratios since it had both lower SN and SC concentrations. The average C:N ratios at Nueces and Guadalupe estuaries were 35 and 49, respectively.

PATTERNS OF BENTHIC NUTRIENT FLUX

Comparison Among Estuaries

Average benthic fluxes of PO_4 ranged from -6 to 17.5 μmol m^{-2} h^{-1} among the different estuaries (Fig. 6-5A; no fluxes are available for Ochlockonee and Apalachicola Bays). The spatial and seasonal variation in PO_4 flux was -8 to 59 μmol m^{-2} h^{-1} in Mississippi River Bight, which had distinctly higher fluxes than those recorded in the shallow estuaries. Large variations in PO_4 fluxes were also observed in Nueces and Guadalupe estuaries and Fourleague Bay. In Nueces estuary, PO_4 uptake by sediment was nearly 36 μmol m^{-2} h^{-1}. Mobile Bay, Trinity-San Jacinto estuary, and Laguna Madre all had small ranges in PO_4 fluxes. Mobile Bay was the only site where all the PO_4 fluxes measured were from the sediment to the water column.

Average NH_4 fluxes were positive in seven of the nine estuaries, with higher averages observed in Fourleague Bay and Mississippi River Bight at about 140 and 125 μmol m^{-2} h^{-1}, respectively (Fig. 6-5B). Ammonium fluxes at Laguna Madre averaged about 2 μmol m^{-2} h^{-1}, while at Guadalupe estuary NH_4 uptake by sediments averaged -155 μmol m^{-2} h^{-1}. There was very little site and seasonal variation in NH_4 regeneration in Trinity-San Jacinto estuary and Laguna Madre, as was also observed in Ochlockonee and Apalachicola Bays. This is in contrast to rates that ranged from 10 to 330 μmol m^{-2} h^{-1} in Fourleague Bay and 10 to -750 μmol m^{-2} h^{-1} in Guadalupe estuary.

There were very distinct differences in average $NO_2 + NO_3$ benthic fluxes among the nine sites (Fig. 6-5C). Four estuaries—Ochlockonee Bay, Mobile Bay, Trinity-San Jacinto estuary, and Laguna Madre—along with the Mississippi River Bight, all had nearly undetectable average fluxes of $NO_2 + NO_3$ across the sediment-water interface, at average rates <20 μmol m^{-2} h^{-1}. This perspective is influenced by the highly variable rates measured in Guadalupe and Nueces estuaries, which had average $NO_2 + NO_3$ fluxes from sediment to the water column at rates of 176 and 85 μmol m^{-2} h^{-1}, respectively. Fluxes in Guadalupe estuary ranged from a slight uptake by sediments at some stations to benthic regeneration of 900 μmol m^{-2} h^{-1} to the water column

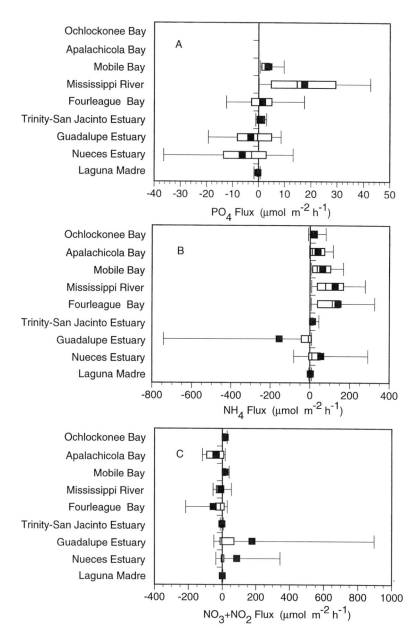

FIG. 6-5. *Average and range of benthic nutrient fluxes for the different stations and sampling dates in the respective estuaries. (A) PO_4 fluxes; (B) NH_4 fluxes; (C) $NO_3 + NO_2$ fluxes. Positive and negative nutrient rates represent fluxes out of and into the sediments, respectively.*

(Fig. 6-5C). In contrast, average $NO_2 + NO_3$ fluxes in Fourleague and Apalachicola Bays were from the water column to the sediment, including rates of -200 μmol m^{-2} h^{-1} observed in the Mississippi River delta during high river discharge in spring.

Average DIN fluxes were from sediment to the water column for all eight estuaries in this survey (Fig. 6-6A). Average benthic DIN release was higher in the Mississippi River Bight at 123 μmol m^{-2} h^{-1}, followed by Fourleague and Mobile Bays at 87 and 77 μmol m^{-2} h^{-1}, respectively. Average rates ranged from 20 to 30 μmol m^{-2} h^{-1} at Ochlockonee Bay, Apalachicola Bay, Guadalupe estuary, and Nueces estuary compared to <10 μmol m^{-2} h^{-1} at Trinity-San Jacinto estuary and Laguna Madre. Extreme variations in rates

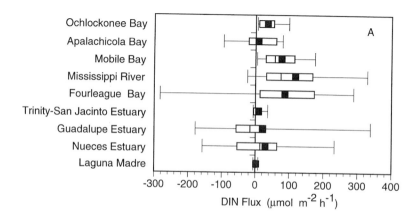

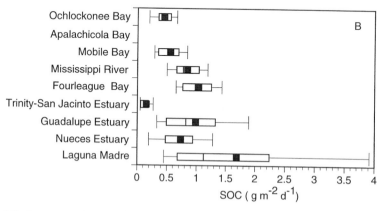

FIG. 6-6. Average and range of benthic nitrogen and dissolved oxygen fluxes for the different stations and sampling dates in the respective estuaries. (A) DIN fluxes. (B) SOC. All SOC measurements represent O uptake by the sediments; positive and negative N rates represent fluxes out of and into the sediments, respectively.

of DIN fluxes were observed for Fourleague Bay, Guadalupe estuary, and Nueces estuary.

Average rates of sediment oxygen consumption (SOC) varied among the nine estuaries from 0.1 to 1.7 g m^{-2} d^{-1}, with the low and high averages found in Texas estuaries, Trinity-San Jacinto estuary and Laguna Madre, respectively (Fig. 6-6B). SOC in Laguna Madre exhibited much variation, with rates from 0.18 to 5.72 g m^{-2} d^{-1} associated with vegetated and unvegetated patches of sediment (see the later section on bioturbation). The low average for Trinity-San Jacinto estuary is associated with low concentrations of SC (Fig. 6-4A). In the coastal plain estuaries of the central Gulf of Mexico, SOC is closely linked to river discharge, with higher rates occurring in Fourleague Bay along with increased SC (Fig. 6-6B). Yet this pattern between SC and SOC does not occur when including the Mississippi River Bight, where much higher concentrations of SC do not correspond to higher SOC. Benthic SOC rates in this survey do not measure anaerobic metabolism directly, and it has been suggested that other electron acceptors may account for much of the metabolism in the Mississippi River Bight region.

The extreme variation in benthic fluxes of nutrients and DO at the sediment-water interface limits any regional comparison among estuaries (Table 6-3). One general trend is that rates of sediment regeneration to the water column of both DIN and PO$_4$ occur in the central Gulf of Mexico estuaries, which are all coastal plain estuaries with significant river discharge. Lower rates of DIN and PO$_4$ flux occurred in all of the estuaries in Texas, particularly in Laguna Madre, where both inorganic N and P fluxes were not significantly different from zero. Comparing benthic fluxes within an estuary based on different studies limits our ability to discern patterns among Gulf estuaries. The results of our survey for Fourleague Bay based on flow-through cores incubated aboard ship are similar to those of a study using in situ chambers at similar stations (Table 6-3). Average uptake of NO$_2$ + NO$_3$ in both studies was 20 to 50 μmol m^{-2} h^{-1}, and NH$_4$ release ranged from 130 to 140 μmol m^{-2} h^{-1}; both studies measured very little exchange of PO$_4$. In contrast, the patchy sediment types in the Texas estuaries resulted in dissimilar patterns of nutrient flux in Guadalupe estuary in this survey compared to a study by Yoon and Benner (1992). In our survey, there are very high releases of NO$_2$ + NO$_3$ and uptake of NH$_4$ compared to very little NH$_4$ or NO$_2$ + NO$_3$ flux across the sediment-water interface in Yoon and Benner (1992). These investigators found that most of the DIN flux in this estuary was N gas (see the discussion of denitrification below). Both our survey and that of Yoon and Benner found similar patterns of N flux in Nueces estuary.

There is no general pattern of benthic N flux in estuaries of the Gulf of Mexico compared to estuaries outside of the Gulf region, as rates fall within ranges reported for other estuaries. There is an interesting trend in PO$_4$ flux associated with the upper range of release from sediments. The highest mean in the estuaries of the Gulf was at the Mississippi River Bight with an upper range of 36 μmol m^{-2} h^{-1}. This peak flux of PO$_4$ from sediments is an order

TABLE 6-3. Systemwide Average of Exchange at the Sediment-Water Interface in Coastal Marine Environments (Results of This Study and Those Reported in Cowan et al. 1996)

Estuary	Ref	Nutrient Fluxes (μmol m^{-2} h^{-1})						SOC (g m^{-2} d^{-1})
		PO$_4$	NN	NO$_3$	NH$_4$	DIN	N$_2$	
Ochlockonee Bay	1		16.0		18.7	34.7	88.4	0.90
Apalachicola Bay	2		−30.3	−37.2	38.0	7.7		0.55
Mobile Bay	3,4	3.9	16.1	14.2	62.8	76.9		0.84
Mississippi River Bight	5,6	17.5	−9.17	−15.8	126.3	117.3		1.03
Fourleague Bay	7,8	1.4	−54.7		141.7	86.9		1.2
Fourleague Bay	9	−8.0	−19		129	110		
Trinity-San Jacinto estuary	10	0.6	−2.7		11.7	9.0	18.5	0.15
Gudalupe estuary	11	−3.1	176.2	183.1	−155.3	20.9		0.98
Gudalupe estuary	12		8–10		30–60		5–30	
Nueces estuary	11	−6.4	106.5	−16.8	54.5	29.1		0.73
Nueces estuary	12		0–10		8–50		4–59	
Laguna Madre-Upper	11	−0.2	0.0	0.1	1.6	1.5		1.68
San Francisco Bay	13	−4.2–54		0–33	17–208			0.35–0.70
Narragansett Bay	14	38–233	10		75–500		59	
Chesapeake Bay	15							
North Bay		−16.3		−117–9	−34–101			0.1–0.65
Mid Bay		0.0–148		−100–12	9–507			0.01–0.86
Lower Bay		−1.5–13		−8–19	15–181			0.3–0.75
Patuxent estuary	16	0–15		12.5–37.5	10–200		133	0.75–2.25
Neuse River estuary	17	−2.3–46		0–6.4	71–454			0.70–1.87
South River estuary	17	−8.3–23		0.0–5.8	0.0–267			0.71–2.72
Bay of Cadiz	18	21–379			258–1525			2.2–7.5

References: 1 (Seitzinger 1987); 2 (Mortazavi unpublished data); 3 (Cowan et al. 1996); 4 (Miller-Way unpublished data); 5 (Twilley and McKee 1996); 6 (Bourgeois 1994); 7 (Miller-Way 1994); 8 (Twilley unpublished data); 9 (Teague et al. 1988); 10 (Zimmerman and Benner 1994); 11 (Montagna unpublished data); 12 (Yoon and Benner 1992); 13 (Hammond et al. 1985); 14 (Elderfield et al. 1981); 15 (Cowan and Boynton 1996); 16 (Boynton et al. 1991); 17 (Fisher et al. 1982); 18 (Forja et al. 1994).

of magnitude less than the peak rates of more eutrophic estuaries of the temperate zone such as the mid-bay region of Chesapeake Bay (148 μmol m^{-2} h^{-1}), Narragansett Bay (233 μmol m^{-2} h^{-1}), and the Bay of Cadiz (379 μmol m^{-2} h^{-1}) (Table 6-3). Fluxes of PO4 from sediments have been used in Chesapeake Bay to evaluate trends in increased P loadings to tributaries in this coastal plain estuary (Boynton et al. 1990). This process in nutrient biogeochemistry may be an important metric to monitor among estuaries in the Gulf of Mexico as a response to nutrient loadings.

Spatial Patterns

Spatial trends in benthic fluxes along salinity gradients have been reported for several estuarine ecosystems such as Chesapeake Bay (Kemp and Boynton 1984; Boynton and Kemp 1985) and Trinity-San Jacinto estuary in the Gulf of Mexico (Zimmerman and Benner 1994). Average benthic fluxes at all the stations in this survey were grouped relative to ranges in average overlying water salinity (Fig. 6-7). Most surveys of benthic fluxes in the shallow estuaries of the Gulf of Mexico observe that water-column inorganic N and P concentrations are higher in the low-salinity zone (<10 g $liter^{-1}$) due to nutrient transport by rivers. This is particularly true for NO_3 and PO_4, especially in estuaries where the watershed is influenced by agricultural land use (Chapter 4 this volume). There is no particular spatial trend among the stations when all the systems are compared; there seems to be as much variation among the different estuaries as there is among the different salinity zones. However, nutrient fluxes of inorganic N and P are consistently lower at the hypersaline sites compared to sites in all the other salinity regimes. All of the sites in the salinity regime >35 g $liter^{-1}$ were from Laguna Madre, and this estuary had much lower rates of benthic fluxes than the other estuaries.

When salinity patterns within an estuary are compared, there are some trends that are different among the estuaries. In Trinity-San Jacinto estuary and Fourleague Bay, PO_4 fluxes are higher within the <10 g $liter^{-1}$ salinity regime and then decrease in the 10–20 g $liter^{-1}$ zone (Fig. 6-7A). However, PO_4 fluxes in Mobile Bay and Guadalupe estuary increase from the low-salinity zone (<10 g $liter^{-1}$) to the mid-salinity zone (10–20 g $liter^{-1}$). Benthic fluxes of PO_4 at stations in Mobile Bay were all from the sediment to overlying water and increased from 2.0 μmol m^{-2} h^{-1} in the low-salinity zone to 4.0 μmol m^{-2} h^{-1} in the mid-salinity zone. In Guadalupe estuary, there was sediment uptake of PO_4 from the water column at the low-salinity station (-13.4 μmol m^{-2} h^{-1}), with a release of PO_4 from sediments at the mid-salinity zone at a rate of about 4 μmol m^{-2} h^{-1}. There was much variation and no consistent pattern in benthic fluxes in the 20–35 g $liter^{-1}$ zone, and all fluxes in the hypersaline zone were $\pm$ 1 μmol m^{-2} h^{-1}.

Benthic exchange of NO_2 + NO_3 also exhibited different patterns along a salinity gradient among the estuaries (Fig. 6-7B). Sediment uptake of NO_2 + NO_3 from overlying water was observed in both Trinity-San Jacinto estuary

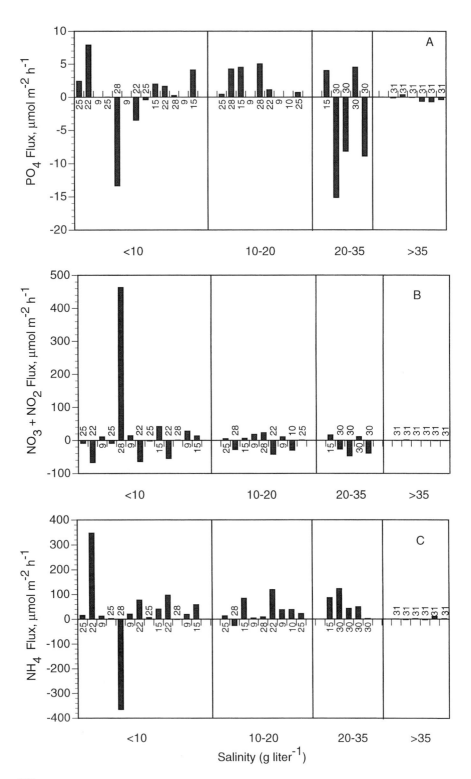

and Fourleague Bay in the low-salinity zone. In the mid-salinity range (10–20 g liter^{-1}), NO_2 + NO_3 fluxes were also from overlying water to sediment in Fourleague Bay, but oxidized inorganic N was released from sediments in Trinity-San Jacinto estuary. Benthic regeneration rates of NO_2 + NO_3 to the water column were observed in Mobile Bay along the entire salinity gradient, with a higher rate at the low-salinity station (41 μmol m^{-2} h^{-1}) and similar rates (6–12 μmol m^{-2} h^{-1}) at the high-salinity stations. The average NO_2 + NO_3 flux to the water column at the low-salinity station in Guadalupe estuary was high, at about 400 μmol m^{-2} h^{-1}, due to rates of 639 and 1154 μmol m^{-2} h^{-1} measured in January and April, respectively. At higher salinities, NO_2 + NO_3 uptake by sediments was observed in this estuary. As observed for PO_4, patterns of benthic flux were inconsistent both among and within estuaries in the 20–35 g liter^{-1} stations, and all rates were low in Laguna Madre with hypersaline waters.

Most of the stations exhibited NH_4 regeneration from sediments to overlying water, and values were highest in Fourleague Bay in the low-salinity region. In Mobile Bay, rates of NH_4 regeneration to the water column increased from the low-salinity to the mid-salinity zone (by a factor of 2) compared to low rates throughout the salinity regime in Trinity-San Jacinto estuary. In Guadalupe estuary, high sediment uptake of NH_4 observed at the low-salinity station coincided with the site of high NO_2 + NO_3 release from sediments. NH_4 benthic regeneration was observed at all of the upper salinity zone stations (20–35 g liter^{-1}). Again NH_4 rates in the hypersaline sites in Laguna Madre were ±10 μmol m^{-2} h^{-1}.

Seasonal Patterns

In Mobile Bay one site was sampled 25 times over 2 years (Cowan et al. 1996). There was a seasonal pattern for most fluxes, with higher rates of nutrient release and O uptake by the sediments in the warmest months of the year (Fig. 6-8). The same significant relation was observed in a seasonal survey of benthic fluxes in Fourleague Bay by Teague et al. (1988). Indeed, temperature was significantly correlated with all fluxes except NO_3 flux in both estuaries. Episodic hypoxic/anoxic events in summer served to diminish this relationship somewhat in Mobile Bay, however, because of their influence on both micro-

---

FIG. 6-7. *Average benthic nutrient fluxes for stations in the respective estuaries in this survey ranked by the average water-column salinity during the different sampling dates. The stations are grouped into four salinity regimes: lower salinity (<10 g liter^{-1}), mid-salinity (10–20), higher salinity (20–35), and hypersaline (>35). (A) PO_4 fluxes. (B) NO_3 + NO_2 fluxes. (C) NH_4 fluxes. Positive and negative nutrient rates represent fluxes out of and into the sediments, respectively. Estuary numbers: 9 = Ochlockonee Bay; 10 = Apalachicola Bay; 15 = Mobile Bay; 22 = Fourleague Bay; 25 = Trinity-San Jacinto estuary; 28 = Guadalupe estuary; 30 = Nueces estuary; 31 = Upper Laguna Madre.*

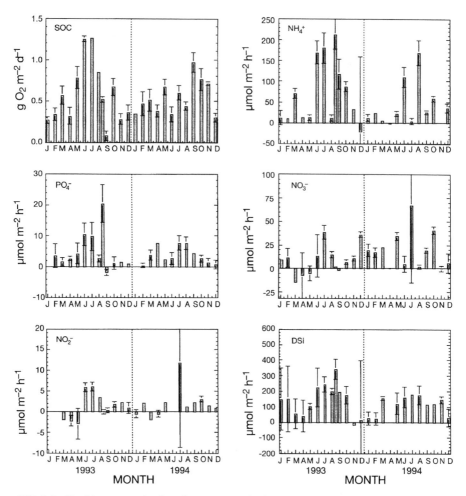

FIG. 6-8. *Monthly averaged values (mean ± standard deviation) of SOC, NH₄, PO₄, NO₃, NO₂, and silicate (DSi) flux at Station DR7 in Mobile Bay from January 1993 to December 1994. All SOC measurements represent O uptake by the sediments; positive and negative nutrient fluxes represent fluxes out of and into the sediments, respectively. Reprinted with permission from Cowan et al. (1996).*

bial and physicochemical processes. As a result, both maximum and, in many cases, minimum rates of nutrient release and SOC were measured in summer. The lack of hypoxia in Fourleague Bay resulted in a more significant regression.

Rates of SOC were highest when temperatures and bottom water DO concentrations were high (Fig. 6-8). However, the lowest rate of SOC was also measured when the temperature was high but the DO concentration was <0.5 mg liter^{-1} (September 1993). This by no means suggests that metabolism was low at this or other times of hypoxia. Anaerobic metabolism (not measured) served to keep DIN fluxes high during warm hypoxic periods, but the speciation of NH_4 and NO_3 fluxes was very different at these times than during oxic periods. In each year, the highest NH_4 release rates and near-zero NO_3 and NO_2 fluxes occurred in the warmest months (Fig. 6-8) when bottom-waters were hypoxic (late August and September 1993, June and August 1994). In contrast, fluxes of NO_3 and NO_2 were highest in the warmest months with non-hypoxic conditions, while NH_4 fluxes tended to be lowest under these conditions. These patterns suggest that nitrification is suppressed under hypoxic conditions, and that nitrification rates are high enough under oxic conditions to oxidize almost all of the regenerated N before it is released from the sediments.

Flux of PO_4 also exhibited a weak positive relationship with temperature in Mobile Bay, but this relationship left much of the variability in PO_4 flux unexplained (Fig. 6-8). Surprisingly, the bottom water DO concentration did not help explain any of the remaining variability. In a related study in Mobile Bay, Fernandez (1995) found that the magnitude of PO_4 released under hypoxic conditions is positively related to the amount of time sediments have been exposed to low DO. This suggests that, upon exposure to hypoxic waters, sediments must first become further reduced before desorption of PO_4 from particulates can occur. Thus, the high periodicity of bottom water re-aeration in Mobile Bay may create a situation in which the sediment response to low-DO water is often slower than the rate of change of the bottom water DO concentration. This scenario would require field samples to be collected relatively late in an extended hypoxic event for a relationship between PO_4 flux and bottom water DO concentration to be detected. Water-column mixing events also likely cause sediment re-suspension in this and other shallow estuaries in the Gulf of Mexico, and may further complicate patterns of PO_4 sorption/desorption onto particles by re-oxidizing sediments and disrupting pore water concentration gradients (Søndergaard et al. 1992; Vidal 1994).

Seasonal patterns of benthic fluxes have also been referred to in other comparisons among estuaries in the Gulf of Mexico (Seitzinger 1987; Teague et al. 1988; Yoon and Benner 1992; Zimmerman and Benner 1994). Denitrification rates are maximum in summer in the Texas estuaries (Galveston, Nueces, and Guadalupe; Zimmerman and Benner 1994) compared to spring in a Florida estuary (Ochlockonee Bay; Seitzinger 1987). The delivery of organic substrates by river discharge and/or phytoplankton blooms is linked to spring peaks in denitrification in several estuaries (Jørgensen and Sørensen 1985; Twilley and Kemp 1986; Seitzinger 1987; Jensen et al. 1988; Binnerup et al. 1992). In the deltaic environments of the central Gulf of Mexico, there is also evidence that spring delivery of NO_3 (Caffrey and Day 1986; Turner and Rabalais 1991) and organic substrates result in higher denitrification rates (Smith et al. 1985; Teague et al. 1988; Miller-Way 1994). In contrast, spatial differences in sediment quality may be more significant than seasonal inputs of organic substrates in patterns of denitrification in dry coastal environments of the Gulf of Mexico that have less seasonal input of fresh water, such as those found along the Texas coast (Zimmerman and Benner 1994).

REGULATION OF BENTHIC FLUXES

Physical Conditions

Temperature has been identified as a major factor influencing sediment nutrient fluxes in many temperate estuaries. Estuaries in the Gulf of Mexico exhibit different seasonal patterns of benthic flux associated with temperature, depending on latitude. Minimum bottom water temperatures generally did not fall below 12°C in the estuaries of this review, and ranges in temperature were generally <15°C. Estuaries in south Florida and along the coast of Mexico are likely to be less influenced by seasonal temperature. The exception is central Gulf estuaries, such as Mobile Bay and Fourleague Bay, where the greatest range in temperatures can occur (region with the lowest number of freeze-free days in the Gulf of Mexico; Table 6-1). In Mobile Bay, bottom water temperatures and/or DO concentrations were found to be the most significant variables explaining variation in benthic nutrient fluxes (Cowan et al. 1996). Temperature was particularly influential in the regulation of SOC and silicate flux. Bottom water DO concentration also influenced the rate of SOC, and together they influenced the rate of DIN flux across the sediment-water interface. Those patterns are related to alterations in microbial activity including rates of organic matter decomposition, nitrification, and denitrification.

Bioturbation

Bioturbation of the benthos can include the effects of both flora and fauna on nutrient exchange across the sediment-water interface. In estuaries with very low turbidity and high light penetration, there are expansive seagrasses (see Table 6-1) in both the Texas and Florida coastal regions. The influence of these communities on benthic fluxes is evident in Florida Bay and Laguna Madre, in contrast to the delta estuaries of coastal Louisiana, where benthic fluxes are not particularly influenced by macrophytes. Benthic fluxes in Laguna Madre included vegetated and unvegetated patches of sediment. There was a significant effect of vegetation on SOC ($p < 0.0001$) and NH_4 flux ($p = 0.0406$), with means of 3.19 and 0.98 g m^{-2} d^{-1} in vegetated and unvegetated patches, respectively, for SOC. Means of NH_4 flux were -1.74 and 1.27 μmol m^{-2} h^{-1} in vegetated and unvegetated patches, respectively. While DO uptake is ecologically significant to the metabolism of the estuary, the difference in DIN flux is minimal in both types of sediment. The effects of seagrasses on patterns of nutrient flux in Florida Bay are presently under investigation and suggest that seagrass can be a significant factor in the fate of nutrients in this oligotrophic estuary.

Macrofauna densities differ spatially among the estuaries of the Gulf of Mexico and may also be linked to patterns of hydrography, geomorphology, and sedimentation of the different estuaries. Macroinfaunal biomass was measured concurrently with nutrient fluxes in Guadalupe estuary, Nueces estuary, and Laguna Madre. There was no significant correlation between benthic biomass and nutrient flux over the 72 samples in these three Texas estuaries. However, there was a significant relationship with SOC and benthic biomass in this survey. The best fit was an exponential decay function where SOC reaches a minimal value with a biomass of about 20 g m^{-2}.

The potential influence of infaunal bioturbation on N cycles has been observed in Galveston (Zimmerman and Benner 1994) and Fourleague (Miller-Way 1994) Bays by clams and oysters, respectively. For example, denitrification rates were elevated by a factor of 10 in a core during experiments of benthic fluxes in Trinity-San Jacinto estuary (Zimmerman and Benner 1994). Over an annual cycle, *Rangia cuneata* and *Crassostrea virginica* significantly increased the rates and altered the patterns of benthic exchange of both particulate and dissolved materials in Fourleague Bay (Miller-Way 1994). These animal communities can enhance substrate quality and sediment oxidation, which promote the coupling of nitrification and denitrification, as has been observed in several other estuarine sediments (Pelegri et al. 1994). Macroinfaunal biomass in Mobile Bay is low and patchy and was not a significant influence on benthic fluxes at Station DR7 (Cowan et al. 1996). But benthic fluxes at other stations in this estuary can be influenced by infaunal communities.

Sediment Quality

The concentration of sediment organic matter has been shown to be an important control on the seasonal and spatial patterns of nutrient regeneration in estuaries (Boynton et al. 1990; Jensen et al. 1990; Cowan et al. 1996). NH_4 and PO_4 fluxes in our survey did vary among the nine coastal systems in relation to the quality of sediment, as indicated by C:N ratios (Fig. 6-9A). As C:N ratios increased, benthic fluxes of NH_4 and PO_4 decreased and were negative at higher C:N ratios of 50. Shifts in C:N ratio were largely due to changes in SC concentration. Higher C:N ratios can cause N immobilization during microbial decomposition of organic matter, resulting in uptake of NH_4 from the water column (Rivera-Monroy and Twilley 1996). However, the uptake of NH_4 can also be associated with release of $NO_2 + NO_3$, indicating the effect of sediment nitrification on NH_4 uptake by sediments (Fig. 6-9B). Thus, dissimilatory processes may be driving NH_4 uptake (such as nitrification) rather than assimilatory uptake (such as immobilization) associated with higher C:N ratios. The former process results in flux of N back to the water column as $NO_2 + NO_3$, while the latter reduces N regeneration from benthic

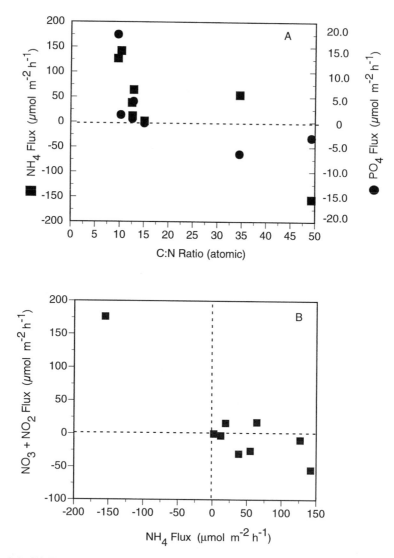

FIG. 6-9. (A) Flux of NH_4 and PO_4 in different estuaries of this survey relative to the average sediment $C:N$ ratio. (B) Average fluxes of $NO_3 + NO_2$ related to ammonium in the different estuaries of this survey.

to pelagic systems. Of course, the coupling of denitrification to nitrification can also reduce the amount of N regenerated from sediments that is available for plankton production (discussed below). However, the low SC content associated with higher C:N ratios may limit denitrification and simply allow the high fluxes of oxidized N back to the water column (Fig. 6-9B).

Sediment quality, as indicated by total chlorophyll in surface sediments (to a depth of 1 cm), has been shown to influence benthic fluxes based on comparisons of Mobile and Chesapeake Bays (Cowan et al. 1996). Sediment chl-a is highly correlated with the magnitude of PO_4, NH_4, and SiO_4 release from the sediments, indicating the importance of sediment quantity and quality on rates of nutrient re-mineralization (Cowan et al. 1996). But we did not observe the effect of sediment chlorophyll on benthic fluxes in comparing Mobile and Fourleague Bays in this survey (data not shown, statistically insignificant regression). Total chlorophyll ranged from 1.9 to 58.8 mg m^{-2} in Fourleague Bay and from 35.8 to 54.8 mg m^{-2} in Mobile Bay compared to values of 60 to 210 mg m^{-2} in Chesapeake Bay (Cowan et al. 1996). The small range in sediment chlorophyll concentration among the two estuaries in the central Gulf of Mexico lacks the gradient to produce a significant relation with benthic fluxes. The lower concentrations in Fourleague Bay can be associated with a high sediment load to the upper regions of this river-dominated estuary during spring discharge. High values are comparable to those in Mobile Bay during periods of high phytoplankton biomass in the water column during summer (Madden et al. 1988). The amount of light that reaches the benthic surface in the shallow estuaries of Gulf of Mexico (<2 m) may also influence the nature of this relationship (Rizzo 1990; Rizzo et al. 1992), although all the fluxes in these surveys were performed under dark conditions. The lower concentrations of sediment chlorophyll among estuaries in the Gulf of Mexico compared to tributaries of the Chesapeake Bay, along with generally lower PO_4 fluxes, indicate the lower condition of eutrophication in the Gulf estuarine ecosystems. Future trends in sediment chlorophyll and benthic fluxes will be an important metric to monitor the trophic conditions of estuaries in the Gulf of Mexico.

Nutrient Loading and Distribution

Unlike studies of many other estuaries (Henriksen et al. 1980; Doering et al. 1987; Jensen et al. 1990) NO_3 fluxes were not correlated with bottom water NO_3 concentration among the Gulf estuaries in this survey. Comparisons of $NO_2 + NO_3$ concentration in overlying water with sediment uptake of $NO_2 + NO_3$ showed no relation at concentrations <20 μM (data not shown). In fact, in Mobile Bay, $NO_2 + NO_3$ flux from sediments to overlying water increased at higher concentrations of $NO_2 + NO_3$, suggesting that the higher concentrations are the result of the flux, rather than the flux responding to the NO_3 concentration. At concentrations >20 μM, $NO_2 + NO_3$ uptake by sediments increases with higher $NO_2 + NO_3$ concentrations in the overlying water, particularly in Fourleague Bay and Mississippi River Bight. Teague et al. (1988) found a significant regression of $NO_2 + NO_3$ flux with $NO_2 + NO_3$ concentration in overlying water in Fourleague Bay associated with higher river discharge in spring. However, this relation did not occur with benthic fluxes in Trinity-San Jacinto estuary, even though this estuary had the highest concentrations of $NO_2 + NO_3$ in the water column. The minimal uptake of $NO_2 + NO_3$ even when concentrations reached >100 μM may be associated with low concentration of SC in this estuary (Fig. 6-4). These contributions of $NO_2 + NO_3$ uptake in river-dominated estuaries with higher SC concentration compared to estuaries with lower SC concentration may influence the fate of N in these coastal ecosystems (discussed below).

Those estuaries with more pronounced seasonal patterns of benthic fluxes in the Gulf of Mexico experience greater seasonal delivery of fresh water and organic matter with river discharge. The cation exchange hypothesis of Gardner et al. (1991) suggests that NH_4 fluxes are normally higher in saline compared to fresh-water sediments because the higher concentration of cations in pore water binds exchange sites that limit the movement of NH_4 ions across the sediment-water interface in fresh-water systems. Among four estuaries in Texas reviewed by Zimmerman and Benner (1994; Table 6-3), there was a direct correlation between bay water salinity and NH_4 benthic efflux ($r = 0.723$, $p < 0.05$). There was no trend in rates of NH_4 flux with increased salinity among the nine coastal systems in this review, since some of the highest NH_4 fluxes were observed at the lower salinities of Fourleague Bay (Fig. 6-7). Lower salinities occur in regions of the estuary with higher loading of organic matter that may contribute to higher rates of nutrient regeneration, overwhelming the cation exchange effect on NH_4 flux.

Loadings of N and P to each of the estuaries in this survey (normalized to the surface area of the respective estuaries) were calculated based on inputs from the fluvial (Dunn 1996) and estuarine (NOAA/EPA 1989) drainage areas (Fig. 6-10). Fluvial inputs of N and P dominate total loadings for most of the estuaries, with the exception of Trinity-San Jacinto and Nueces estuaries, where most of the inputs are from the estuarine drainage area. Total loadings of N and P are much higher in the two estuaries associated with the Mississippi River delta, where inputs are more than twice those of the other estuaries. The exception is total P loading to the Trinity-San Jacinto estuary, which has a per unit area loading rate similar to that of the Mississippi River Bight and Fourleague Bay (Fig. 6-10). The lowest rates of N and P loading occur in the most eastern and western estuaries of this survey in Ochlockonee Bay and Laguna Madre, respectively. Benthic regeneration of N and, to a lesser extent of P, to the water column generally followed the pattern of nutrient loading (Fig. 6-10, lower panel). Fluxes of DIN were significantly correlated with respective rates of N input, with a slope that was about 1 μmol m^{-2} h^{-1} of DIN regenerated for each Mg km^{-2} yr^{-1} that was loaded to an estuary (approximately 12% of the loading is remineralized per year). For P, there was only one estuary that followed a pattern of regeneration linked to loading, the Mississippi River estuary, while rates of P regeneration in the other estuaries were low. No net flux of P from sediments to the overlying water was observed at loading rates <4 Mg P km^{-2} yr^{-1} (Fig. 6-10). Above this loading rate, only one estuary, the Mississippi, had P regeneration rates above 5 μmol

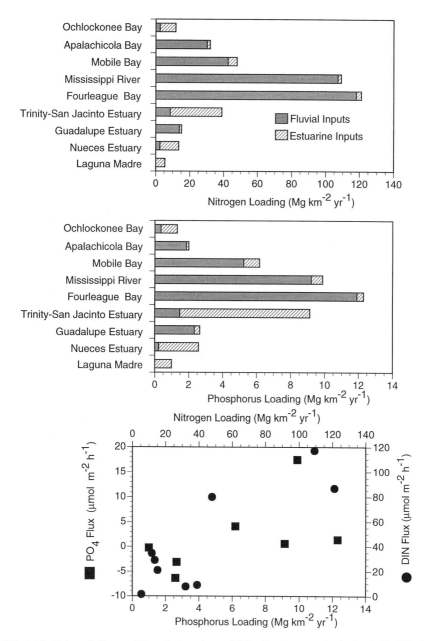

FIG. 6-10. *Rates of nitrogen (N) and phosphorus (P) loading to estuaries in the Gulf of Mexico based on fluvial and estuarine drainage estimates normalized to the per square kilometer surface area of the respective estuary (NOAA/EDA 1989; Dunn 1996). The lower panel shows the relationship between the respective loading rates of N and P compared to the average rates of DIN and PO₄ flux measured in each estuary.*

m^{-2} h^{-1}. Thus, P loading to these estuaries seems to be more conserved in sediments than for N, which exhibited a response to rates of N loading.

ECOLOGICAL SIGNIFICANCE OF BENTHIC FLUXES

Denitrification and N Loss

Denitrification has a significant effect on the fate of N in several estuaries in the Gulf of Mexico (Seitzinger 1987; Zimmerman and Benner 1994), as has been found in several other global regions (reviewed by Seitzinger 1988). Much of the benthic denitrification in estuaries in Texas and Florida is coupled to nitrification in the sediments rather than to NO_3 uptake from the water column. In general, about 30% of the SOC in these four estuaries can be attributed to nitrification, and most N_2 gas produced is from denitrification of NO_3 produced within the sediments. The importance of nitrification to denitrification rates is indicated by the significant correlation between SOC and N_2 production in Ochlockonee Bay [N_2 (μg-at N m^{-2} hr^{-1}) = 0.083 * O_2 (μg-at O m^{-2} hr^{-1}) + 2, r = 0.65; Seitzinger 1987]. In contrast, the deltaic bays of coastal Louisiana are characterized by high concentrations of NO_3, and here denitrification in slurries of sediments (Y) is controlled by the availability of NO_3 (X) in the water column ($Y = 2.9 + 29.8*X$; Smith et al. 1985), as observed by Teague et al. (1988). Rates ranged from 2 to 70 ng N gdw^{-1} h^{-1} and were generally higher in the upper bay stations, especially from May to August. Annual rates of NO_3 loss via denitrification in Fourleague Bay account for about 50% of the loading of this nutrient to this river-dominated estuary (Smith et al. 1985). In the more distal portions, there is evidence that coupled nitrification/denitrification results in N loss from Fourleague Bay (Teague et al. 1988; Day et al. 1994; Miller-Way 1994).

Rates of denitrification in four estuaries of Gulf of Mexico range from 4.5 to 9.0 g m^{-2} yr^{-1}, based on seasonal and spatial measures of N_2 production from sediments (Zimmerman and Benner 1994). Compared to total N loading rates to these respective estuaries, denitrification can account for the loss of 14–136% of the N input compared to an average loss of 40–50% for other estuarine ecosystems (Seitzinger 1988). The high estimate is due to low N loading to estuaries in drier coastal environments of Texas, which causes denitrification to be fueled mainly by nitrification rather than by NO_2 + NO_3 uptake by sediments. Because of low N loading, denitrification is a significant process in N budgets of these lagoon estuaries. The low estimate of N loss via denitrification is for an estuary in Texas with extremely low denitrification rates that may be associated with the low C content of sediments. Even with low inputs, denitrification does not control the fate of N in this lagoon. Although system comparisons are not available, the strong non-conservative behavior of NO_3 in the turbid deltaic estuaries of central Gulf of Mexico suggests that direct utilization of this N from the water column by sediments accounts for much of this nutrient loss (Smith et al. 1985; Madden et al. 1988; Teague et al. 1988).

The shallow nature of estuaries in the Gulf of Mexico, infrequent occurrence of hypoxia, and generally higher temperatures are excellent conditions to enhance the coupling of nitrification and denitrification. In Ochlockonee Bay, 69% of the inorganic N exchange at the sediment-water interface is associated with denitrification, indicating the significance of this process to N cycling in this estuary (Seitzinger 1987). Although there are no direct measures of denitrification in Mobile Bay, Cowan et al. (1996) also suggest that NH_4 fluxes are low under oxic conditions due to high rates of nitrification that are coupled to denitrification, limiting DIN regeneration across the sediment-water interface. In more deltaic environments with higher loadings of N from river discharge, more denitrification and N loss may be accounted for by NO_3 uptake by sediments.

Stoichiometry of Benthic Fluxes

The stoichiometry of SOC and NH_4 release from sediments in the selected estuaries of this survey indicates the relative role of nitrification to the fate of N in these Gulf of Mexico estuaries. There should be 13.25 moles of O taken up by sediments per mole of NH_4 released, and variations from this ratio are indicative of specific N transformations (notice the use of μmol O instead of O_2). Elevated ratios are indicative of nitrification, since the process increases O uptake and reduces NH_4 release. If there is no denitrification, the O:DIN ratio should be about 17, with the release of $NO_2 + NO_3$ for each mole of NH_4 that is oxidized (4 μmol O consumed per μmol of NO_3 produced). Usually, O:NH_4 ratios are used to estimate the ecological significance of nitrification to patterns of benthic N flux, but they do not include negative ratios caused by NH_4 uptake. An analysis of all the SOC and NH_4 fluxes (positive and negative) can be evaluated for evidence of nitrification by calculating an expected NH_4 flux based on SOC rates and comparing those to measured fluxes (Fig. 6-11A). A line is included that marks the 1:1 ratio, and values that fall below that line are indicative of nitrification. The majority of points are below the theoretical ratio, indicating the prevalence of nitrification among the estuaries. A comparison of estuaries based on O:NH_4 ratios (dropping the negative ratios) shows a regional trend in the amount of nitrification in the Gulf of Mexico (Fig. 6-11B). Those estuaries in river-dominated environmental settings in the central Gulf of Mexico have lower rates of nitrification than eastern and western regions with less terrigenous inputs. Direct measures of denitrification with estimates of nitrification have been made of estuaries in the eastern and western regions of the Gulf of Mexico (described above), where significant N loss is associated with this process. Nitrification has less influence on the fate of N in river-dominated estuaries, where NO_3 supplied to the overlying water from river discharge may fuel denitrification.

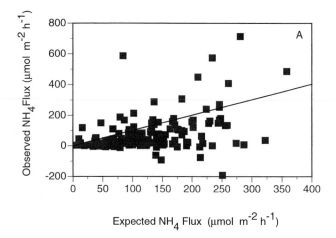

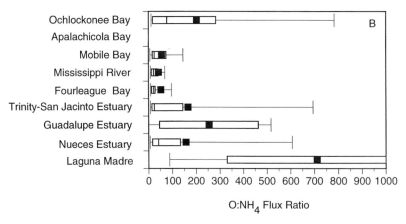

FIG. 6-11. *(A) Expected flux of NH₄ based on sediment O consumption plotted against the actually measured flux of NH₄. (B) Average and range in O : NH₄ flux ratios (O is based on SOC) for each of the estuaries in our survey.*

The stoichiometry of N and P flux can be evaluated based on the expected flux of 16 μmol of DIN per μmol of PO_4 (Fig. 6-12A). Again, the complete data set can be observed by plotting the raw values rather than just using ratios that exclude negative values. Most of the points fall below the theoretical value of 16, indicating that N is regenerated at lower rates relative to the flux of P. A plot of the positive ratios for each of the estuaries in our survey indicates much similarity among the estuaries, with mean values from five of the seven estuaries falling near the expected ratio of 16 (Fig. 6-12B). There does not seem to be a proportionate loss of N relative to P among these systems based on this analysis, but again, that conclusion is biased by omitting N or P uptake. Fourleague and Mobile Bays, on the other hand, both have

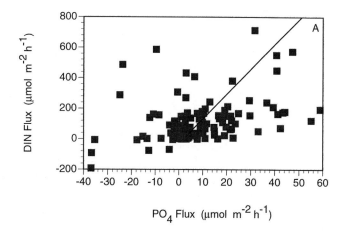

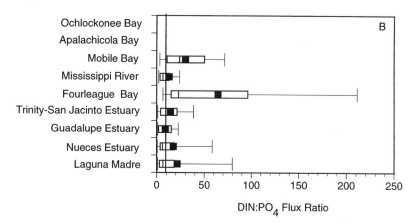

FIG. 6-12. *(A) DIN flux (= NH_4 + NO_3 + NO_2) against PO_4 flux measured in the flux chambers used for eight of the nine estuaries in our survey. (B) Average and range in DIN : PO_4 ratios of respective fluxes among the different estuaries in our survey of the Gulf of Mexico.*

mean ratios >16, indicating the reduced regeneration of P relative to N. The distinct difference between Fourleague Bay and Mississippi River Bight suggests that this shift in N : P regeneration is not related to the loading of terrigenous nutrients, as observed for O : NH_4 ratios.

Contribution of Benthic Flux to Pelagic Productivity

Benthic regeneration of N and P has been linked to the elevated rates of primary productivity in shallow coastal ecosystems. Monthly average measurements of benthic fluxes together with water-column productivity in Mobile

Bay suggest that sediments supplied overlying waters with about 36% of the N and about 23% of the P required by phytoplankton (Cowan et al. 1996). However, the percent contribution varied during the year from 0 to 94% for N and 0 to 83% for P. The lack of a seasonal pattern in primary productivity in contrast to seasonal benthic fluxes (associated with temperature) caused the huge variation in contribution of benthic nutrients to plankton demand. Cowan et al. (1996) suggest that benthic nutrient regeneration supports the greatest amount of primary productivity during periods of low river discharge, thus maintaining rates above that supported strictly by allochthonous nutrient input. Average DIN flux in Fourleague Bay can provide about 15–20% of the N required for annual productivity (Madden et al. 1988). Most of the other estuaries in the Gulf of Mexico have much lower estimates of benthic contribution to the plankton N demand, based on productivity estimates (Zimmerman and Benner 1994). The contribution of benthic N regeneration in Ochlockonee Bay varies from 10 to 19%, depending on annual estimates of primary productivity. The contribution of Texas estuaries is much less, ranging from only 3 to 7%, because of low benthic N flux. In addition, more of the N flux from the Florida and Texas estuaries is associated with coupled nitrification/denitrification. For example, during summer in Ochlockonee Bay, the amount of N lost from the benthic recycling pathway via denitrification is equivalent to approximately 40% of the N requirement of the phytoplankton. This represents a loss of reduced N that may have otherwise been regenerated back to the water column to support primary productivity. Terrigenous inputs and a lower fraction of N loss via nitrification/denitrification in the river-dominated estuaries of central Gulf of Mexico may contribute to the elevated plankton productivity in these estuaries (Chapter 5 this volume).

SUMMARY

Rates of benthic regeneration of nutrients to the water column in selected estuaries of the Gulf of Mexico show some very distinguishing characteristics. Rates in general are lower than those observed in more eutrophic estuaries along the east coast of the United States and in Europe. This is particularly true for P, with rates generally <5 μmol m^{-2} h^{-1}, with the exception of the Mississippi River Bight, where rates are about 18 μmol m^{-2} h^{-1}. But even this higher rate in the Mississippi River is well below rates of benthic P regeneration that are >100 μmol m^{-2} h^{-1} in estuaries that receive high inputs of this nutrient. This is a function of several factors, including lower rates of N and P loading to estuaries in the Gulf of Mexico and the low residence time of water in Gulf estuaries, particularly in river-dominated estuaries in the central region. Low rates of P regeneration from sediments and significant coupling of nitrification to denitrification both combine to limit the contribution of benthic nutrients to water-column primary productivity. In addition, the shallowness of the estuaries in the eastern and western regions with low turbidity results in seagrass communities that apparently reduce nutrient exchange at the sediment-water interface. The possible ecological significance of these plant and animal communities warrants further studies in the estuaries of these regions. Missing from this survey are good estimates of benthic nutrient fluxes in more carbonate-dominated estuaries in south Florida and the Yucatan region of Mexico (although there are a few estimates that these rates are low) and information on the exchange of dissolved organic N and P from sediments to overlying water. But the patterns associated with this survey do demonstrate the diverse response of benthic nutrient fluxes to deltaic compared to more coastal plain and lagoon-type estuaries as a function of nutrient loading, water residence time, sediment quality, coupled nitrification-denitrification, and effects of bioturbation. Eutrophication of these estuaries will lead to changes in these patterns of benthic nutrient regeneration.

REFERENCES

Aller, R. C., J. E. Mackin, W. J. Ullman, C. H. Wang, S. M. Tsai, J. C. Jin, Y. N. Sui, and J. Z. Hong. 1985. Early chemical diagenesis, sediment-water solute exchange, and storage of reactive organic matter near the mouth of the Chang Jiang, East China Sea. Cont. Shelf Res. **4:** 227–251.

Banta, G. T., A. E. Giblin, J. E. Hobbie, and J. Tucker. 1995. Benthic respiration and nitrogen release in Buzzards Bay, Massachusetts. J. Mar. Res. **53:** 107–135.

Berner, R. A. 1978. Sulfate reduction and the rate of deposition of sediments. Earth Planetary Sci. Lett. **37:** 492–498.

Billen, G., C. Lancelot, and M. Meybeck. 1991. N, P, and Si retention along the aquatic continuum from land to ocean, p. 19–44. *In:* R. F. C. Mantoura, J. M. Martin, and R. Wollast [eds.], Ocean margin processes in global change. Wiley.

Binnerup, S., K. Jensen, N. P. Revsbech, M. H. Jensen, and J. Sørensen. 1992. Denitrification, dissimilatory reduction of nitrate to ammonium, and nitrification in a bioturbated estuarine sediment as measured with ^{15}N and microsensor techniques. Appl. Environ. Microbiol. **58:** 303–313.

Blackburn, T. H. 1991. Accumulation and regeneration: Processes at the benthic boundary layer, p. 181–195. *In:* R. F. C. Mantoura, J. M. Martin, and R. Wollast [eds.], Ocean margin processes in global change. Wiley.

Bourgeois, J. 1994. Patterns of benthic nutrient fluxes on the Louisiana continental shelf. M.S. thesis, University of Southwestern Louisiana.

Boynton, W. R., and W. M. Kemp. 1985. Nutrient regeneration and oxygen consumption by sediments along an estuarine salinity gradient. Mar. Ecol. Prog. Ser. **23:** 45–55.

———, W. M. Kemp, J. M. Barnes, J. L. W. Cowan, S. E. Stammerjohn, L. L. Matteson, F. M. Rohland, M. Marvin, and J. H. Garber. 1990. Long-term characteristics and trends of benthic oxygen and nutrient fluxes in the Maryland portion of the Chesapeake Bay, p. 339–354. *In:* J. H. Mihursky [ed.], New perspectives in the Chesapeake system: A research and management partnership. Chesapeake Research Consortium.

————, L. L. Matteson, J. L. Watts, S. E. Stammerjohn, D. A. Jasinski, and F. M. Rohland. 1991. Maryland Chesapeake Bay water quality monitoring program: Eco-system processes component level 1 interpretive report no. 8. UMCEES, CBL Ref. No. 91-110.

————, W. M. Kemp, and C. W. Keefe. 1982. An analysis of nutrients and other factors influencing estuarine phytoplankton, p. 69–90. *In:* V. S. Kennedy [ed.], Estuarine comparisons. Academic Press.

————, W. M. Kemp, and C. G. Osborne. 1980. Nutrient fluxes across the sediment-water interface in the turbid zone of a coastal plain estuary, p. 93–109. *In:* V. Kennedy [ed.], Estuarine perspectives. Academic Press.

Caffrey, J. M., and J. Day. 1986. Control of the variability of nutrients and suspended sediments in a Gulf coast estuary by climatic forcing and spring discharge of the Atchafalaya River. Estuaries **9:** 295–300.

————, and W. M. Kemp. 1990. Nitrogen cycling in sediments with estuarine popula-tions of *Potamogeton perfoliatus* and *Zostera marina.* Mar. Ecol. Prog. Ser. **66:** 147–160.

Canfield, D. E. 1989. Sulfate reduction and oxic respiration in marine sediments: Implications for organic preservation in euxinic environments. Deep-Sea Res. **36:** 121–138.

Conley, D. J., and R. W. Johnstone. 1995. Biogeochemistry of N, P and Si in Baltic Sea sediments: Response to a simulated deposition of a spring diatom bloom. Mar. Ecol. Prog. Ser. **122:** 265–276.

Corredor, J. E., and J. M. Morell. 1989. Assessment of inorganic nitrogen fluxes across the sediment-water interface in a tropical lagoon. Estuar. Coast. Shelf Sci. **28:** 339–345.

Cowan, J. L. W., and W. R. Boynton. 1996. Sediment water oxygen and nutrient exchanges along the longitudinal axis of Chesapeake Bay: Seasonal patterns, control-ling factors and ecological significance. Estuaries **19:** 562–580.

————, J. R. Pennock, and W. R. Boynton. 1996. Seasonal and interannual patterns of sediment-water nutrient and oxygen fluxes in Mobile Bay, Alabama (USA): Regulating factors and ecological significance. Mar. Ecol. Prog. Ser. **141:** 229–245.

Cratsley, D. W. 1975. Recent deltaic sedimentation, Atchafalaya Bay Louisiana. M.S. thesis, Louisiana State University.

Dagg, M., C. Grimes, S. Lohrenz, B. McKee, R. R. Twilley, and W. Wiseman, Jr. 1991. Continental shelf food chains of the northern Gulf of Mexico, p. 67–106. *In:* K. Sherman, L. M. Alexander, and B. D. Gold [eds.], Food chains, yields, models, and management of large marine ecosystems. Westview Press.

Day, J. W., Jr., C. J. Madden, R. R. Twilley, R. F. Shaw, B. A. McKee, M. J. Dagg, D. L. Childers, R. C. Raynie, and L. J. Rouse. 1994. The influence of Atchafalaya River discharge on Fourleague Bay, Louisiana (USA), p. 151–160. *In:* K. R. Dyer and R. J. Orth [eds.], Changes in fluxes in estuaries. Olsen and Olsen.

Deegan, L. A., J. W. Day, Jr., J. G. Gosselink, A. Yañez-Arancibia, G. S. Chavez, and P. Sanchez-Gil. 1986. Relationships among physical characteristics, vegetation distribution and fisheries yield in Gulf of Mexico estuaries, p. 83–100. *In:* D. A. Wolfe [ed.], Estuarine variability. Academic Press.

DeMaster D. J., B. A. McKee, C. A. Nittrouer, J. Qian, and G. Cheng. 1985. Rates of sediment accumulation and particle reworking based on radiochemical measurements from continental shelf deposits in the East China Sea. Cont. Shelf Res. **4:** 143–155.

————, and C. A. Nittrouer. 1983. Uptake, dissolution and accumulation of silica near the mouth of the Changjiang River, p. 235–240. *In:* Proceedings of the International Symposium in Sedimentation on the Continental Shelf, with Special Reference to the East China Sea, China Ocean Press, Beijing.

Denes, T. A. 1983. Seasonal transports and circulation of Fourleague Bay, Louisiana. M.S. thesis, Louisiana State University.

Doering, P. H., J. R. Kelly, C. A. Oviatt, and T. Sowers. 1987. Effect of the hard clam *Mercenaria mercenaria* on benthic fluxes of inorganic nutrients and gases. Mar. Biol. **94:** 337–383.

Dunn, D. D. 1996. Trends in nutrient inflows to the Gulf of Mexico from streams draining the conterminous United States, 1972–93. U.S. Geological Survey, Water-Resources Investigations Report 96-4113, Austin, Texas.

Elderfield, H., N. Luedtke, R. J. McCaffrey, and M. Bender. 1981. Benthic flux studies in Narragansett Bay. Am. J. Sci. **281:** 768–787.

Fernandez, F. 1995. Nitrogen and phosphorous fluxes across the sediment-water interface during summer oxic and hypoxic/anoxic periods in Mobile Bay, Alabama. M.S. thesis, University of South Alabama.

Fisher, R. R., P. R. Carlson, and R. T. Barber. 1982. Sediment nutrient regeneration in three North Carolina estuaries. Estuar. Coast. Shelf Sci. **14:** 101–116.

Forja, J. M., J. Blasco, and A. Gomez-Parra. 1994. Spatial and seasonal variation of in situ benthic fluxes in the Bay of Cadiz (south-west Spain). Estuar. Coast. Shelf Sci. **39:** 127–141.

Gardner, W. S., S. P. Seitzinger, and J. M. Malczyk. 1991. The effects of sea salts on the forms of nitrogen released from estuarine and freshwater sediments. Estuaries **14:** 157–166.

Glibert, P. M. 1982. Regional studies of daily, seasonal and size fraction variability in NH_4 remineralization. Mar. Biol. **70:** 209–222.

Hammond, D. E., C. Fuller, D. Harmon, B. Hartman, M. Korsec, L. G. Miller, R. Rea, S. Warren, W. Berelson, and S. W. Hagar. 1985. Benthic fluxes in San Francisco Bay. Hydrobiologia **129:** 69–90.

Hargrave, B. R. 1969. Similarity of oxygen uptake by benthic communities. Limnol. Oceanogr. **14:** 801–805.

————. 1973. Coupling carbon flow through some pelagic and benthic communities. J. Fish. Res. Bd. Can. **30:** 1317–1326.

Harrison, W. G., D. Douglas, P. Falkowski, G. Rowe, and J. Vidal. 1983. Summer nutrient dynamics of the Middle Atlantic Bight: Nitrogen uptake and regeneration. J. Plankton Res. **5:** 539–556.

Henriksen, K., J. I. Hansen, and T. H. Blackburn. 1980. The influence of benthic infauna on exchange of inorganic nitrogen between sediment and water. Ophelia **1:** 294–256.

————. 1981. Rates of nitrification, distribution of nitrifying bacteria, and nitrate fluxes in different types of sediment from Danish waters. Mar. Biol. **61:** 299–304.

————, and W. M. Kemp. 1986. Nitrification in estuarine and coastal marine sediments: Methods, patterns and regulating factors, p. 207–249. *In:* J. Sørensen, T. H. Blackburn, and T. Rosswall [eds.], Nitrogen cycling in coastal marine sediments. Wiley.

————, M. B. Rasmussen, and A. Lensen. 1983. Effect of bioturbation on microbial nitrogen transformations in the sediment and fluxes of ammonium and nitrate to the overlying water. Ecol. Bull. (Stockholm) **35:** 193–205.

Jensen, M. H., T. K. Andersen, and J. Sørensen. 1988. Denitrification in coastal bay sediment: Regional and seasonal variation in Aarhus Bight, Denmark. Mar. Ecol. Prog. Ser. **48:** 155–162.

————, E. Lomstein, and J. Sørensen. 1990. Benthic NH_4^+ and NO_3^- flux following sedimentation of a spring phytoplankton bloom in Aarhus Bight, Denmark. Mar. Ecol. Prog. Ser. **61:** 87–96.

Jørgensen, B. B., and J. Sørensen. 1985. Seasonal cycles of O_2, NO_3^-, and SO_4^{2-} reduction in estuarine sediments. The significance of an NO_3^- reduction maximum in spring. Mar. Ecol. Prog.Ser. **24:** 65–74.

Kelly, J. R., and S. W. Nixon. 1984. Experimental studies of the effect of organic deposition on the metabolism of a coastal marine bottom community. Mar. Ecol. Prog. Ser. **17:** 157–169.

Kemp, W. M., and W. R. Boynton. 1984. Spatial and temporal coupling of nutrient inputs to estuarine primary production: The role of particulate transport and decomposition. Bull. Mar. Sci. **35:** 522–535.

————, P. Sampou, J. Caffrey, M. Mayer, K. Henriksen, and W. R. Boynton. 1990. Ammonium recycling versus denitrification in Chesapeake Bay sediments. Limnol. Oceanogr. **35:** 1545–1563.

————, R. L. Wetzel, W. R. Boynton, D. C. F., and J. C. Stevenson. 1982. Nitrogen cycling and estuarine interfaces: Some current concepts and research directions, p. 209–230. *In:* V. S. Kennedy [ed.], Estuarine comparisons. Academic Press.

Klump, J. V., and C. S. Martens. 1983. Benthic nitrogen regeneration, p. 411–457. *In:* E. J. Carpenter and D. G. Capone [eds.], Nitrogen in the marine environment. Academic Press.

MacIntyre, H. L., and J. J. Cullen. 1988. Primary production in San Antonio Bay, Texas: Contribution by phytoplankton and microphytobenthos. A report to the Texas Water Development Board. The University of Texas Marine Science Institute, Port Aransas.

Madden, C. J. 1986. Distribution and loading of nutrients in Fourleague Bay, a shallow Louisiana estuary. M.S. thesis, Louisiana State University.

————. 1992. Factors controlling phytoplankton production in a shallow, turbid estuary. Ph.D. thesis, Louisiana State University.

————, J. Day, and J. Randall. 1988. Coupling of freshwater and marine systems in the Mississippi deltaic plain. Limnol. Oceanogr. **33:** 982–1004.

Malone, T. C., W. M. Kemp, H. W. Ducklow, W. R. Boynton, J. H. Tuttle, and R. B. Jonas. 1986. Lateral variation in the production and fate of phytoplankton in a partially stratified estuary. Mar. Ecol. Prog. Ser. **32:** 149–160.

McKee, B. A., C. A. Nittrouer, and D. J. DeMaster. 1983. The concepts of sediment deposition and accumulation applied to the continental shelf near the mouth of the Yangtze River. Geology **11:** 631–633.

McRoy, C., R. Barsdate, and M. Nebert. 1972. Phosphorus cycling in an eelgrass (*Zostera marina* L.) ecosystem. Limnol. Oceanogr. **17:** 58–67.

Middelburg, J. J., G. Klver, J. Nieuwenhuize, A. Wielemaker, W. de Haas, T. Vlug, and J. F. W. A. van der Nat. 1996. Organic matter mineralization in intertidal sediments along an estuarine gradient. Mar. Ecol. Prog. Ser. **132:** 157–168.

Miller-Way, T. 1994. The role of infaunal and epifaunal suspension feeding macrofauna on rates of benthic-pelagic coupling in a southeastern estuary. Ph.D. thesis, Louisiana State University.

———, G. Boland, G. Rowe, and R. R. Twilley. 1994. Sediment oxygen consumption and benthic nutrient fluxes on the Louisiana continental shelf: A methodological comparison. Estuaries **17:** 809–815.

———, and R. R. Twilley. 1996. A comparison of batch and continuous flow methodologies for determining benthic fluxes. Mar. Ecol. Prog. Ser. **142:** 257–269.

Montagna, P. A., and R. D. Kalke. 1992. The effect of freshwater inflow on meiofaunal and macrofaunal populations in the Guadalupe and Nueces Estuaries, Texas. Estuaries **15:** 266–285.

———. 1995. Ecology of infaunal Mollusca in south Texas estuaries. Amer. Malacol. Bull. **11:** 163–175.

Morisawa, M. 1968. Streams, their dynamics and morphology. McGraw-Hill.

Nixon, S. W. 1981. Remineralization and nutrient cycling in coastal marine ecosystems, p. 111–138. *In:* B. J. Neilson and L. E. Cronin [eds.], Estuarine and nutrient. Humana Press.

———, J. R. Kelly, B. N. Furnas, C. A. Oviatt, and S. S. Hale. 1980. Phosphorus regeneration and the metabolism of coastal marine bottom communities. *In:* K. R. Tenore and B. C. Coull [eds.], Marine benthic dynamics. University of South Carolina Press.

———, C. A. Oviatt, and S. S. Hale. 1976. Nitrogen regeneration and the metabolism of coastal marine bottom communities, p. 269–283. *In:* J. M. Anderson and A. Macfadyed [eds.], The role of terrestrial and aquatic organisms in decomposition. Proceedings of the 17th symposium of the British Ecological Society. Blackwell Scientific.

———, and M. Pilson. 1983. Nitrogen in estuarine and coastal marine ecosystems, p. 565–648. *In:* E. Carpenter and D. Capone [eds.], Nitrogen in the marine environment. Academic Press.

NOAA/EPA. 1989. Strategic assessment of near coastal waters: Susceptibility and status of Gulf of Mexico estuaries to nutrient discharges. Strategic Assessment Branch, NOS/NCA.

Orlando, S. P., Jr., L. P. Roxas, G. H. Ward, and C. J. Klein. 1993. Salinity characteristics of Gulf of Mexico estuaries. National Oceanic and Atmospheric Administration, Office of Ocean Resources Conservation and Assessment.

Patrick, W. H., Jr., and R. A. Khalid. 1974. Phosphate release and sorption by soils and sediments: Effect of aerobic and anaerobic conditions. Science **186:** 53–55.

Pelegri, S. P. 1995. Effect of bioturbation by *Nereis* sp., *Mya arenaria* and *Cerastoderma* sp. on nitrification and denitrification in estuarine sediments. Ophelia **42:** 289–299.

———, N. P. Nielsen, and T. H. Blackburn. 1994. Denitrification in estuarine sediment stimulated by the irrigation activity of the amphipod *Corophium volutator.* Mar. Ecol. Prog. Ser. **105:** 285–290.

Pennock, J. R. 1987. Temporal and spatial variability in phytoplankton NH_4 and nitrate uptake in the Delaware estuary. Estuar. Coast. Shelf Sci. **24:** 841–857.

Redfield, A. C. 1958. The biological control of chemical factors in the environment. Am. Sci. **46:** 205–221.

Rivera-Monroy, V. H., and R. R. Twilley. 1996. The relative role of denitrification and immobilization on the fate of inorganic nitrogen in mangrove sediments of Terminos Lagoon, Mexico. Limnol. Oceanogr. **41:** 284–296.

Rizzo, W. M. 1990. Nutrient exchanges between the water column and a subtidal benthic microalgal community. Estuaries **13:** 219–226.

————, G. J. Lackey, and R. R. Christian. 1992. Significance of euphotic, subtidal sediments to oxygen and nutrient cycling in a temperate estuary. Mar. Ecol. Prog. Ser. **86:** 51–61.

Rudnick, D. T., and C. A. Oviatt. 1986. Seasonal lags between organic carbon deposition and mineralization in marine sediments. J. Mar. Res. **44:** 815–837.

Schroeder, W. W., and W. J. Wiseman, Jr. 1986, Low-frequency shelf-estuarine exchange processes in Mobile Bay and other estuarine systems on the northern Gulf of Mexico, p. 355–367. *In:* D. A. Wolfe [ed.], Estuarine variability. Academic Press.

Seitzinger, S. P. 1987. Nitrogen biogeochemistry in an unpolluted estuary: The importance of benthic denitrification. Mar. Ecol. Prog. Ser. **41:** 177–186.

————. 1988. Denitrification in freshwater and coastal marine ecosystems: Ecological and geochemical significance. Limnol. Oceanogr. **33:** 702–724.

————, S. Nixon, M. E. Q. Pilson, and S. Burke. 1980. Denitrification and N_2O production in near-shore marine sediments. Geochim. Cosmochim. Acta. **44:** 1853–1860.

Shokes, R. F. 1976. Rate-dependent distributions of lead-210 and interstitial sulfate in sediments of the Mississippi River delta. Ph.D. thesis, Texas A & M. University.

Smith, C. J., R. D. DeLaune, and W. H. Patrick, Jr. 1985. Fate of riverine nitrate entering an estuary: I. Denitrification and nitrogen burial. Estuaries **8:** 15–21.

Smith, S. V. 1991. Stoichiometry of C : N : P fluxes in shallow-water marine ecosystems, p. 259–286. *In:* J. J. Cole et al. [eds.], Comparative analyses of ecosystems: Patterns, mechanisms, and theories. Springer-Verlag.

Søndergaard, M., P. Kristensen, E., and E. Jeppsen. 1992. Phosphorus release from resuspended sediment in the shallow and wind-exposed Lake Arreso, Denmark. Hydrobiologia **228:** 91–99.

Sundby, B., C. Gobeil, N. Silverberg, and A. Mucci. 1992. The phosphorus cycle in coastal marine sediments. Limnol. Oceanogr. **37:** 1129–1145.

Taghon, G. L., A. R. M. Nowell, and P. A. Jumars. 1984. Transport and breakdown of fecal pellets: Biological and sedimentological consequence. Limnol. Oceanogr. **29:** 64–72.

Teague, K., C. Madden, and J. Day. 1988. Sediment oxygen uptake and net sediment-water nutrient fluxes in a river-dominated estuary. Estuaries **11:** 1–9.

Turner, R. E., and N. N. Rabalais. 1991. Changes in Mississippi River water quality this century. BioScience **41:** 140–147.

Twilley, R. R., and W. M. Kemp. 1986. The relation of denitrification potentials to selected physical and chemical factors in sediments of Chesapeake Bay, p. 277–293. *In:* D. A. Wolfe [ed.], Estuarine variability. Academic Press.

————, and B. McKee. 1996. Ecosystem analysis of the Louisiana Bight and adjacent shelf environments. Volume I. The fate of organic matter and nutrients in the sediments of the Louisiana Bight. OCS study/MMS No. U.S. Department of the Interior, Minerals Management Service, Gulf of Mexico OCS Regional Office, New Orleans.

Vidal, M. 1994. Phosphate dynamics tied to sediment disturbances in Alfacs Bay (NW Mediterranean). Mar. Ecol. Prog. Ser. **110:** 211–221.

Ward, L. G., W. M. Kemp, and W. R. Boynton. 1984. The influence of waves and seagrass communities on suspended particulates in an estuarine embayment. Mar. Geol. **59:** 85–103.

Westrich, J. T., and R. A. Berner. 1984. The role of sedimentary organic matter in bacterial sulfate reduction: The G model tested. Limnol. Oceanogr. **29:** 236–249.

Yoon, W. B., and R. Benner. 1992. Denitrification and oxygen consumption in sediments of two south Texas estuaries. Mar. Ecol. Prog. Ser. **90:** 157–167.

Zeitzschel, B. 1980. Sediment water interactions in nutrient dynamics, p. 195–218. *In:* K. R. Tenore and B. C. Coull [eds.], Marine benthic dynamics. University of South Carolina Press.

Zimmerman, A. R., and R. Benner. 1994. Denitrification, nutrient regeneration and carbon mineralization in sediments of Galveston Bay, Texas, USA. Mar. Ecol. Prog. Ser. **114:** 275–288.

Chapter *7*

Wetland-Water Column Interactions and the Biogeochemistry of Estuary-Watershed Coupling Around the Gulf of Mexico[1]

Daniel L. Childers, Stephen E. Davis III, Robert Twilley, and Victor Rivera-Monroy

INTRODUCTION

Historically, the Gulf of Mexico had one of the most wetland-rich coasts of any land margin in the Western Hemisphere. Even today, after significant losses of wetlands in certain areas (Wells and Coleman 1987; Templet and Meyer-Arendt 1988; Turner 1991), there are over 14,500 km² of estuarine wetlands bounding the Gulf of Mexico. Of these, approximately 476,841 ha are forested mangrove wetlands and 974,500 ha are herbaceous marsh (Dahl

[1]Publication No. 73 of the Southeast Environmental Research Program Contribution Series.

Biogeochemistry of Gulf of Mexico Estuaries, Edited by Thomas S. Bianchi, Jonathan R. Pennock, and Robert R. Twilley.
ISBN 0-471-16174-8 © 1999 John Wiley & Sons, Inc.

and Johnson 1991; Field et al. 1991; Snedaker 1993). The U.S. coast is a mixture of intertidal marsh (974,471 ha) and mangrove forest (225,447 ha), though most of the marsh is in Louisiana, which contains roughly 25% of all tidal wetlands in the 48 contiguous U.S. states (Templet and Meyer-Arendt 1988). Florida's peninsular Gulf Coast is dominated by mangroves, with approximately 219,610 ha of mangrove compared to 2606, 2956, 250, and 25 ha of mangroves in Texas, Louisiana, Mississippi, and Alabama, respectively (Dahl and Johnson 1991; Field et al. 1991; Snedaker 1993). Virtually all of the Mexican coast is mangrove forest (254,000 ha; Snedaker 1993). Notably, the Laguna de Terminos estuarine basin in southeastern Mexico (Number 38 on Fig. 7-1) contains more mangrove wetlands (250,000 ha) than does the entire U.S. Gulf coast (Snedaker 1993). Given this distribution of wetlands, it is easy to understand why the coasts of south Florida, Louisiana, and southern Mexico have been the geographical foci of intertidal wetland research in the Gulf of Mexico.

Wetland processes and exchanges of sediments, nutrients, and organic matter are intuitively important to estuarine and near-shore coastal ecosystems along the Gulf of Mexico by virtue of the sheer extent of these estuarine wetlands. While numerous researchers have studied biogeochemical processes in these wetlands, far fewer studies have *directly* quantified wetland-water column interactions in these systems. In this chapter, we review the data from all such studies conducted in Gulf of Mexico estuaries, and for the first time we compare and contrast wetland-water column exchange data from mangrove forests and intertidal marshes. Our objectives are: (1) to compile and summarize these wetland flux datasets; (2) to analyze these datasets to elucidate general patterns in the way wetlands are coupled to the greater estuarine ecosystem; (3) to synthesize these patterns into a broader picture of how the degree of upstream-downstream coupling (between estuaries, their watersheds, and the adjacent Gulf of Mexico) affects wetland-water column interactions, and; (4) to integrate this review of wetland fluxes in Gulf of Mexico estuaries with other biogeochemical processes summarized in this volume.

A key challenge with any comparison of data from a number of studies is separating natural variability from variability induced by differences in methodologies. Datasets generated from a wide application of the same methodology are particularly conducive to synthesis using broad-scale integrative factors. As an example, several authors have reported that aboveground primary production of *Spartina alterniflora* is negatively related to latitude (Keefe 1972; Turner 1976; Dame 1989) and positively related to local tidal range (Steever et al. 1976). In all cases, these large-scale syntheses were possible because of the large number of *S. alterniflora* productivity datasets generated over a wide geographic area using comparable techniques. In this review, we only consider studies that have used the wetland flume technique (Table 7-1).

The wetland flume methodology, reviewed by Childers (1994), was originally introduced by Lee (1979). Flumes used in intertidal marsh settings have used the same basic approach: The flumes consist of vertical walls that are

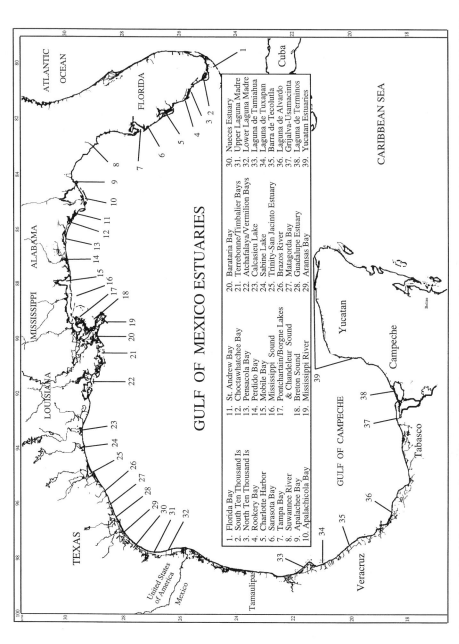

FIG. 7-1. Map of the Gulf of Mexico showing the locations of all major estuaries. Those represented in the flume flux datasets listed in Tables 7-1 and 7-2 are: Barataria Basin, LA (20), Fourleague Bay, LA (between 21 and 22), Estero Pargo and Boca Chica, Mexico (both in Laguna de Terminos, 38), Shark River, Florida (2), Taylor River, Florida, (north-central Florida Bay, 1), and Rookery Bay, Florida (4).

GULF OF MEXICO ESTUARIES

1. Florida Bay
2. South Ten Thousand Is
3. North Ten Thousand Is
4. Rookery Bay
5. Charlotte Harbor
6. Sarasota Bay
7. Tampa Bay
8. Suwannee River
9. Apalachee Bay
10. Apalachicola Bay

11. St. Andrew Bay
12. Choctawhatchee Bay
13. Pensacola Bay
14. Perdido Bay
15. Mobile Bay
16. Mississippi Sound
17. Pontchartrain/Borgne Lakes
   & Chandeleur Sound
18. Breton Sound
19. Mississippi River

20. Barataria Bay
21. Terrebonne/Timbalier Bays
22. Atchafalaya/Vermilion Bays
23. Calcasieu Lake
24. Sabine Lake
25. Trinity-San Jacinto Estuary
26. Brazos River
27. Matagorda Bay
28. Guadalupe Estuary
29. Aransas Bay

30. Nueces Estuary
31. Upper Laguna Madre
32. Lower Laguna Madre
33. Laguna de Tamiahua
34. Laguna de Tuxapan
35. Barra de Tecolutla
36. Laguna de Alvarado
37. Grijalva-Usamacinta
38. Laguna de Terminos
39. Yucatan Estuaries

TABLE 7-1. Listing of All Wetland Flume Studies Conducted in Gulf of Mexico Estuaries by Estuary Name and Location

Estuary	Location	Reference	Wetland Type (Ecological)	Geology (Geochemical)	Morphology (Geomorphological)	Watershed Coupling
Barataria brackish	Louisiana, USA	Childers and Day (1990b)	Marsh	Terrigenous	Proximal to open water	Minimal
Fourleague brackish	Louisiana, USA	Childers and Day (1990b)	Marsh	Terrigenous	Proximal to wetlands	Intense
Barataria saline	Louisiana, USA	Childers and Day (1990b)	Marsh	Terrigenous	Proximal to open water	Minimal
Estero Pargo	Campeche, Mexico	Rivera-Monroy et al. (1995)	Mangrove	Carbonate	Proximal to open water	Minimal
Boca Chica	Campeche, Mexico	Rivera-Monroy (unpubl. data)	Mangrove	Terrigenous	Proximal to open water	Intensive
Shark River	South Florida, USA	Rivera-Monroy (unpubl. data)	Mangrove	Carbonate	Proximal to wetlands	Intensive
Taylor River dwarf	South Florida, USA	Davis and Childers (unpubl. data)	Mangrove	Carbonate	Proximal to wetlands	Minimal
Taylor River fringe	South Florida, USA	Davis and Childers (unpubl. data)	Mangrove	Carbonate	Proximal to wetlands	Minimal
Rookery Bay basin	South Florida, USA	Twilley (1985)	Mangrove	Carbonate	Proximal to wetlands	Minimal
Rookery Bay fringe	South Florida, USA	Twilley (1985)	Mangrove	Carbonate	Proximal to open water	Minimal

Notes: The four classifications for each refer to the hypothetical matrices shown in Fig. 7-2. See the text for details. The Shark River, Florida flux study was conducted using a core technique and not the flume technique.

generally 0.5–1 m tall and 1.5–3 m apart; boardwalks are generally built on either side of the flume walls to prevent disturbance of the marsh surface. The walls are removed after each sampling to prevent long-term wall effects, such as shading, edge scouring, and wrack accumulation. The flumes are designed to prevent lateral water movement, without altering normal flow, as the flooding tide inundates the marsh. Flumes used in Gulf Coast estuaries, both marshes and mangroves, were all open at both ends. As water flowed through the experimental wetland within the flume walls, tidal water was sampled at both ends simultaneously, generating a time series of fluxes into and out of the flume.

The statistical design for this "throughflow" flume technology is based on a known volume of tidal water being treated by a known area of wetland (see Childers and Day 1988 for computations). Constituent flux is based on water flux, which is volumetrically derived by combining the wetland microtopography in the flume with water level change over the course of the tidal cycle. When water flow through the flume is detectable, advective water flux must also be included. Samples are drawn from the tidal water at both ends of the flume at regular intervals throughout the tidal cycle. Sample concentration multiplied by water flux generates a time series of instantaneous constituent fluxes. Net tidal cycle exchange is based on the balance of flux into the flume (across one end) and flux out of the flume (across the other end). All of the marsh and mangrove flux data used in our review (Table 7-1) were based on exactly this flume methodology except for an ongoing study in which we were forced to modify the methodology slightly (Davis and Childers in prep.). In this chapter, we quantify wetland-water column interactions in a narrow creekside fringe of red mangrove along Taylor River, Florida, using flume walls set in the tidal creek itself. These walls are parallel to water flow and completely separate 14 m of the *Rhizophora* fringe forest on both creekbanks from the main creek channel. For these duplicate "half-flumes," we use the high vertical banks of the creek itself as the second flume wall. The half-flumes on either side of the creek are thus experimental replicates, and the creek between the two (which contains no mangrove trees) is the control. We use the same upstream-downstream sampling approach, but virtually all water flux is driven by advection rather than volume change. Secondly, we quantify wetland-water column interactions in dwarf red mangrove wetlands by isolating mangrove clumps 3–5 m in diameter with walls. Once we have "corraled" our triplicate dwarf mangrove islands, we sample the water both inside and outside each corral every 1–2 h for 24–48 h. Flux is thus based on concentration change within the dwarf *Rhizophora* corrals relative to the exterior water column. In all of the studies shown in Table 7-1—including this ongoing Taylor River, Florida, study—the design involved isolating a portion of intact, undisturbed wetland in situ and quantifying the way it modifies constituent concentrations in the overlying water column. It is this unifying theme that allows us to compare and contrast these studies while assuming consistent sampling techniques.

APPROACH

Details of the study sites are listed in Table 7-1, and Fig. 7-1 shows the geographic location of each. All raw flux data were converted to milligrams of constituent per square meter of wetland (marsh or mangrove) per hour inundated. We present this complete dataset in Table 7-2. Our overall goal is to identify broad-scale controls underlying wetland-water column interactions in Gulf of Mexico estuaries. To address this goal, we categorized each wetland flux study site using several schemes based on wetland type, geological setting, morphology, and upstream-downstream interactions (Table 7-1). We then synthesized the data in several combinations to test hypothetical similarities in wetland flux behavior across categories. Details of the categories shown in Table 7-1 are: (1) Ecological—whether a flux study was conducted in an herbaceous marsh or a mangrove forest; (2) Geomorphological—whether a flux study was conducted with the flume adjacent to an open estuary or in an area proximal to extensive wetlands. Alternatively, we classified sites as "proximal to open water" when the wetland in question was relatively small and contiguous with open estuarine waters or as "proximal to wetlands" when the wetland in question dominated the local setting; (3) Geochemical— whether a flux study took place in a terrigenous-clastic geochemical setting or in a semi-tropical, carbonate-dominated geochemical setting (because of the Boca Chica, Mexico, site, this is not a surrogate for latitude), and; (4) Degree of watershed-offshore coupling—whether a flux study site was located in an estuary experiencing extensive runoff from its upland watershed or one that had only minimal coupling with an upland watershed (because of the Barataria marsh sites and Boca Chica mangrove site, this is not a surrogate for the geochemical classification).

Given these classifications, we grouped the flux datasets into matrices of paired categories based on where we hypothesized that similarities and differences in patterns of wetland flux would be revealed. These matrices are shown in Fig. 7-2. The hypotheses we tested are listed below. In all cases, these hypotheses were tested with unbalanced ANOVA. Where significant differences were found, we compared mean fluxes graphically ($\pm$ standard error).

In the Ecological-Geochemical matrix, we hypothesized that wetland-water column fluxes in mangrove forests and intertidal marshes would not be significantly different. Unless a wetland is in a growth phase (showing net biomass increase) or a progradational phase (showing net deposition of material in soils), a quasi-steady state balance should exist in the whole system nutrient budget—regardless of wetland type. However, we expected that wetlands in terrigenous-clastic geochemical settings would have a higher magnitude of flux than carbonate-based geochemical settings because of nutrients and organic matter supplied by the upstream watershed (Twilley 1988; Yañez-Arancibia and Day 1988; Rivera-Monroy et al. 1995).

In the Ecological-Geomorphological matrix we hypothesized that wetland-water column fluxes would be greater in both marshes and mangroves located proximal to open estuarine settings than in those in wetland-dominated settings. A key component of flux calculated using wetland flumes is the initial composition of water entering the flume. In areas where wetlands are dominant, water column concentrations should strongly reflect wetland influence. Where open water is dominant, however, water column characteristics are presumably not as greatly affected by wetland interactions; the more extensive water column should dilute the effects of the less extensive wetlands (Ridd et al. 1990; Childers and Day 1991). Thus, the water entering a marsh flume that is proximal to other wetlands should be more similar in constituent concentration to water leaving that flume than to water entering a flume that is proximal to an open estuary. The result of this greater contrast between open water concentrations (before treatment by the wetland within the flume) and wetland-influenced concentrations (after treatment) would be greater flux rates.

In the Watershed Coupling-Geochemical matrix we hypothesized that wetland-water column fluxes would be greater in both terrigenous-clastic and carbonate estuaries with tight coupling between the estuary and the watershed because of nutrients and organic matter supplied by the upstream watershed (Madden et al. 1988; Yañez-Arancibia and Day 1988). Within this comparison, we expected that wetlands in terrigenous-clastic geochemical settings would have a higher magnitude of flux than carbonate-based geochemical settings because the terrigenous-clastic sediments must have an upstream source, suggesting a more dynamic watershed-estuary connection. The stronger an estuary's connection to a fertile upstream watershed, the greater the exchange of materials between that estuary and its wetlands.

Finally, in the Ecological-Watershed Coupling matrix, we hypothesized that wetland-water column fluxes would be greater in both marshes and mangroves located in systems with close coupling between the estuary and the watershed because of nutrients and organic matter supplied by the upstream watershed. In this case, the degree of wetland-watershed coupling was positively related to the magnitude of watershed inputs to the estuary. Thus, estuaries that are most tightly coupled with their watersheds will have the largest fresh-water inputs from upstream.

FLUX DATASETS

The wetland-water column flux datasets from Gulf of Mexico estuaries presented some challenges. Geographically, datasets from several studies in Louisiana marshes and in mangrove wetlands of south Florida and southeast Mexico do not adequately represent the entire Gulf of Mexico. However, the study locations represented happen to be the areas with the largest concentrations of these types of intertidal wetlands (Fig. 7-1). This flux dataset also presents several statistical challenges. As Table 7-2 clearly shows, these data are biased toward mangrove studies: Of the 52 tidal flux values shown, only 9 were from

TABLE 7-2. All Flux Data Available from the Studies Shown in Table 7-1

Estuary	Wetland Type	Date Sampled	Season	Constituent											
				NH4	NN	SRP	DOC-1	DOC-2	DON	POC-1	POC-2	PON	TN	TP	TSS
Barataria	Marsh	15 Sep 86	F	-18.00	-1.10	-0.30	-460.0	-460.0		98.0	98.0	16.0			2500.0
Barataria	Marsh	23 Jun 87	SU	0.40	-0.50	0.40	-1200.0	-1200.0		-2.3	-2.3	1.8			180.0
Fourleague	Marsh	27 Sep 86	F	-5.40	0.10	-1.50	54.0	54.0		-48.0	-48.0	-5.1			62.0
Fourleague	Marsh	7 Mar 87	SP	3.40	-2.20	-1.30	3000.0	3000.0		-78.0	-78.0	0.0			-1800.0
Fourleague	Marsh	10 Jun 87	SU	-1.70	-2.10	-23.00	5100.0	5100.0		-140.0	-140.0	-35.0			980.0
Barataria	Marsh	2 Jun 87	SU	-2.10	-1.20	-0.10	-60.0	-60.0		-26.0	-26.0	-6.5			-510.0
Barataria	Marsh	10 Aug 87	SU	11.00	1.20	2.10	1200.0	1200.0		-15.0	-15.0	7.0			-460.0
Barataria	Marsh	15 Nov 87	F	-0.30	3.00	-0.80	-1200.0	-1200.0		97.0	97.0	13.0			700.0
Barataria	Marsh	18 Apr 88	SP	-15.00	-7.30	-0.10	-1100.0	-1100.0		160.0	160.0	23.0			1100.0
Estero Pargo	Mangrove	Aug 90	Wet	3.48	-0.29				-0.91			-8.7			
Estero Pargo	Mangrove	Sep 90	Wet	0.89	0.39				-1.28			0.1			
Estero Pargo	Mangrove	Oct 90	Wet	-0.31	0.38				-0.35			0.8			
Estero Pargo	Mangrove	Nov 90	Wet	14.20	0.10				-1.38			-5.0			
Estero Pargo	Mangrove	Dec 90	Dry	-0.70	0.40				0.35			-19.2			
Estero Pargo	Mangrove	Feb 91	Dry	1.62	0.10				-1.72			2.1			
Estero Pargo	Mangrove	Apr 91	Dry	11.20	-0.10				0			2.2			
Estero Pargo	Mangrove	Jun 91	Wet	0.15	2.54				-0.24			0.2			
Estero Pargo	Mangrove	Jul 91	Wet	-0.16	0.10				0.05			0.3			
Estero Pargo	Mangrove	Aug 91	Wet	2.59	0.11				0			1.6			
Estero Pargo	Mangrove	Jan 92	Dry	1.62	0.10				0.58			-4.6			
Boca Chica	Mangrove	Nov 90	Wet	2.46	2.43				0			0.0			
Boca Chica	Mangrove	Dec 90	Wet	-4.43	0.26				4.07						
Boca Chica	Mangrove	Jun 91	Dry	-5.56	0.15				-9.02						
Shark River	Mangrove	May 95	Dry	0.79	0.45	-0.02									
Shark River	Mangrove	May 95	Dry	0.38	0.15	-0.23									
Shark River	Mangrove	May 95	Dry	0.78	0.27	0.00									
Taylor R. dwarf	Mangrove	Aug 96	Wet	0.70	-3.59	0.16	60.0			-104.0			2.9	-0.2	
Taylor R. dwarf	Mangrove	Nov 96	Wet	0.25	-0.80	-0.01				-2.6			0.2	0.0	

Site	Wetland type	Date	Season	DOC-1	POC-1	DOC-2	POC-2
Taylor R. dwarf	Mangrove	Jan 97	Dry				
Taylor R. fringe	Mangrove	Nov 96	Wet				
Taylor R. fringe	Mangrove	Jan 97	Dry				
Rook. Bay basin	Mangrove	Jan 79	Dry	-5.3	-6.7	-1.6	-2.8
Rook. Bay basin	Mangrove	Feb 79	Dry	-7.2	-8.4	-5.0	-2.5
Rook. Bay basin	Mangrove	Mar 79	Dry	-1.2	-1.5	-0.4	-1.8
Rook. Bay basin	Mangrove	Apr 79	Dry	-5.7	-6.5	-2.0	-1.9
Rook. Bay basin	Mangrove	May 79	Dry	-3.2	-4.6	-1.1	-3.2
Rook. Bay basin	Mangrove	Jun 79	Dry	-3.6	-4.7	-1.4	-3.1
Rook. Bay basin	Mangrove	Jul 79	Wet	-4.5	-6.0	-2.0	-3.8
Rook. Bay basin	Mangrove	Aug 79	Wet	-0.5	-4.0	-1.5	-9.1
Rook. Bay basin	Mangrove	Sep 79	Wet	-9.0	-13.7	-2.9	-5.9
Rook. Bay basin	Mangrove	Oct 79	Wet	-4.6	-4.6	-1.5	-1.2
Rook. Bay basin	Mangrove	Nov 79	Wet	-2.4	-2.8	-1.1	-2.0
Rook. Bay basin	Mangrove	Dec 79	Wet	-1.8	-3.4	-0.4	-4.4
Rook. Bay fringe	Mangrove	Jan 79	Dry	-5.2	-6.8	-1.4	-2.9
Rook. Bay fringe	Mangrove	Feb 79	Dry	-5.3	-6.7	-1.5	-2.0
Rook. Bay fringe	Mangrove	Mar 79	Dry	-2.4	-2.7	-1.1	-1.5
Rook. Bay fringe	Mangrove	Apr 79	Dry	-1.2	-2.1	-1.1	-4.2
Rook. Bay fringe	Mangrove	May 79	Dry	-3.3	-5.0	-0.1	-3.0
Rook. Bay fringe	Mangrove	Jun 79	Dry	-2.8	-4.2	-1.0	-3.4
Rook. Bay fringe	Mangrove	Jul 79	Dry	-7.3	-9.0	-0.7	-2.3
Rook. Bay fringe	Mangrove	Aug 79	Wet	-3.4	-7.6	-1.1	-5.7
Rook. Bay fringe	Mangrove	Sep 79	Wet	-13.5	-19.1	-4.5	-6.9
Rook. Bay fringe	Mangrove	Oct 79	Wet	-4.3	-4.3	-1.5	-1.2
Rook. Bay fringe	Mangrove	Nov 79	Wet	-3.0	-3.5	-0.2	-1.3
Rook. Bay fringe	Mangrove	Dec 79	Wet	-2.2	-4.2	-0.6	-4.3

Note: Each estuarine site is classified here by wetland type, and a season is shown for each sampling date (note that the subtropical mangrove wetlands have two seasons and the herbaceous marshes have four seasons). All flux values are in mg constituent m^{-2} wetland hr^{-1} inundated, with positive flux representing uptake by the experimental wetland and negative flux representing release. DOC-1 AND POC-1 are tidal fluxes only for all sites; DOC-2 and POC-2 are tidal fluxes plus rainfall inputs at Rookery Bay, Florida, but tidal fluxes only for all other sites.

A. Ecological - Geochemical Matrix

	Herbaceous (marsh)	Forested (mangrove)
Terrigenous-clastic based system	Barataria Basin Fourleague Bay	Boca Chica
Carbonate based system	??????	Estero Pargo Rookery Bay Taylor R., Shark R.

B. Ecological - Geomorphological Matrix

	Herbaceous (marsh)	Forested (mangrove)
Wetlands proximal to open estuary	Barataria Basin	Boca Chica Estero Pargo Rookery Bay (fringe)
Wetlands proximal to other wetlands	Fourleague Bay	Shark River Taylor River Rookery Bay (basin)

C. Watershed Coupling - Geochemical Matrix

	Intensive upstream, downstream coupling	Minimal upstream, downstream coupling
Terrigenous-clastic based system	Fourleague Bay Boca Chica	Barataria Basin
Carbonate based system	Shark River	Taylor River Estero Pargo Rookery Bay

D. Ecological - Watershed Coupling Matrix

	Herbaceous (marsh)	Forested (mangrove)
Intensive upstream, downstream coupling	Fourleague Bay	Boca Chica Shark River
Minimal upstream, downstream coupling	Barataria Basin	Estero Pargo Rookery Bay Taylor River

(a)

FIG. 7-2. Hypothetical matrices used in flux analysis, including (a) examples of the four comparisons tested using two-way ANOVA (all sites and classifications from Table 7-1) and (b) the same matrices showing all constituent flux datasets applicable to tests of each of the four hypothetical matrices. See text for definitions of constituent abbreviations.

A. Ecological - Geochemical Matrix

	Herbaceous (marsh)	Forested (mangrove)
Terrigenous-clastic based system	NH_4, NO_3+NO_2	
Carbonate based system		

B. Ecological - Geomorphological Matrix

	Herbaceous (marsh)	Forested (mangrove)
Wetlands proximal to open estuary	NH_4, NO_3+NO_2, SRP, DOC-1, POC-1 DOC-2, POC-2	
Wetlands proximal to other wetlands		

C. Watershed Coupling - Geochemical Matrix

	Intensive upstream, downstream coupling	Minimal upstream, downstream coupling
Terrigenous-clastic based system	NH_4, NO_3+NO_2, SRP, DOC-1, POC-1 DOC-2, POC-2	
Carbonate based system		

D. Ecological - Watershed Coupling Matrix

	Herbaceous (marsh)	Forested (mangrove)
Intensive upstream, downstream coupling	NH_4^+, NO_3+NO_2, SRP; DOC, POC same as watershed-geochem.	
Minimal upstream, downstream coupling		

FIG. 7-2. (Continued).

(b)

herbaceous marshes. All of these nine flux values were from studies conducted in Louisiana estuaries (Table 7-1). Notably, the Rookery Bay basin and fringe mangrove sites dominate the mangrove flux dataset, as 24 of the 42 fluxes are from this site. The most significant challenge, however, comes from the fact that not all researchers quantified fluxes of the same constituents; in fact, no single constituent was represented in every data set (Table 7-2; Fig. 7-2b). For example, ammonium (NH_4), and nitrate + nitrite ($NO_3 + NO_2$) flux data were available for all sites except Rookery Bay, Florida, while dissolved organic carbon (DOC) and particulate organic carbon (POC) flux data were available for all marsh sites and the Rookery Bay mangrove sites. Fluxes of soluble reactive phosphorus (SRP), total nitrogen (TN), total phosphorus (TP), and total suspended sediments (TSS) were available only for the marsh sites, ruling out even basic marsh-mangrove comparisons for exchanges of these constituents (Fig. 7-2b). Finally, we were unable to use sampling time or even sampling season as an explanatory variable because no two studies were conducted at the same time and because the mangrove sites typically followed a wet-dry tropical seasonality, while the marsh sites followed a winter-summer temperate seasonality.

The nine sites used in this analysis included (1) estuarine wetlands experiencing a wide range of coupling with their upstream watersheds; (2) wetlands representing both terrigenous-clastic and carbonate geochemical settings; (3) wetlands that we classified as being either major or minor components of their estuaries, and; (4) flux data from January 1979 through the present (Table 7-2). Except for the Rookery Bay study, wetland-water column exchange data were all reported in the literature in units of mg m^{-2} hr^{-1}. We converted the Rookery Bay flux data from units of g C m^{-2} mo^{-1} to mg C m^{-2} hr^{-1} using monthly inundation data from Twilley (1985). The Rookery Bay dataset was also unique in that Twilley (1985) presented carbon flux for both tidal exchange and export mediated by rainfall runoff. For our analyses, we separated these into (1) tidal exchanges of DOC and POC and (2) organic C fluxes for tidal exchange plus rainfall runoff and treated these as separate variables in our analysis (shown as DOC-1 and POC-1, and DOC-2 and POC-2, respectively, in Table 2). In all cases, positive flux represented wetland uptake, while negative flux represented release to the inundating water column.

RESULTS OF FLUME FLUX RELATIONSHIPS

Single-Factor Comparisons

We used single-factor ANOVA to test for differences between (1) marsh and mangrove wetland fluxes; (2) wetland fluxes in terrigenous-clastic versus carbonate geochemical settings; (3) fluxes in wetland-dominated versus open water-dominated estuaries; and (4) wetland exchanges in estuaries with intense versus minimal watershed coupling (note that in the last three analyses, marsh and mangrove data were combined). Fluxes were the dependent variables for these tests (Table 7-2). In all cases, we included only those constituent fluxes for which data were available for both factors being tested.

Given the many obvious differences (structural, latitudinal, etc.) between herbaceous marshes and mangrove forests, one might expect to see some difference in the way these two wetland types interact with the inundating water column. In fact, several researchers have warned about extrapolating estuarine exchange concepts from temperate marshes to tropical mangroves (Boto and Wellington 1984; Alongi 1990; Rivera-Monroy et al. 1995). To the contrary, we found no significant difference for any of the constituents for which this comparison was possible (Table 7-3). We also investigated simple differences between those Gulf coast wetlands characterized by terrigenous-clastic geochemistry and those that are carbonate-based. Notably, herbaceous marsh is found only in the former category, while mangrove forest is found in both categories (Table 7-1, Fig. 7-2). Ammonium exchanges were significantly different (Table 7-3), with NH_4 release from terrigenous-clastic wetlands and NH_4 uptake by carbonate wetlands (see Table 7-3 for mean fluxes from each). The release of reduced inorganic N from intertidal marshes has been well documented (Nixon 1980; Welsh 1980; Wolaver et al. 1983; Whiting et al. 1989; Childers and Day 1990b; Childers et al. 1993a). The geochemical characterization did not generate significant differences in any other viable flux constituents tested (Table 7-3; note that tests using DOC and POC flux were identical to the ecological characterization test because of the nature of the dataset).

Using geomorphology as the distinguishing characteristic, we found simple differences in organic fluxes but not in dissolved inorganic nutrient exchanges (Table 7-3). Wetlands in proximity to open water took up particulate organics and exported DOC (though the difference in DOC flux was marginal; Table 7-3). When particulate organics were taken up, the C:N molar ratio of flux was about 14:1 compared to a ratio of less than 2:1 when exported. Wetlands that were proximal to other wetlands appeared to export preferentially. It is also intriguing that these wetlands imported large amounts of DOC, though this value may be somewhat skewed by large uptake rates measured in the Fourleague Bay, Louisiana, study (Table 7-2). Notably, however, the sampling design of the Rookery Bay mangrove study precluded the possibility of any positive (import) fluxes in this dataset (Twilley 1985; Table 7-2).

We also investigated how the degree of watershed coupling affected wetland-water column exchanges in Gulf estuaries. In this simple ANOVA comparison, the significant differences were again seen only in organic fluxes, whereas mean particulate and dissolved exchanges were in the opposite direction. Wetlands in estuaries with intensive (upstream) watershed interactions exported POC and took up DOC, while mean flux from wetlands with only minimal upstream interaction were negligible (Table 7-3). Particulate organic N (PON) flux showed a weak response to watershed coupling ($p = 0.08$). For both the morphological and watershed coupling comparisons, the addition of

TABLE 7-3. Results of the Simple ANOVA Tests of Differences in Flux Between Wetland Type, Geochemical Setting, Geomorphological Proximity to Open Water or Other Wetlands, and Watershed Coupling

Comparison	Mean ± SE	NH4	NO3 + NO	SRP	DOC-1	DOC-2	DON	POC-1	POC-2	PON	TN	TP	TSS
Ecological	Significance	NS	NS	NS	NS	NS	NA	NS	NS	NS	NA	NA	NA
	Marsh												
	Mangrove												
Geochemical	Significance	$p = 0.03$	NS	NS	NA	NA	NA	NA	NA	NS	NA	NA	NA
	Terrigenous-clastic	−2.94 ± 2.26											
	Carbonate	2.34 ± 1.05											
Geomorphological	Significance	NS	NS	NS	$p = 0.076$	$p = 0.072$	NA	$p = 0.016$	$p = 0.031$	$p = 0.042$	NA	NA	NA
	Proximal to open water				−160 ± 131	−161 ± 131		18.1 ± 11.4	15.2 ± 11.7	1.34 ± 2.23			
	Proximal to wetlands				510 ± 359	539 ± 382		−20.7 ± 10.8	−20.5 ± 10.2	−13.4 ± 10.9			
Watershed coupling	Significance	NS	NS	NS	$p < 0.0001$	$p < 0.0001$	NA	$p = 0.0009$	$p = 0.0003$	$p = 0.08$	NA	NA	NA
	Minimal coupling				−92 ± 77	−98.7 ± 79.2		5.3 ± 7.5	7.71 ± 7.12				
	Intensive coupling				2720 ± 1460	2720 ± 1460		−88.7 ± 27.1	−88.7 ± 27.1				

Note: For each significant finding, the mean flux (± SE) is shown for the two classifications of that category. NS = ANOVA test results were not significant ($p > 0.10$); NA = data were not available to allow this particular test.

rain-induced flux from Rookery Bay (DOC-2 and POC-2; Table 7-2) made no difference in the ANOVA results and only minimal differences in the mean fluxes (Table 7-3).

Comparisons within Wetland Type

Before testing the hypotheses conceptualized in Fig. 7-2, we also checked for seasonal effects on flux patterns. Gulf of Mexico mangrove wetlands are found in sub-tropical and tropical climates, where there are effectively only two seasons—a summer-fall wet season and a winter-spring dry season. This clearly contrasts with the four-season temperate climatic regime where most Gulf herbaceous marshes are found. Thus, we split the dataset into marsh versus mangrove fluxes to test for seasonal effects. Using one-way ANOVA, we found no simple seasonal control on wetland-water column exchanges in mangrove forests. When season and morphology were tested together, how-ever, we did see a significant morphology effect on mangrove $NO_3 + NO_2$ flux ($p = 0.016$) and a significant interaction with season ($p = 0.008$), though the season effect was weak ($p = 0.068$). The most interesting pattern seen in the mean $NO_3 + NO_2$ fluxes from this comparison was strong release of $NO_3 + NO_2$ from mangroves in wetland-dominated estuaries during the wet season compared with low-magnitude $NO_3 + NO_2$ uptake in all other situations (Fig. 7-3). Wetlands, with their reducing soils, have long been thought of as consistent transformers of inorganic N from oxidized ($NO_3 + NO_2$) to reduced (NH_4) forms (e.g., Nixon 1980). In fact, fluxes measured at the Estero Pargo, Mexico, site (which we classified as proximal to open water) followed this pattern (Rivera-Monroy et al. 1995; Table 7-2). Recent data from south Florida sites (which make up the "proximal to other wetlands" classification for the DIN data) suggest the opposite, though (Table 7-2), and it is these data that appear to be controlling this statistical outcome.

Wetland-water column exchange data from herbaceous marshes were avail-able only for the spring, summer, and fall seasons (Table 7-2). The only seasonal effect observed in our simple comparisons was with $NO_3 + NO_2$ flux, and it was weak ($p = 0.079$). In this case, spring and summer were characterized by $NO_3 + NO_2$ export (-0.65 ± 0.7 and -4.75 ± 2.55 mg N $m^{-2} h^{-1}$, respectively), while uptake occurred in the fall (0.67 ± 1.22 mg N $m^{-2} h^{-1}$). It was fortunate that we found no simple, significant effects of season in either the marsh or mangrove flux datasets. This "complication of time" would have rendered our more complex hypotheses either impossible to test (since marshes and mangroves have different categorizations of seasons, we could not add a season interaction term to any test across wetland types) or, at best, difficult to interpret.

Two-way marsh flux comparisons using season as an independent varia-ble were more restricted because (1) all fluxes were from terrigenous-clastic marshes and (2) the morphological and watershed characterizations exactly overlapped (Table 7-1). Given this limitation, we found significant season and

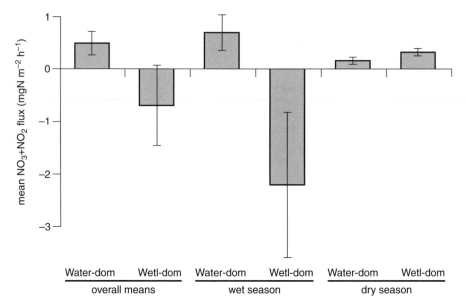

FIG. 7-3. *$NO_3 + NO_2$ flux means from mangrove studies generated by the test of season versus morphology. Water-dom = $NO_3 + NO_2$ flux in wetlands proximal to open water; Wetl-dom = $NO_3 + NO_2$ flux in wetlands proximal to similar wetlands. Error bars represent ± 1 SE; positive values represent uptake of $NO_3 + NO_2$ by the mangrove forest, while negative fluxes represent $NO_3 + NO_2$ export.*

morphology effects on SRP flux ($p = 0.017$ and 0.012, respectively), POC flux ($p = 0.014$ and 0.0002, respectively), and PON flux ($p = 0.021$ and 0.008, respectively), as well as a significant season-morphology interaction in the SRP ($p = 0.001$) and POC ($p = 0.017$) tests. Gulf Coast marshes appeared to release SRP in all three seasons, though the mean summer flux was not significantly different from zero (Fig. 7-4). In contrast, salt marshes in South Carolina and Georgia consistently took up SRP during non-winter months (Wolaver and Spurrier 1988; Childers et al. 1993a). Particulate organics were taken up in the spring and fall and exported in the summer, though, again, several of these mean fluxes were indistinguishable from zero (Fig. 7-4). It is interesting that marshes proximal to open water released all three constituents, while those proximal to other wetlands took up all three constituents (Fig. 7-4).

Ecological-Geochemical Matrix

The results we predicted from this comparison included that (1) inorganic nutrient fluxes should be greatest in mangrove wetlands located in terrigenous-clastic regimes and (2) organic matter fluxes should be greatest in mangrove wetlands, regardless of the geochemical setting. Unfortunately, only the NH_4 and $NO_3 + NO_2$ flux datasets were adequate to address this question (our

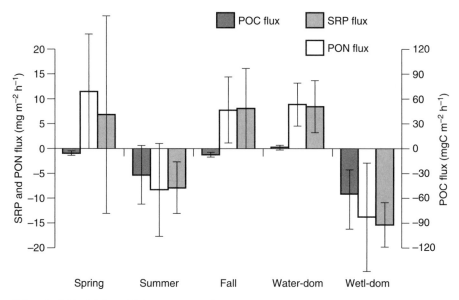

FIG. 7-4. *SRP, POC, and PON flux means from marsh studies generated by the test of season versus morphology. Water-dom = flux in wetlands proximal to open water; Wetl-dom = flux in wetlands proximal to similar wetlands. Error bars represent ± 1 SE; positive values represent uptake of SRP, POC, or PON by the marsh, while negative fluxes represent export.*

criterion being that all matrix combinations shown in Fig. 7-2a must be represented with flux data). Neither the type of wetland nor the geochemical setting of that wetland played a significant role in explaining the observed patterns in wetland-water column exchange for either form of dissolved inorganic N. We were thus unable to support our hypotheses for this interaction.

Ecological-Geomorphological Matrix

We expected this comparison to show that fluxes in open estuarine wetlands should be greater than those in locations dominated by the wetlands in question. This somewhat counter-intuitive hypothesis was based on the idea that, in both cases, the water column inundating a wetland (where wetland flume studies are conducted) will be affected by wetland fluxes and that waterborne concentrations will reflect this. In open estuarine wetlands, flooding tide concentrations represent open water conditions, and concentrations are likely quite different from those driven by wetland influences. In contrast, where wetlands dominate, the water column should reflect this wetland influence nearly everywhere and most of the time. Because flux is calculated from concentration differences before and after water has been exposed to a given area of wetland, the flux should be greater where the concentration differences are greater. This is the open estuary case. Naturally, if wetland-water column interactions, measured per m^2 wetland, are normalized to the whole-system area of wetland, wetland influence will be greater in the wetland-dominated system. However, we hypothesized greater fluxes on a per m^2 flux basis in the open water-dominated systems (for more details, see Childers and Day 1991).

Neither wetland type nor geomorphological setting affected dissolved inorganic N flux, but both showed an effect with SRP flux ($p = 0.05$ and 0.04, respectively). Soluble reactive P was exported from both marsh and mangrove wetlands, but mean flux was two orders of magnitude greater from herbaceous marshes (marsh $= -2.73 \pm 2.56$ mg P m^{-2} h^{-1}; mangrove $= -0.02 \pm 0.06$ mg P m^{-2} h^{-1}). Mean SRP flux in water-dominated settings was not different from zero (0.20 ± 0.41 mg P m^{-2} h^{-1}), while wetland-dominated systems showed a mean SRP export of -3.24 ± 2.83 mg P m^{-2} h^{-1}.

Wetland-water column exchanges of both dissolved and particulate organic C were affected by wetland type and geomorphology. For DOC, both characteristics showed significant effects ($p = 0.001$ and $p < 0.0001$, respectively), as did the interaction of the two ($p < 0.0001$). For POC, the morphology effect was significant ($p < 0.0001$), as was the interaction term ($p < 0.0001$). Wetlands in water-dominated settings exported DOC, both as a mean flux, regardless of wetland type, and as separate mangrove and marsh means (Fig. 7-5A). Mean DOC fluxes in wetland-dominated settings, however, consistently showed uptake (Fig. 7-5A). The pattern was the opposite for POC flux, with the exception of the mean export from water-dominated mangrove settings (Fig. 7-5B). The results of this comparison highlight a major difference in flux behavior in marsh versus mangrove wetlands: DOC flux measured in marshes is consistently several orders of magnitude greater than in mangroves, while POC flux is generally of similar magnitude (Fig. 7-5). Much of the DOC being exchanged by intertidal wetlands is presumably a product of macrophyte exudation (Pomeroy et al. 1977; Chalmers et al. 1985). It is possible that herbaceous macrophytes simply exude more dissolved organic matter than the woody trees that dominate mangrove wetlands.

Geochemical-Watershed Coupling Matrix

We predicted that this comparison would show greater mean fluxes in wetlands found in estuaries that are tightly coupled to their upstream watersheds. This close connection between upland and riverine inputs and biological activity has been demonstrated for many estuarine components, including wetland fluxes (Boto and Bunt 1981; Madden et al. 1988; Twilley 1988; Childers and Day 1990a, 1990b; Childers et al. 1993b). In our two-way ANOVA tests using geochemical and watershed coupling classifications as effects, NH_4 showed a weak effect of geochemical setting ($p = 0.095$). Both organic C fluxes showed strong responses to watershed coupling (DOC: $p < 0.0001$; POC: $p < 0.0001$). In both cases, the greatest mean organic matter flux occurred in wetlands with intense coupling to their watersheds, with strong mean export of POC and uptake of DOC (Fig. 7-6). We hypothesized that estuarine wetlands, being

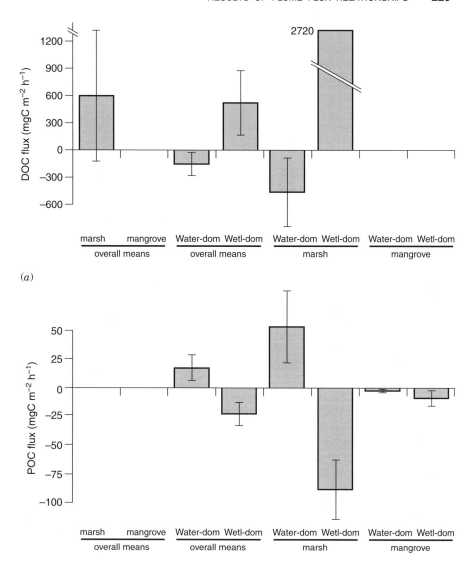

FIG. 7-5. DOC (A) and POC (B) flux means generated by the test of wetland type versus morphology. Water-dom = flux in wetlands proximal to open water; Wetl-dom = flux in wetlands proximal to similar wetlands. Note the break in (a) for Wetl-dom. marsh DOC flux (actual mean flux = 2720 ± 1460 mg C m^{-2} hr^{-1}) and note that in (b) the wetland-type component of this model was not significant (p > 0.05); thus no means are shown for marsh versus mangrove overall means. Error bars represent ± 1 SE; positive values represent uptake by the wetland, while negative fluxes represent export.

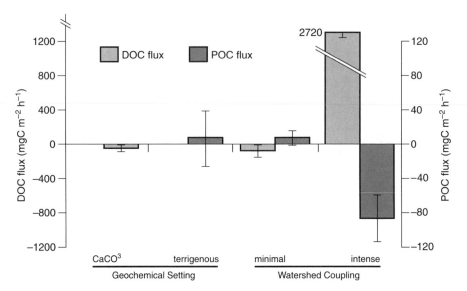

FIG. 7-6. *DOC and POC flux means generated by the test of geochemical setting versus watershed coupling. Note the break in DOC flux in wetlands with intense watershed coupling (actual mean flux = 2720 ± 1460 mg C m⁻² hr⁻¹). Due to limitations and overlaps in the dataset, note that the geochemical setting classification here is identical to the wetland-type classification for this test. Thus, CaCO₃ = mangrove forest and terrigenous-clastic = marsh. Error bars represent ± 1 SE; positive values represent uptake by the wetland, while negative fluxes represent export.*

strongly influenced by upstream influences—and presumably by upstream particulate inputs—would respond to this large estuarine sediment load by importing POC. The mean POC export we found here, however, was based on flux data from the Fourleague Bay, Louisiana, site only (the only site with intense watershed coupling that also had POC and DOC flux data available). With one exception, on every sampling date on which a POC export was measured from this marsh, TSS uptake was also observed (Childers and Day 1990a; Table 7-2). Thus, this marsh was importing inorganic sediment being provided by the Atchafalaya drainage basin while, at the same time, releasing particulate organics. Only POC also showed a geochemical setting effect ($p = 0.0017$); mean POC flux was highly variable in terrigenous-clastic wetland systems, while carbonate systems showed a low-magnitude mean POC export (Fig. 7-6). Neither ANOVA model had interaction terms because of our limited dataset.

Ecological-Watershed Coupling Matrix

We expected this comparison to show greater mean fluxes in wetlands found in estuaries that are tightly coupled to their upstream watersheds, regardless of whether they are mangrove or herbaceous wetlands. Our logic was the same as for Matrix C in Fig. 7-2. In our two-way ANOVA tests of wetland type and watershed coupling, the only inorganic nutrient to show an effect was $NO_3 + NO_2$—where we found only a weak effect of wetland type ($p = 0.087$). Both organic C fluxes showed strong responses to watershed coupling (DOC: $p < 0.0001$; POC: $p < 0.0001$), and POC also showed a significant effect of wetland type ($p = 0.0017$). Because of limitations posed by the DOC and POC flux datasets, however, wetland type and geochemical setting were identical in these two-way tests. All marsh flux data were from terrigenous-clastic geochemical settings, and we had DOC and POC flux data only from carbonate-based mangrove sites (Table 7-1). Thus, in Fig. 7-6, the $CaCO_3$ geochemical setting also represents mangroves and the terrigenous geochemical setting also represents marsh.

SUMMARY

Our objectives for this chapter included summarizing all wetland flux studies from Gulf of Mexico estuaries that have directly quantified wetland-water column biogeochemical exchanges in situ. Our presentation of these data in Table 7-2 is the first compilation of all such data. Given the wide range of time represented by these data (1979 to the present) and the wide range of wetland types, the inorganic nutrient flux values were surprisingly similar. Exchanges of organics and particulates, however, varied widely. Most wetland-water column exchange data were from mangrove forests, and that dataset was dominated by the intensive Rookery Bay study by Twilley (1985). This situation, coupled with the fact that no single constituent was quantified in all Gulf of Mexico flux studies, greatly complicated our statistical analyses and interpretations. In spite of having to use widely unbalanced ANOVA designs and being able to test only some hypothetical interactions, we were able to discern some interesting patterns in wetland-water column flux. This kind of broad, all-encompassing search for commonalities in the way disparate systems behave can often be as instructional as a detailed look at the specific processes responsible for the dynamics of a given system.

Another of our objectives was to use this flux dataset to search for general patterns in wetland flux behavior across the Gulf of Mexico region. We found several interesting similarities, including no significant difference in wetland-water column exchange in mangrove forests compared to herbaceous marshes—for any constituent (Table 7-3). In fact, fluxes were very similar in magnitude in some cases, such as with inorganic nutrients and PON. In our geochemical classification, only NH_4 flux showed a significant difference, with wetland uptake in carbonate systems and export in terrigenous-clastic systems (Table 7-3). With the organic constituents, the marsh fluxes were larger but quite variable in direction (and, notably, the only DOC and POC flux data available from mangroves did not include uptake or import because of the sampling procedures; see Twilley 1985). We also found that season did not significantly explain the variance in wetland fluxes in either mangroves or marshes (though marsh $NO_3 + NO_2$ flux showed a weak response to season at $p = 0.079$; Table 7-3). Similarities and predictable patterns in ecosystem-level behaviors across geographically and morphologically different estuarine systems have been reported (Childers 1994).

As with any integrative analysis, however, we learn more from statistical differences than from similarities. When simple ANOVA tests were run for geomorphological and watershed coupling effects on wetland fluxes, we found significant differences in organics fluxes (Table 7-3). In this case, the pattern was quite consistent. Wetlands proximal to open water and with minimal coupling to their upstream watersheds exported DOC while taking up POC and PON. Wetlands proximal to other wetlands and having tight communication with their watersheds took up DOC but released POC and PON (Table 7-3).

Next, we separated the dataset into marsh fluxes and mangrove fluxes, then categorized exchanges by wetland geomorphology. In doing so, we found that SRP, POC, and PON tend to be taken up by marshes proximal to open water but to be released by marshes proximal to other wetlands (Fig. 7-4). The mangrove flux results were slightly more confusing: We found large differences in $NO_3 + NO_2$ flux based on the geomorphological classification but only during the wet season (Fig. 7-3). Mangroves took up $NO_3 + NO_2$ in the dry season, regardless of geomorphological type. These types of relationships are difficult to explain, though. For this reason, we addressed specific hypotheses about the way major features of Gulf wetlands might control wetland-water column interactions.

Only dissolved inorganic N data were applicable in addressing our first hypothesis, regarding the interaction of wetland type and geochemical setting (Fig. 7-2). We found no significant differences based on these classifications, confirming our hypothesis about the similarity of mangrove and marsh flux patterns but rejecting our hypothesis that wetlands in terrigenous-clastic geochemical settings would show greater rates of exchange (Fig. 7-2a).

For all three of our remaining matrices of hypothesized comparisons, the most significant results were seen in the DOC and POC data. Wetland type and geomorphology had strong independent and interactive effects on DOC and POC flux. Wetlands proximal to open water tended to release DOC and take up POC; fluxes in wetlands proximal to other wetlands were consistently in the opposite direction (Fig. 7-5). Most interestingly, marsh DOC fluxes were large, variable, and often orders of magnitude greater than mangrove DOC fluxes (wetland type did not play a significant role in explaining POC flux variability; Fig. 7-5). Marshes and mangrove forests have similar rates of net C fixation (Day et al. 1989; Twilley et al. 1992; Mitsch and Gosselink 1993). Standing crop biomass is very different in these two systems, however, as typical values for marshes are 500–2000 g dry wt m^{-2} (Day et al. 1989) compared to 10,000–40,000 g dry wt m^{-2} in mangrove forests (Twilley et al.

1992). Thus marshes and mangroves have similar rates of C input to dramatically different pools of C and quite different DOC flux behaviors, yet there is no real difference in POC exchange rates between them. Our explanation for this divergence in DOC flux rates and directions is that much of the net production in marshes incorporates C into biomass that senesces every fall and is replaced every spring, while most net production in mangroves fixes C that is utilized to maintain the extant structure of the trees and forest. At this scale, marsh productivity may thus be viewed as more labile and more readily exchanged with the estuarine water column, and much of this exchange is via dissolved organic matter.

The interactions among wetland type, geochemical setting, and degree of watershed coupling followed our hypotheses fairly closely with DOC and POC data. Wetlands in terrigenous-clastic geochemical settings exchanged both forms of organic matter at higher and more variable rates than did carbonate wetlands (Fig. 7-6). This finding is closely tied to the fact that wetlands in estuaries that are tightly coupled to their upstream watersheds and rivers had far greater DOC and POC exchanges than those with minimal watershed coupling (Fig. 7-6). This occurs because the watershed is usually the source of terrigenous-clastic sediments, while autochthonous sources are generally responsible for carbonate sediments. Wetlands in estuaries with intense watershed coupling appeared to take up DOC and release POC, but the mean rate of DOC uptake was nearly 30 times the mean rate of POC release (Fig. 7-6).

REFERENCES

Alongi, D. M. 1990. Effects of detrital outwelling on nutrient fluxes in coastal sediments of the central Great Barrier Reef Lagoon. Estuar. Coast. Shelf Sci. **31:** 581–598.

Boto, K. G., and J. S. Bunt. 1981. Tidal export of particulate organic matter from a northern Australian mangrove forest. Estuar. Coast. Shelf Sci. **13:** 247–255.

————, and J. T. Wellington. 1984. Soil characteristics and nutrient status in a northern Australian mangrove forest. Estuaries. **7(1):** 61–69.

Chalmers, A. G., R. G. Wiegert, and P. L. Wolf. 1985. Carbon balance in a salt marsh: Interactions of diffusive export, tidal deposition, and rainfall-caused erosion. Estuar. Coast. Shelf Sci. **21:** 757–771.

Childers, D. L. 1994. Fifteen years of marsh flumes—A review of marsh-water column interactions in Southeastern USA estuaries, p. 277–294. *In:* W. Mitsch [ed.], Global wetlands. Elsevier.

Childers, D. L., S. Cofer-Shabica, and L. Nakashima. 1993a. Spatial and temporal variability in marsh-water column interactions in a Georgia salt marsh. Mar. Ecol. Prog. Ser. **95(1,2):** 25–38.

————, and J. W. Day, Jr. 1988. A flow-through flume technique for quantifying nutrient and materials fluxes in microtidal estuaries. Estuar. Coast. Shelf Sci. **27(5):** 483–494.

————. 1990a. Marsh-water column interactions in two Louisiana estuaries. I. Sediment dynamics. Estuaries **13(3):** 393–403.

————. 1990b. Marsh-water column interactions in two Louisiana estuaries. II. Nutrient dynamics. Estuaries **13(3):** 404–417.

————. 1991. The dilution and loss of wetland function associated with conversion to open water. Wetl. Ecol. Manag. **3:** 163–171.

————, H. N. McKellar, Jr., R. Dame, F. Sklar, and E. Blood. 1993b. A dynamic nutrient budget of subsystem interactions in a salt marsh estuary. Estuar. Coast. Shelf Sci. **36:** 105–131.

Dahl, T. E., and C. E. Johnson. 1991. Wetlands status and trends in the conterminous United States mid-1970s to mid-1980s. U.S. Department of Interior, FWS.

Dame, R. F. 1989. The importance of Spartina alterniflora to Atlantic coast estuaries. Aquat. Sci. **1(4):** 639–660.

Davis, S. E., and D. L. Childers. In prep. Wetland-water column interactions in two types of south Florida mangrove forests.

Day, J. W., Jr., C. A. S. Hall, W. M. Kemp, and A. Yañez-Arancibia. 1989. Estuarine ecology. Wiley.

Field, D. W., A. J. Reyer, P. V. Genovese, and B. D. Shearer. 1991. Coastal wetlands of the United States. NOAA and USFWS.

Keefe, C. W. 1972. Marsh production: A summary of the literature. Contrib. Mar. Sci. **16:** 163.

Lee, V. 1979. Net nitrogen flux between the emergent marsh and tidal waters. M. S. Thesis, University of Rhode Island.

Madden, C. J., J. W. Day, Jr., and J. M. Randall. 1988. Freshwater and marine coupling in estuaries of the Mississippi River Deltaic Plain. Limnol. Oceanogr. **33(4.2):** 982–1004.

Mitsch, W. J., and J. G. Gosselink. 1993. Wetlands, 2nd ed. Van Nostrand Reinhold.

Nixon, S. W. 1980. Between coastal marshes and coastal waters: A review of 20 years of speculation and research on the role of saltmarshes in estuarine productivity and water chemistry, p. 437–525. *In:* P. Hamilton and K. B. MacDonald [eds.], Estuarine and wetland processes. Plenum.

Pomeroy, L. R., K. Bancroft, J. Breed, R. R. Christian, D. Frankenberg, J. R. Hall, L. G. Maurer, W. J. Wiebe, R. G. Wiegert, and R. L. Wetzel. 1977. Flux of organic matter through a saltmarsh. Estuar. Processes **2:** 270.

Ridd, P. V., E. Wolanski, and Y. Mazda. 1990. Longitudinal diffusion in mangrove-fringed tidal creeks. Estuar. Coast. Shelf Sci. **31:** 541–554.

Rivera-Monroy, V. H., J. W. Day, R. R. Twilley, F. Vera-Herrera, and C. Coronado-Molina. 1995. Flux of nitrogen and sediment in a fringe mangrove forest in Terminos Lagoon, Mexico. Estuar. Coast. Shelf Sci. **40:** 139–160.

Snedaker, S. C. 1993. Impact on mangroves, p. 282–299 *In:* G. A. Maul [ed.], Climatic change in the intra-Americas Sea. Routledge, Chapman, and Hall.

Steever, E. Z., R. S. Warren, and W. A. Niering. 1976. Tidal energy subsidy and standing crop production of *Spartina alterniflora.* Estuar. Coast. Shelf Sci. **4:** 473–490.

Templet, P. H., and K. J. Meyer-Arendt. 1988. Louisiana wetland loss: A regional water management approach to the problem. Environ. Manag. **12(2):** 181–192.

Turner, R. E. 1976. Geographic variations in salt marsh macrophyte production: A review. Contrib. Mar. Sci. **20:** 47–69.

————. 1991. Tide gauge records, water level rise, and subsidence in the northern Gulf of Mexico. Estuaries **14(2):** 139–147.

Twilley, R. R. 1985. The exchange of organic carbon in basin mangrove forests in a SW Florida estuary. Estuar. Coast. Shelf Sci. **20:** 543.

————. 1988. Coupling of mangroves to the productivity of estuarine and coastal waters, p. 155–180. *In:* B. O. Jansson [ed.], Coastal-offshore ecosystem interactions. Springer-Verlag.

————, R. H. Chen, and T. Hargis. 1992. Carbon sinks in mangroves and their implications to carbon budget of tropical coastal ecosystems. Water Air Soil Poll. **64:** 265–288.

Weber, M., R. Townsend, and R. Bierce. 1992. Environmental quality in the Gulf of Mexico. Center for Marine Conservation.

Wells, J. T., and J. M. Coleman. 1987. Wetland loss and the subdelta life cycle. Estuar. Coast. Shelf Sci. **25:** 111–125.

Welsh, B. L. 1980. Comparative nutrient dynamics of a marsh-mudflat ecosystem. Estuar. Coast. Mar. Sci. **10:** 143–164.

Whiting, G. J., H. N. McKellar, and T. G. Wolaver. 1989. Nutrient exchange between a portion of vegetated saltmarsh and the adjoining creek. Limnol. Oceanogr. **34(2):** 463–473.

Wolaver, T. G., and J. D. Spurrier. 1988. The exchange of phosphorus between a euhaline vegetated marsh and the adjacent tidal creek. Estuar. Coast. Shelf Sci. **26:** 203–214.

————, R. L. Wetzel, J. C. Zieman, and K. L. Webb. 1983. Tidal exchange of nitrogen and phosphorus between a mesohaline vegetated marsh and the surrounding estuary in the Lower Chesapeake Bay. Estuar. Coast. Shelf Sci. **16:** 321–332.

Yañez-Arancibia, A., and J. W. Day, Jr. 1988. Ecología de los ecosistemas costeros en el sur del Golfo de México: La región de la Laguna de Términos. UNAM Press.

Section III

Organic Matter Cycling

Chapter *8*

Particulate Organic Matter in Gulf of Mexico Estuaries—Implications for Net Heterotrophy

Luis A. Cifuentes, Richard B. Coffin, Jeff Morin, Thomas S. Bianchi, and Peter M. Eldridge

INTRODUCTION

Estuaries lie at the junction between land and sea and receive allochthonous organic matter from riverine inflow and exchange with coastal ecosystems and from autochthonous primary production (Burton and Liss 1976). The degree to which these inputs influence the living resources and biogeochemical cycling at the fresh-water–sea-water interface depends on the residence time of the organic matter and on its reactivity. As discussed by Deegan et al. (1986), Gulf of Mexico estuaries are excellent for comparisons among systems because (1) the number of estuaries is large (Fig. 8-1); (2) the climate in this region is diverse, ranging from tropical to temperate and from humid to arid; (3) the fresh-water influence varies dramatically; and (4) the size of estuarine areas ranges from very small to the largest in the United States. Consequently, these

Biogeochemistry of Gulf of Mexico Estuaries, Edited by Thomas S. Bianchi, Jonathan R. Pennock, and Robert R. Twilley.
ISBN 0-471-16174-8   © 1999 John Wiley & Sons, Inc.

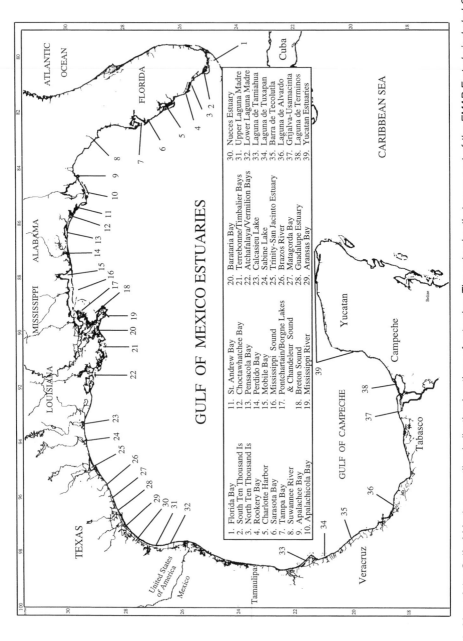

FIG. 8-1. *Map of the Gulf of Mexico coastline indicating locations of estuaries. The estuaries that were part of the EMAP-E study included Suwanee River, Apalachee Bay, Apalichicola Bay, St. Andrew Bay, Choctawhatchee Bay, Pensacola Bay, Perdido Bay, Mobile Bay, Mississippi Sound, Calcasieu Lake, Galveston Bay, Brazos River, Matagorda Bay, Antonio Bay, Upper Laguna Madre, and Lower Laguna Madre.*

systems have a wide range of water and particle residence times—from days to about a year (Baskaran and Santschi 1993; Baskaran et al. 1997; Chapter 2 this volume). Moreover, it is evident that the composition (terrestrial versus marine) of the organic matter within these estuaries is variable (e.g., Trefry et al. 1994; Qian et al. 1996; Argyrou et al. 1997; Bianchi et al. 1997; Goni et al. 1997), implying different organic matter reactivity (Henrichs 1993) among estuaries. Although there is information on the nature and cycling of particulate organic matter (POM) in Gulf estuaries (e.g., Hedges and Parker 1976; Twilley 1985; Day et al. 1988, Flores-Verduge et al. 1988; Twilley 1988; Trefry et al. 1994; Argyrou et al. 1997; Bianchi et al. 1997; Goni et al. 1997), our knowledge of the impact that riverine inputs have on land-sea exchange processes is more limited. Thus, while some theoretical generalizations about these effects can be made, quantifying the absolute rate, or even the relative magnitude, of *specific* processes is difficult and often time-consuming owing to the heterogeneity observed in these systems.

Net heterotrophy occurs when the ratio of respiration (R) of allochthonous and autochthonous organic carbon to autochthonous primary production (P) is greater than 1, that is, $R/P > 1$. This phenomenon, occurring over long (annual or inter-annual) time scales, is thought to be fueled by inputs of reactive allochthonous organic matter (Smith and Hollibaugh 1993). However, net heterotrophy is also considered to occur over shorter (diel or seasonal) cycles if there is de-coupling of autotrophic and heterotrophic production in time and/or space. Ittekkot (1989), however, suggested that terrestrial organic material is refractory and reacts over long time scales. Others (Imberger et al. 1983; Henrichs 1992; Smith and Hollibaugh 1993) argue that terrestrial organic matter is composed of several different components with distinct reactivities that could, therefore, contribute to heterotrophy in estuarine systems (Smith and Hollibaugh 1997). Moreover, anthropogenically derived materials (i.e., sewage) may have significantly higher reactivities. The issue of organic matter reactivity is crucial to our understanding of the autotrophic-heterotrophic balance in estuaries. In turn, comprehension of the nature of this balance and factors that influence this balance is important to determining how carbon (energy) flows in these systems.

A diverse array of allochthonous and autochthonous POM sources (terrestrial, submersed macrophytes, plankton) enter the food webs of riverine and estuarine ecosystems (Wetzel 1984; Valiela 1995). Temporal variations in these sources are expected to vary seasonally, with maximum phytoplankton production in summer and peak allochthonous inputs in spring and winter or late fall. Describing the patterns of cycling of these POM sources is crucial in understanding how seasonal changes in the quality and reactivity of POM affect the overall rate at which POM is cycled in estuaries. For example, there are usually high inputs of recalcitrant terrestrially derived carbon sources into many of the shallow fresh-water–dominated estuaries in the northern Gulf of Mexico in winter months (when rainfall is high), with more labile phytoplankton-derived sources in summer months (Bianchi and Agyrou 1997;

Bianchi et al. 1997). These types of seasonal changes can affect the availability and residence time of POM in these systems. To fully understand the implications of these changes on net heterotrophy, it is necessary to identify sources of POM, which in many cases is a heterogeneous mixture of living and decomposed materials primarily derived from both terrestrial plant and algal sources.

Stable isotopes, often in combination with elemental analyses, have been used traditionally to distinguish allochthonous from autochthonous sources of organic matter in estuarine systems (Coffin et al. 1994). In particular, isotopic analyses of carbon ($\delta^{13}C$) can often discriminate among C3 and C4 terrestrial plants, macrophytes, and benthic and pelagic algae (Fry and Sherr 1984; Coffin et al. (1993). Stable nitrogen isotopes ($\delta^{15}N$) combined with $\delta^{13}C$ offer an additional level of discrimination, including identification of anthropogenic contributions in estuarine systems (Fogel and Cifuentes 1993; Coffin et al. 1994). Elemental ratios (C:N) can also be used to suggest dominant organic sources, owing to differences among algae (C:N of 7 to 10; e.g., Holligan et al. 1984), bacteria (C:N of 3 to 5; Lee and Furhman 1987), and vascular plant material (e.g., C:N >50; Hedges and Mann 1979), provided that decomposition does not alter these values significantly (e.g., Twilley et al. 1986).

Through our participation in the Louisiana Province Environmental Monitoring and Assessment Program-Estuaries (EMAP-E) during the summers of 1991 and 1993, we had a unique opportunity to sample one or more stations in each of 16 diverse estuaries stretching from the southern coast of Texas to the western Florida coast. Although site selection was based on historical evidence of hypoxia and evidence of contamination in bottom sediment, the sites included combinations of these effects and were not biased toward either terrestrial or marine inputs. Therefore, we asked whether allochthonous or autochthonous sources of organic matter dominated in these systems.

We measured $\delta^{13}C$ and $\delta^{15}N$ in suspended particulate matter, humic acids, dissolved inorganic carbon, and bacterial bioassays at most of these 16 stations. Elemental analysis and other selected chemical and biological analyses were conducted to aid in the interpretation of the data. This combination of data provided two levels of information. First, the principal source of organic matter (carbon and nitrogen) was determined qualitatively. Second, the degree to which this primary source of organic matter and microheterotrophic activity was linked was established by comparing isotopic measurements of suspended particulate matter and humic acids with those in the bacterial bioassays (Coffin et al. 1989). In this chapter, we compare results of the EMAP-E study with those of other reports describing the nature of organic matter in Gulf estuaries. In turn, we describe a simple model that considers how composition (i.e., reactivity) and residence time influence the potential for net heterotrophy. We then discuss the implications of this preliminary formulation for Gulf estuaries using the Sabine Lake and Lake Ponchartrain systems as examples.

EMAP-E FIELD SAMPLING AND ANALYTICAL TECHNIQUES

Sixteen estuaries were sampled in the summers of 1991 and 1993. Both particulate and dissolved samples were obtained from a depth of 1 m at all locations. Suspended particulate matter (SPM) samples were isolated for isotopic analyses by pumping 1 to 5 liters of water through 47-mm GF/F filters (pre-treated by heating at 450°C for 2 h), which have a nominal pore size of 0.7 μm and were housed in a stainless steel filter holder with a Masterflex peristaltic pump. Humic acids were collected in the summer of 1991 by a modification of the method of Fox (1983). First, 1 liter of water was pre-filtered through a 47-mm GF/F filter, as described for SPM samples. The filtrate was then acidified to pH 2 with 8 N H_2SO_4 to precipitate humic acids. This precipitate was retained on a 47-mm GF/F filter.

Bacterial bioassays were conducted in the summer of 1991 following the procedures described in Coffin et al. (1989) and Coffin and Cifuentes (1993). Water was pumped through 1.0-μm and 0.2-μm serially connected Millipore cartridge filters. Approximately 4.5 liters of filtrate passing through both filters served as the growth medium. Filtrate from the 1.0-μm filter (250 ml) was used as an inoculum. Bioassays were incubated in cubitainers in the dark at room temperature. Five-milliliter samples were removed and preserved with glutaraldehyde for bacterial abundance at $t = 0$, 24 and 48 h. Bacteria were counted by epifluorescence microscopy following the technique of Hobbie et al. (1977). The remaining growth medium was collected on a 47-mm GF/F filter for isotopic analysis. All filters (containing SPM, humic acid, or bacterial bioassays) were immediately frozen on dry ice and stored at -20°C. In the laboratory, these filters were dried in a 50°C oven and flushed with high-grade N_2 gas.

Approximately 30 ml of unfiltered water was collected in Quorpak bottles and preserved with 2% $HgCl_2$ for isotopic analysis of dissolved inorganic carbon. These samples were refrigerated at 4°C prior to isolation of CO_2 for isotopic analysis by the in vacuo acidification and purging technique described in Grossman (1984).

Nitrate was reduced to NH_4 with Devarda's alloy, distilled, and exchanged onto zeolite (Velinkski et al. 1989). All samples were analyzed isotopically by a modified Dumas combustion that converts organic carbon and organic nitrogen to CO_2 and N_2 gas, respectively, for mass spectral analysis (Macko 1981). CO_2 gas was analyzed on a Finnigan MAT 251, isotope ratio mass spectrometer and dinitrogen gas was analyzed on a Nuclide 3-60-RMS. We present the results in standard notation:

$$\delta^h X = \left[\frac{\left(\dfrac{^hX}{^1X} \right)_{SAM}}{\left(\dfrac{^hX}{^1X} \right)_{STD}} - 1 \right] \times 1000 \tag{8-1}$$

where X is either carbon or nitrogen, h is the heavier isotope, l is the lighter isotope, SAM is the sample, and STD is the standard. The standard for nitrogen was atmospheric N_2, and the standard for carbon was PeeDee Belemnite. The reproducibility of the measurement for $\delta^{13}C$ of particulate matter was $\pm$ 0.2‰. For $\delta^{15}N$, the precision was $\pm$ 0.3‰ for particulate samples, and $\pm$0.8‰ for NH_4 and NO_3.

Salinity was measured with a refractometer. Filtrate (47-mm GF/F) was reserved for measurements of nutrient, dissolved organic carbon (DOC), and total dissolved nitrogen (TDN) concentration. These water samples were immediately frozen on dry ice and stored at $-20°C$ prior to analysis. Nutrient concentrations were determined with a Technicon AA-II Autoanalyzer following procedures outlined by Biggs et al. (1982). Dissolved organic carbon was measured by high-temperature combustion at 680°C in the presence of Pt catalyst (0.5%) with a Shimadzu TOC-5000 analyzer. To remove dissolved inorganic carbon (DIC), 25 μl of Ultrex HCl was added to the samples followed by purging for 5 min with high-purity N_2 gas. Dissolved organic nitrogen (DON) was measured with a Shimadzu TOC-5000 analyzer, which was modified by coupling to a Sievers Model 279 Nitric Oxide Analyzer (Lopez-Veneroni and Cifuentes 1994). DON was calculated by subtracting total dissolved inorganic nitrogen (TDIN) from the TDN measured by the analyzer.

Samples for elemental carbon and nitrogen analyses were collected by passing approximately 40 ml (depending upon the particle load) of water through a 13-mm GF/F filter. A stainless steel filter holder and a 60-ml glass syringe were used for this purpose. Particulate organic carbon and nitrogen were measured by combustion and thermal conductivity detection in a Carlo-Erba 1400 CNS Analyzer. Finally, pigment samples were collected by filtering 100–150 ml of water with a 60-ml glass syringe through 25-mm GF/F filters (pre-treated by heating at 450°C for 2 h). Chlorophyll-a measurements were made by the standard method of Parsons et al. (1985).

EMAP-E FIELD DATA

Salinity

A large range in salinity was measured, such that the stations sampled in the 16 estuaries ranged from fresh water to coastal waters (Fig. 8-2A). Specifically, salinity varied from 0 in the Choctawhatchee River to 41 in Southern Laguna Madre. Eighteen of the 47 stations sampled for salinity had values below 5 and above 25; thus, a large proportion of stations sampled were estuarine sites of intermediate salinity.

Isotopic and Elemental Analyses

Previous data for Gulf of Mexico estuaries suggest that values span the range from seagrasses (i.e., Laguna Madre: McMillan et al. 1980) to algal (i.e.,

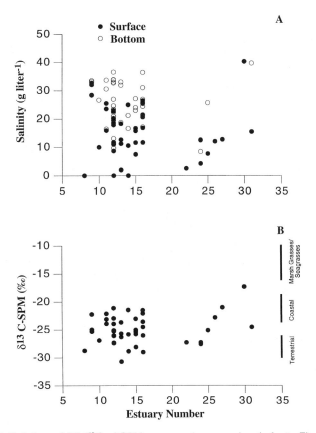

FIG. 8-2. (A) Salinity and (B) $\delta^{13}C$ of SPM versus estuary number (refer to Fig. 8-1). Typical $\delta^{13}C$ ranges for marsh grasses/seagrasses, terrestrial matter, and coastal organic matter are included. (Fry and Sherr 1984; Fogel and Cifuentes 1993; Coffin et al. 1994).

Galveston Bay: Calder and Parker, 1968, Cifuentes unpublished data) and terrestrial C3 and C4 plants (i.e., Mississippi River: Eadie et al. 1978, 1994; Goni et al. 1997) material. We measured carbon isotope ratios for SPM in 16 Gulf of Mexico estuaries ranging from −30.7 to −17.3‰ (Fig. 8-2B), similar to those reported for other extensively studied estuaries (e.g., Delaware estuary: Cifuentes et al. 1988; Fogel et al. 1992). Specifically, the range of values we measured in Texas estuaries was also similar to that reported more recently by Qian et al. (1996). Without further analysis, it would appear that we also sampled waters spanning the range of terrestrial and algal sources of organic matter, consistent with the fact that these samples originated from both fresh and coastal-dominated waters. The most positive value (−17.3‰) was from southern Laguna Madre, where seagrasses, which are relatively enriched in ^{13}C, are known to contribute significant quantities of organic matter (Fry et al. 1987). Pensacola Bay SPM had the most negative $\delta^{13}C$, −30.7, which is

[13]C-depleted compared with values generally reported for terrestrially derived organic matter in estuaries.

When the stations are ordered from most negative to positive $\delta^{13}C$ of SPM, the corresponding and expected change (based on classical terrestrial-marine distinctions) from terrestrial or sewage-derived nitrogen ($\delta^{15}N$ of -2 to $+4$; Fogel and Cifuentes 1993) to coastal nitrogen ($\delta^{15}N$ of $+8$ to $+12$; Fogel and Cifuentes 1993) was not observed (Fig. 8-3). In fact, approximately 10 stations had values that were not consistent with terrestrial, coastal, and/or seagrass organic matter. Lake Calcasieu, Arroyo Colorado, and Galveston Bay all had significantly enriched ^{15}N values. The degree of enrichment was similar to that noted by others (Mariotti et al. 1984; Cifuentes et al. 1988), which was explained by algae assimilating NO_3 or NH_4 enriched in [15]N.

Elemental ratios (C:N) are not as discriminating as stable isotope ratios, but the former can be used to confirm the presence of allochthonous (C:N ratio ~45) and autochthonous (C:N ratio ~6) sources (Wetzel 1983), except in situations where extensive degradation or nitrogen immobilization alters the original values sufficiently to erase these distinctions. Considering the range of $\delta^{13}C$ measured in these estuarine stations, it was surprising to find that the C:N ratios of SPM were generally limited to the range reported for autochthonous sources (Fig. 8-4), with only a few values associated with allochthonous materials. Higher values (>15) have been reported in more

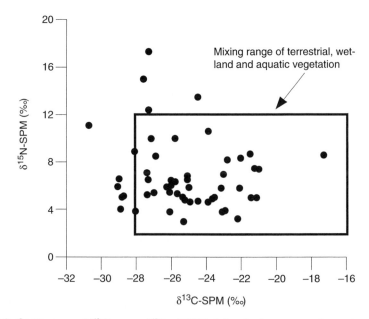

FIG. 8-3. *Scattergram of $\delta^{13}C$ versus $\delta^{15}N$ of SPM. A box is drawn around samples that fall within the mixing range of terrestrial, wetland, and aquatic organic matter (Fry and Sherr 1984; Fogel and Cifuentes 1993; Coffin et al. 1994).*

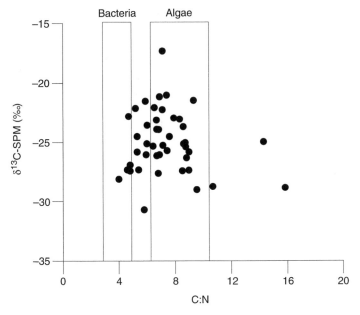

FIG. 8-4. *Scattergram of $\delta^{13}C$ versus $C:N$ (atomic). Boxes are drawn to designate typical $C:N$ ranges for bacteria (Lee and Furhman 1987) and algae (Holligan et al. 1984).*

extensive studies conducted in the Sabine-Neches estuary (Bianchi et al. 1997) and Lake Pontchartrain (Argyrou et al. 1997). However, the EMAP-E study was conducted in summer. Typically, higher $C:N$ ratios were observed during the late fall and winter months in the Sabine-Neches and Lake Pontchartrain studies. At the other extreme, some values were highly enriched in nitrogen ($C:N$ ratio <5). These values could result from algae growing in nitrogen-enriched conditions or from extensive bacterial colonization of particles. Rice (1982) argued, however, that the nitrogen increase observed during degradation was due to humification processes. Therefore, the combination of low $C:N$ ratios and negative $\delta^{13}C$ in SPM is more likely the result of algae growing on isotopically light CO_2 (Fogel et al. 1992).

More than half of the stations sampled had chlorophyll-a concentrations >2.5 μg liter^{-1} (not shown), suggesting that algae contributed to the carbon in the SPM. The isotopic discrimination during CO_2 fixation by algae is usually smaller than that of terrestrial C3 plants owing to CO_2 transport constraints in aqueous media (Fogel and Cifuentes 1993). Fractionation by C3 plants can be as high as 29‰ when the availability of CO_2 is not limited and as low as 4‰ when CO_2 transport into the cells is limiting. If the organic matter sampled at the estuarine sites was influenced to a large degree by algal inputs and the isotopic discrimination did not change significantly from site to site, then there should be a correspondence among the $\delta^{13}C$ of SPM and that of DIC. This behavior was observed in the estuarine stations we sampled (Fig. 8-5A.

Under ideal conditions, the linear regression between $\delta^{13}C$ of SPM and DIC should have a slope of 1.0 and the intercept should be the isotopic discrimination. The slope and intercept for the data shown in Fig. 8-5A are 0.6 and $-22.3‰$ ($r^2 = 0.54$), respectively. The latter value provides a rough estimate for isotopic discrimination between SPM and DIC, which is taken to be representative of the mean algal fractionation and is in accord with published values (e.g., Spiker and Schemel 1979; Fry and Wainright 1991; Fogel et al. 1992; Goericke et al. 1994).

Another explanation for this relationship is simple mixing of terrestrial end members with marine end members. Dividing mean values for terrestrial ($-26‰$) and coastal ($-20‰$) SPM by our measured range of $\delta^{13}C$ of DIC

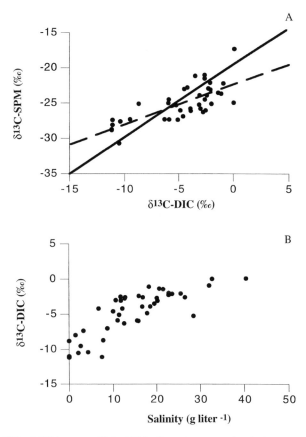

FIG. 8-5. (A) $\delta^{13}C$ of SPM versus $\delta^{13}C$ of DIC. Also shown are the 1:1 correspondence (solid line) and the linear regression (dashed line; $\delta^{13}C\text{-}SPM = 0.6 \times \delta^{13}C\text{-}DIC - 22.3$; $r^2 = 0.54$) for the data (B) $\delta^{13}C$ of DIC versus salinity.

(11.2‰) gives a slope of about 0.5. The measured slope of 0.6 is closer to the slope predicted by end-member mixing (0.5) compared to that required by a simple model for algal fractionation (1.0). Moreover, isotopic discrimination between SPM and DIC was larger than expected for algae at low chlorophyll-a values (Fig. 8-6), suggesting that detrital material (low chlorophyll−a) was not derived primarily from dead algal material, which would have no chlorophyll-a, but would retain the isotopic signature obtained during fixation. Thus, we have identified two pools of organic matter in the SPM, terrestrial material and algae growing on isotopically varying DIC. These results do not contradict those of earlier reports that Gulf estuaries are carrying terrestrial organic matter that is diluted to varying degrees of algal production (e.g., Hedges and Parker 1976; Trefry et al. 1994; Argyrou et al. 1997; Bianchi et al. 1997).

The range of $\delta^{13}C$ values for DIC (-11.1 to 0.1‰) was similar to that found in other estuaries (see Coffin et al. 1994). A conservative mixing line for the $\delta^{13}C$ of the DIC-salinity diagram (e.g., Spiker and Schemel 1979) cannot be described because it would require values for end members in each estuary. Upon visual inspection, however, the isotopic ratio of DIC appears to be enriched in ^{12}C in some estuarine stations (Fig. 8-5B). In order to generate isotopically negative DIC, the respiratory production of CO_2, which is enriched in ^{12}C, must be greater than its incorporation by algae (Coffin et al. 1994). In other words, isotopic values of DIC that are more negative than predicted by conservative mixing imply net heterotrophy.

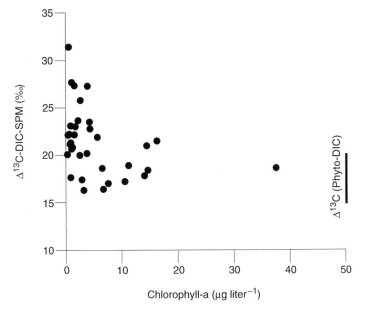

FIG. 8-6. *Isotopic discrimination between SPM and DIC ($\Delta^{13}C$-DIC-SPM) versus chlorophyll a. Typical $\Delta^{13}C$-DIC-SPM ranges for estuaries and coastal waters are included (Spiker and Schemel 1979; Fry and Wainright 1991; Fogel et al. 1992; Goericke et al. 1994).*

Bacterial Substrate Sources

Another piece of information needed to understand the balance of autotrophy and heterotrophy in estuaries is whether bacteria are consuming more allochthonous or autochthonous organic matter. Bacteria generally have a C:N ratio of 3 to 5 (Lee and Fuhrman 1987). When growth occurs on substrates with high C:N ratios, bacteria either supplement their nitrogen requirement with NH_4 or respire much of the assimilated carbon, lowering their growth efficiency. In contrast, very low C:N ratios coupled with high organic matter concentrations indicate a rich medium for bacterial growth. In most aquatic systems, the C:N ratio does not necessarily represent the elemental ratio of bacterial substrate sources because parts of the organic matter pool (e.g., urea, amino acids, nucleic acids) are selectively assimilated and because bacteria also utilize inorganic nitrogen. Extreme values (high or low), however, can be used as a first-order indicator of the lability of the organic matter pool.

As stated earlier, C:N ratios of SPM were typically in the range expected for autochthonous organic matter (Fig. 8-4). During 1991, the C:N ratio of dissolved organic matter was $\geq$ the Redfield ratio at stations sampled from southern Laguna Madre to Galveston Bay, whereas values were nitrogen-depleted to the east (data for dissolved C:N ratios not shown). The absence of a correlation between $\delta^{13}C$ of humic acids and the C:N ratio of dissolved organic matter ($r^2 = 0.18$) reflects the fact that humic acids contribute <20% of the DOC (Fox 1983). Thus, humic acids, as operationally defined, do not appear to be good indicators of terrestrial organic matter in these estuaries. Instead, the $\delta^{13}C$ of humic acids was within ±2‰ of SPM isotopic ratios, with the exception of a few estuarine sites (Fig. 8-7). This result suggests a link between sources of SPM and humic acids among these estuaries. It must be noted, however, that substances such as proteins and mucopolysaccharides, possibly associated with SPM, also precipitate with humic acids at low pH (Thurman 1985).

Bacteria have isotopic ratios similar to those of the assimilated substrates (Coffin et al. 1989, 1994). The greater similarity among the isotopic ratios of bacterial bioassays and humic acids compared with those of bacterial bioassays and SPM (Fig. 8-8A) could imply uptake of acid-precipitable organics, but not necessarily humic acids, which are commonly believed to be refractory. With the exception of the Brazos River, two stations in Galveston Bay and Choctawhatchee River, the $\delta^{15}N$ of bacterial bioassays was within the range of SPM and humic acids (Fig. 8-8B). Although the $\delta^{13}C$ of bacterial bioassays was similar to that of humic acids and SPM in the Brazos River (Fig. 8-8A), the $\delta^{15}N$ of humic acids or SPM was different from that of bacterial bioassays (Fig. 8-8B). This discrepancy could be explained by bacterial assimilation of another organic source, which has an $\delta^{13}C$ similar to that of humic acids and SPM but does not precipitate at low pH.

High C:N ratios of dissolved organic matter were measured at the Chochtawatchee River, and the DOC concentration was in excess of 1000 μM. Here,

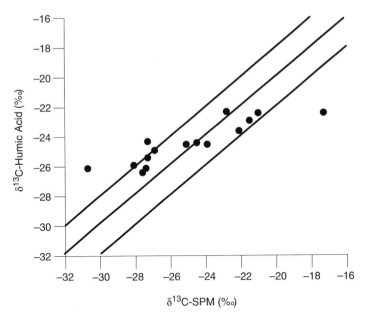

FIG. 8-7. *Scattergram of δ¹³C of humic acid versus δ¹³C of SPM. Lines are drawn to designate the 1 : 1 correspondence and ±2‰.*

both $\delta^{13}C$ and $\delta^{15}N$ of bacterial bioassays were enriched in the heavier isotope compared with humic acids and SPM (Fig. 8-8A,B). At this site, the major nitrogenous nutrient was NO_3, and its isotopic ratio was even more depleted in ^{15}N (3.3‰). Nitrate, however, is not known to be a nitrogen source for bacteria in aquatic systems (D. R. Kirchman personal communication). Some other nitrogen sources, such as free amino acids and nucleic acids that do not precipitate at low pH, probably supported bacterial production at this site.

In the organic-rich Houston Ship Channel portion of Galveston Bay, the $\delta^{15}N$ of bacterial bioassays was 28.5‰; here, both organic carbon and nitrogen sources for bacterial growth were different from that of bacterial bioassays (Figs. 8-8A,B). The low C:N ratio of dissolved organic matter would imply that there was sufficient organic nitrogen to balance bacterial growth; however, the similarity between $\delta^{15}N$ of NH_4 (32.6‰) and bacterial bioassays strongly indicates bacterial assimilation of NH_4 in the Houston Ship Channel. Bacterial uptake of NH_4 has been observed in coastal waters (Wheeler and Kirchman 1986) and estuaries (Keil and Kirchman 1991) and would be expected in environments where available substrates had low nitrogen content. The study of Kirchman et al. (1990) was conducted in Delaware Estuary, which, like the Houston Ship Channel, is heavily influenced in the upper reaches by petroleum transport and refining activity.

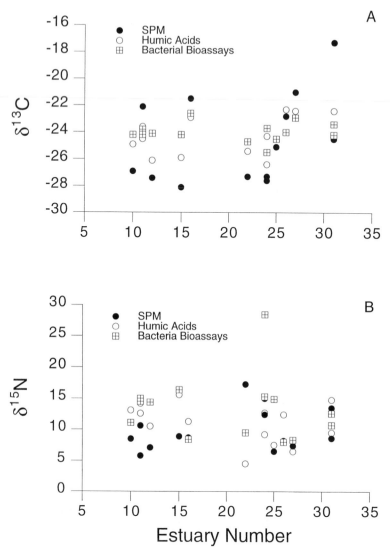

FIG. 8-8. (A) $\delta^{13}C$ and (B) $\delta^{15}N$ of SPM, humic acid, and bacterial bioassays versus estuary number (refer to Fig. 8-1).

IMPLICATIONS FOR NET HETEROTROPHY

Anthropogenic activity often leads to increased nutrient, organic, and sediment loading and to decreased fresh-water inflow to estuaries (i.e., longer residence time of nutrients). In addition, sediment loading, which is influenced by inflow, is coupled with the amount of carbon carried by particles (Fig. 8-9). Thus, changes in inflow will alter both the absolute and relative concentrations of dissolved and particulate organic matter. This will impact the reactivity of allochthonous inputs. In turn, variation in the absolute and relative magnitude of these inputs, coupled with interactions among physical, chemical, and biological processes, influence the balance between production (P) and respiration (R) in estuarine systems.

The suggested outcome of anthropogenic activity is a shift toward net heterotrophy, that is, the ratio of respiration to production (R/P) > 1 (Smith and Mackenzie 1987; Smith and Hollibaugh 1993). For example, net heterotrophy has been reported in the Georgia Bight (Hopkinson 1985), Tomales Bay, California (Smith et al. 1987, 1991), the Newport River Estuary, North Carolina (Kenney et al. 1988), and the upper Delaware Estuary (Kirchman and Hoch 1989). High R/P values have also been reported for Gulf of Mexico estuaries (Table 8-1). However, recent work with radionuclides (Baskaran and Santschi 1993; Baskaran et al. 1997) suggests that particle residence times are low in many of the Gulf of Mexico estuaries, as are water residence times (Chapter 2 this volume). Using a simple model described below, we ask: does the organic matter remain in these estuaries long enough for net heterotrophy to occur?

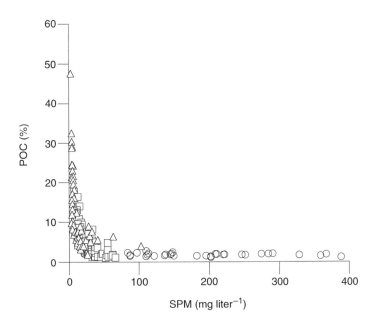

FIG. 8-9. *Percent particulate organic carbon (% POC) versus SPM (mg liter⁻¹) in the Sabine-Neches estuary (Bianchi et al. 1997; squares), Lake Ponchartrain (Argyrou et al. 1997; triangles) and Mississippi River (Trefry et al. 1994; circles).*

TABLE 8-1. Primary Production (P), Respiration (R) and R/P for Gulf of Mexico Estuaries

System	P	R	R/P	Reference
Texas Bays	872	1122	1.29	Odum and Wilson (1961)
El Verde Lagoon, Mexico	521	599	1.15	Flores-Verdugo et al. (1988)
Fourleague Bay,	517	543	1.05	Randall and Day (1987);
Louisiana				Teague et al. (1988)
Estero Pargo, Mexico	345	405	1.17	Day et al. (1988)
Terminos Lagoon, Mexico	219	219	1.00	Day et al. (1988)

Note: All rates are in g C m^{-2} yr^{-1}.

Source: Modified from Smith and Hollibaugh (1993).

Model Description

Net heterotrophy occurs when the ratio of respiration of allochthonous (R_{al}) and autochthonous (R_{au}) organic carbon to autochthonous primary production (P) (Eq. 8-2)

$$\frac{R}{P} = \frac{R_{al} + R_{au}}{P} \tag{8-2}$$

is greater than 1, that is, $R/P > 1$. This cannot take place unless there is an allochthonous source of carbon (Eq. 8-3):

$$\frac{I_{au} + I_{al}}{I_{au}} \geq 1 \tag{8-3}$$

where I_{al} is the input of allochthonous carbon to the estuary and I_{au} is the in situ production in the estuary (same as P) (both in mmol C d^{-1}). For simplicity, we assume that allochthonous inputs are carried by rivers and streams such that (Eq. 8-4)

$$I_{au} = QC \tag{8-4}$$

where Q (m^3 d^{-1}) is flow and C (mmol C m^{-3}) is the total (DOC and POC) carbon concentration. Using the simple first-order decay equation (Eq. 8-5)

$$C = C_0 e^{-kt} \tag{8-5}$$

where C_0 is the concentration entering the estuary, we can estimate the rate at which this input dissipates. However, for our purposes, it is more appropriate to consider the amount of allochthonous carbon degraded over its estuarine residence time (t_r) $- C_0(1 - e^{-kt}R)$. If we consider that (1) microbial organisms are primarily responsible for degradation of carbon in the estuary and

(2) generation times for the microbial populations are much shorter than estuarine residence time, then we can assume that all carbon lost is eventually respired. Combining Eqs. 8-2 and 8-4 with the modified version of Eq. 8-5 and normalizing to autochthonous primary production, we arrive at the following expression (Eq. 8-6):

$$\frac{R_{al}}{P} = \frac{I_{al}(1 - e^{-kt})}{I_{au}} \tag{8-6}$$

It is convenient to simplify this expression by defining $\alpha = I_{al}/I_{au}$ as the ratio between allochthonous and autochthonous estuarine inputs (Eq. 8-7):

$$\frac{R_{al}}{P} = \alpha(1 - e^{-k_r}) \tag{8-7}$$

However, it is more realistic to consider that allochthonous carbon contains pools of differing reactivity. Taking a simplistic approach and assuming that C_i can be divided into a labile (C_{rl}) and a refractory pool (C_{rr}), Eq. 8-7 can be expanded to Eq. 8-8

$$\frac{R_{al}}{P} = \alpha(\psi(1 - e^{-k_{all}\tau}) + (1 - \psi)(1 - e^{-k_{alr}\tau})) \tag{8-8}$$

where k_{lr} and k_{rr} are the first-order coefficients for labile and refractory carbon and ψ is the fraction of labile to total riverine carbon (Eq. 8-9):

$$\psi = \frac{[C_{rl}]}{[C_{rr}] + [C_{rl}]} \tag{8-9}$$

An obvious but often ignored question is whether autochthonous organic matter is respired completely before exiting the estuary. If $R_{au}/P < 1$, then more respiration of allochthonous organic matter is required to reach net heterotrophy. An analogous approach is used to formulate the autochthonous component of respiration. We assume that most of the TOC produced autochthonously is partitioned between either phytoplankton in the form of dissolved material released (ϕ) from phytoplankton (exudate) or structural material $(1 - \phi)$, each with characteristic first-order decay coefficients—[k_{er} and k_{el}]. The expression is then normalized by the autochthonous input (Eq. 8-10)

$$\frac{R_{au}}{P} = \frac{I_{au}(\phi(1 - e^{-k_{el}\tau}) + (1 - \phi)(1 - e^{-k_{er}\tau}))}{I_{au}} \tag{8-10}$$

and simplified to (Eq. 8-11)

$$\frac{R_{au}}{P} = \phi(1 - e^{-k_{el}\tau}) + (1 - \phi)(1 - e^{-k_{er}\tau}) \qquad (8\text{-}11)$$

Equations 8-8 and 8-11 are then combined to give the expanded form of Eq. 8-2. Model equations were formulated with MATLAB software (The Math Works, Inc.) and runs were performed on a Macintosh Power PC.

R/P, Organic Matter Reactivity, and Residence Times

The fractional lability (ψ) of allochthonous organic matter is generally thought to be small (refer to the discussion in Smith and Hollibaugh 1993), with a range from about 0.1 to 0.4. Estimates of algal DOC release as a fraction (ϕ) of primary production in estuaries range from 0.1 to 0.4 (Lignell 1990). Residence times of water (τ) in Gulf estuaries range from days to about a year (Table 8-2; Chapter 2 this volume). However, particle residence times may be significantly lower (Baskaran and Santschi 1993; Baskaran et al. (1997) due to degradation, predation and settling. In turn, the range in primary productivity (P) in these estuaries varies from 160 to 500 g C m^{-2} yr^{-1} (Smith and Hollibaugh 1993). Finally, the reactivities of various types of organic matter, expressed as first-order decay coefficients, are found in Table 8-3.

Given the above ranges of data, our model predicts a rather interesting relationship between residence time, fractional lability of allochthonous material, and net heterotrophy ($\alpha = 1.5$; Fig. 8-10). Estimates of R/P by Smith and Hollibaugh (1993) and others suggest that estuaries are generally net heterotrophic ($R/P = 1.2$). In contrast, our model predicts that this degree of net heterotrophy does not occur in systems with residence times of less than a year. If we assume higher fractional labilities, net heterotrophy could occur under residence times calculated for Gulf of Mexico estuaries (Table 8-2). However, compositional determinations, including lignin oxidation product and pigment analyses (e.g., Hedges and Parker 1976; Trefry et al. 1994; Argyrou et al. 1997; Bianchi et al. 1997; this Chapter), suggest that a significant fraction of the allochthonous carbon is of vascular as opposed to algal origin and therefore would be expected to have low reactivity (low k; see Henrichs 1993). Unless the TOC entering most of the Gulf of Mexico estuaries has a much greater residence time than that of the water, which is in contrast to estimates derived with radionuclides (Baskaran and Santschi 1993; Baskaran et al. (1997), the results of our simple model contradict those found in published reports (summarized in Smith and Hollibaugh 1993).

In the case discussed above, α was set at 1.5, which is an average value reported for estuarine systems by Smith and Hollibaugh (1993). Many of the Gulf estuaries, particularly those in the south Texas coast, have little river input and, consequently, have much lower estimated α (see estimates in Table 8-2). The relationship between residence time, α, and net heterotrophy (fractional lability of 0.15; Fig. 8-11) suggests that net heterotrophy can occur in

TABLE 8-2. Allochthonous and Autochthonous Fluxes and α in Various Gulf of Mexico Estuaries

River/Estuary	Estuary Number	Residence Time[a] (d)	Allochthonous Flux (10^{10} g C yr^{-1})	Autochthonous Flux (10^{10} g C yr^{-1})	α	Reference
Corpus Christi Bay	29	356	0.7	43.6	.02	D. Brock (unpublished data)
Aransas Bay	28	360	1.1	30.8	.04	D. Brock (unpublished data)
San Antonio Bay	27	39	2.8	29.7	.10	D. Brock (unpublished data)
Matagorda Bay	26	81	4.2	200	.02	D. Brock (unpublished data)
Galveston Bay	24	41	17.3	58.1	.30	D. Brock (unpublished data)
Sabine Lake	23	9	9.4–10.8	N.A.	—	Benoit et al. (1994); Bianchi et al. (1997)
Lake Ponchartrain	—	136	3.3–3.5	N.A.	—	Argyrou et al. (1997)
Mobile Bay[b]	15	10	23–38	N.A.	—	Pennock et al. (submitted)

[a] From Chapter 2, this volume.
[b] DOC only.

TABLE 8-3. Estimates of First-Order Reaction Rate Coefficients

Component	Rate Constant (yr⁻¹)
Refractory riverine TOC	.001
Labile riverine TOC	0.1
Structural algal material	>1
Soluble algal material	>50

Source: Henricks (1993).

Gulf estuaries only if $\alpha > 1.5$. At least for most Texas estuaries, estimated α values are <1. Therefore, in contrast to the observations of Odum and Wilson (1961), the model predicts net autotrophic and not net heterotrophic systems in these waters. The model formulation, however, accounts only for primary production and bacterial degradation such as might occur in estuarine food-webs that are dominated by microbial processes. The role of grazer organisms is not explored here but could account for the differences in model results

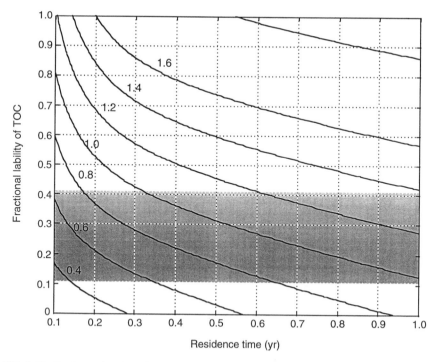

FIG. 8-10. *Contours showing R/P as a function of estuarine residence time and fractional lability of allochthonous carbon ($\alpha = 1.5$, an average value—Smith and Hollibaugh 1993). Shaded box indicates accepted ranges of TOC lability (Smith and Hollibaugh 1993) and residence times for the estuaries in Table 8-2.*

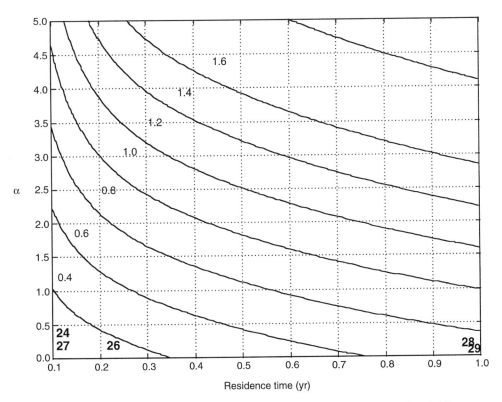

FIG. 8-11. *Contours showing* R/P *as a function of estuarine residence time and* α. *Boldface values refer to estuary number as used in Fig. 8-1 and Table 8-2.*

and measured *R/P* ratios. Thus, much more detail on the composition of organic matter, both allochthonous and autochthonous, in Gulf estuaries is needed before we can better understand how biogeochemical cycling at the fresh-water–sea-water interface depends on the residence time and reactivity of organic matter.

EXAMPLES: LAKE PONTCHARTRAIN AND SABINE-NECHES ESTUARIES

In shallow, turbid estuaries, which tend to be well mixed and usually un-stratified, gross and net primary production are primarily controlled by light availability to phytoplankton (Pennock 1985; Cole et al. 1992). Shallow, turbid estuaries are usually more productive since phytoplankton POC is governed by light availability, along with high nutrients from riverine and terrestrial inputs and from remineralization processes in sediments (Kremer and Nixon 1978; Nixon 1981; Kemp and Boynton 1984). The Lake Pontchartrain and Sabine-Neches Estuaries, located in the northern Gulf of Mexico (Fig. 8-1), are shallow systems with very different hydraulic residence times, 125 days and 10 days, respectively (Chapter 2 this volume). The similarities of POM inputs, water column depth, and geographic region, along with their distinct differences in residence times, provide a unique opportunity to compare the effects of residence time on the cycling of POM and net heterotrophy of these two estuaries.

Seasonal variations in fresh-water discharge to the Lake Pontchartrain and the Sabine-Neches estuaries have been shown to contribute significantly to the spatial and temporal distribution of POM (Argyrou et al. 1997; Bianchi and Argyrou 1997; Bianchi et al. 1997). The highest observed POC and SPM concentrations in both systems generally occurred during winter and early summer, when fresh-water discharges were at their maximum. POC concentrations in both systems, which ranged from 0.2 to 4.2 mg C liter^{-1}, were considerably lower than those of other shallow river-dominated estuaries (Boynton et al. 1982). During periods of low fresh-water input, SPM concentrations decreased below 10–20 mg liter^{-1} in both estuaries, while the POC and PON content of the particles increased (Bianchi et al. 1997). In general, this trend is consistent with patterns observed in large rivers (Meybeck 1982). In these highly turbid systems, the increase in % POC and % PON typically occurs at SPM concentrations of 50 mg liter^{-1} (Meybeck 1982; Milliman et al. 1984; Trefry et al. 1994). This relationship can be attributed to a dilution effect due to an increase in sediment load at high fresh-water discharges. Another important mechanism controlling SPM concentrations in this shallow system is resuspension of bottom sediments. For example, it has been shown that sediments in the Lake Pontchartrain estuary are capable of being resuspended by fairly low wind speeds (~10 mph) (Swenson 1980).

The phytoplankton communities in the Lake Pontchartrain and Sabine-Neches estuaries frequently experience light limitation due to high SPM concentrations (Argyrou et al. 1997; Bianchi and Argyrou 1997; Bianchi et al. 1997). Based on concentrations of the carotenoids fucoxanthin and zeaxanthin, it was determined that diatoms and cyanobacteria were the dominant classes of phytoplankton throughout most of the year in both estuaries. The range of light extinction coefficients found in both estuaries (ca. 0.6 to 3.7 m^{-1}) was similar to that found in other shallow, turbid systems along the east coast (e.g., Cole et al. 1992; Mallin and Paerl 1992). Light extinction coefficients in both estuaries increased to a maximum during winter, when high river inflow and SPM concentrations occurred. It was during the winter months that the lowest average chlorophyll −a concentrations in the Lake Pontchartrain (1.3 ± 0.6 μg liter^{-1}) and Sabine-Neches (1.0 ± 0.5 μg liter^{-1}) estuaries were observed. Conversely, the highest average chlorophyll-a concentrations were observed in summer months for the Lake Pontchartrain (3.3 ± 1.7 μg liter^{-1})

and Sabine-Neches (8.2 ± 5.6 μg liter^{-1}) estuaries, when light extinction coefficients were generally <1 m^{-1}.

High C:N and carbon to chlorophyll−a ratios (C:Chla) in the upper and mid-regions of the Lake Pontchartrain and Sabine-Neches estuaries suggested that terrigenous inputs of organic carbon from rivers and surrounding wetlands are likely to be important components of the carbon cycle. The regional range of C:N ratios for both systems was between 7.4 and 30.6. These ratios were similar to those often found in fresh-water systems that typically result from a mixing of organic matter derived from allochthonous (C:N ratio = 45.0) and autochthonous (C:N = 6.0) sources (Wetzel 1983). While it is well established that vascular plant inputs are characterized by high C:N ratios, preferential utilization of nitrogen by bacteria may have also contributed to these high ratios. High C:Chla ratios further suggest that allochthonous inputs of organic carbon (i.e., riverine input, wetlands exchange) are likely to have been substantial since the C:Chla ratio of estuarine phytoplankton is typically 60.0 (Tantichodok 1989). Concentrations of ligninphenols in a limited number of POC samples collected in winter months from Lake Pontchartrain ($\Lambda = 0.39 \pm 0.07$ and $\Lambda = 0.57 \pm 0.14$) (Argyrou 1996) and Sabine-Neches estuaries ($\Lambda = 0.41 \pm 0.09$ and $\Lambda = 0.87 \pm 0.27$) (Bianchi, unpublished) were similar to those found in sediment trap materials from Dabob Bay, Washington (Hedges et al. 1988). This similarity further supports that terrestrial inputs are important in the Lake Pontchartrain and Sabine-Neches estuaries. Moreover, lignin-phenol concentrations in sediments from the upper regions of both estuaries indicated that the Lake Pontchartrain and Sabine-Neches estuaries had received higher inputs of vascular plant materials than other regional estuaries (Bianchi, unpublished data). These extremely high C:Chla ratios were primarily due to low chlorophyll-a concentrations. The annual average chlorophyll −a concentrations in the Lake Pontchartrain (2.3 ± 0.9 μg liter^{-1}) and Sabine-Neches (3.0 ± 0.7 μg liter^{-1}) estuaries were relatively low, particularly during high fresh-water inflow periods (Argyrou et al. 1997; Bianchi et al. 1997). For example, during a high-flow period with low chlorophyll−a concentrations (in January 1996) in the Lake Pontchartrain estuary, phytoplankton comprised 45% of the total chlorophyll−a, while the other 55% was mostly derived from inputs of degraded vascular plant detritus (Bianchi and Argyrou 1997).

Net heterotrophic characteristics have been described for both Sabine-Neches and Lake Pontchartrain estuaries (Argyrou et al. 1997; Bianchi and Argyrou 1997; Bianchi et al. 1997). The model discussed earlier would argue that, given that both estuaries receive similar allochthonous inputs, *R/P* would not be expected to be similar in both systems owing to large differences in hydraulic residence times. It is likely, though, that net heterotrophy occurs over various time scales, particularly if there is de-coupling of autotrophic and heterotrophic production in time and/or space. In the case of the Sabine-Neches system, particles are moved rapidly through a shallow, turbid estuary with large inputs of recalcitrant terrestrial materials and with low primary production of labile phytoplankton material due to high SPM and light extinction coefficients. This estuary essentially serves as a "digester" for allochthonous materials that are passing through at a rapid rate (ca. 10 d); (Baskaran et al. 1997). At certain times of the year, when "digestion" is high and productivity is low, net heterotrophy would be measured. Conversely, the Lake Pontchartrain system, which is also shallow, with high SPM and light extinction coefficients, and similar POM sources, has a very long hydraulic residence time (ca. 125 d) that allows for more complete breakdown of carbon sources (allochthonous and autochthonous) introduced into the system. In fact, net losses of DOC (primarily derived from terrestrial sources) in the Lake Pontchartrain estuary appeared to be primarily controlled by heterotrophic consumption (conversion to CO_2)—which may be amplified by the long residence time (Argyrou et al. 1997). Thus, it appears that in spite of very different hydraulic residence times, organic matter processing in these two estuaries produced net heterotrophy on certain time and spatial scales. However, it may not be correct to use data taken on these scales and to report that these estuaries are net heterotrophic over their entire space and residence times. This last point highlights the importance of conducting studies on more extensive spatial and seasonal scales.

ACKNOWLEDGMENTS

We thank Maria-Lourdes Arazate, Diego Lopez-Veneroni, Jeff Kovacs, and Lynn Roelke for their assistance with analyses. Personnel from EPA's Gulf Breeze Environmental Research Laboratory and Texas A&M's Geochemical Environmental Research Group provided support during sampling. Funding for this research was provided in part by the EMAP-E Program through U.S. EPA Cooperative Agreement Project CR-816736.

REFERENCES

Argyrou, M. E., 1996. Spatial and temporal variability of particulate and dissolved organic carbon in the Lake Pontchartrain estuary: The use of chemical biomarkers. Master's thesis, Tulane University.

————, T. S. Bianchi, and C. D. Lambert. 1997. Transport and fate of dissolved organic matter in the Lake Pontchartrain estuary, Louisiana, U.S.A. Biogeochemistry **38:** 207–226.

Baskaran, M., M. Ravichandran, and T. S. Bianchi. 1997. Cycling of [7]Be and ^{210}Pb in a high DOC, shallow, turbid estuary of Southeast Texas. Estuar. Coastal Shelf Sci. **45:** 165–176.

————, and P. H. Santschi. 1993. The role of particles and colloids in the transport of radionuclides in coastal environments of Texas. Mar. Chem. **43:** 95–114.

Benoit, G., S. Otkay, A. Cantu, M. O. Hood, C. H. Coleman, O. Corapcioglu, and P. H. Santschi. 1994. Partitioning of Cu, PB, Ag, Zn, Fe, Al, Mn between filter-retained particles, colloids and solution in 6 Texas estuaries. Mar. Chem. **45:** 307–336.

Bianchi, T. S., and M. E. Argyrou. 1997. Temporal and spatial dynamics of particulate organic carbon in the Lake Pontchartrain Estuary, southeast Louisiana, U.S.A. Estuar. Coast. Shelf Sci. **45:** 165–176.

———, M. Baskaran, J. DeLord, and M. Ravichandran. 1997. Carbon cycling in a shallow turbid estuary of southeast Texas I. The use of plant pigment biomarkers and water quality parameters. Estuaries **20:** 404–415.

———, C. Lambert, P. H. Santschi, M. Baskaran, and L. Guo. 1995. Plant pigments as biomarkers of high-molecular-weight dissolved organic carbon. Limnol. Oceanogr. **40:** 422–428.

Biggs, D. C., M. A. Johnson, R. B. Bidigare, J. D. Guffy, and O. Holm-Hansen. 1982. Shipboard autoanalyzer studies of nutrient chemistry. *In:* Technical Report 82-11-T. Department of Oceanography, Texas A&M University.

Boynton, W. R., W. M. Kemp, and C. W. Keefe. 1982. A comparative analysis of nutrients and other factors influencing estuarine phytoplankton production, p. 69–90. *In:* V. S. Kennedy [ed.], Estuarine comparisons, Academic Press.

Burton, J. D., and P. S. Liss. 1976. Estuarine chemistry, Academic Press.

Calder, J. A., and, P. L. Parker. 1968. Stable carbon isotope ratios as indices of petrochemical pollution of aquatic systems. Environ. Sci. Tech. **2:** 536–539.

Cifuentes, L. A., J. H. Sharp, and M. L. Fogel. 1988. Stable carbon and nitrogen isotope biogeochemistry in the Delaware estuary. Limnol. Oceanogr. **33:** 1102–1115.

Coffin, R. B., and L. A. Cifuentes. 1993. Approaches for measuring stable carbon and nitrogen isotopes in bacteria, p. 663–676. *In:* P. F. Kemp, B. F. Sherr, E. B. Sherr, and J. J. Cole [eds.], Current methods in aquatic microbial ecology. Lewis.

———, L. A. Cifuentes, and P. Eldridge. 1994. The use of stable carbon isotopes to study microbial processes in estuaries, p. 222–240. *In:* K. Lajtha and R. H. Michener [eds.], Stable isotopes in ecology, Blackwell Scientific.

———, B. Fry, B. J. Peterson, and R. T. Wright. 1989. Identification of bacterial carbon sources with stable isotope analysis. Limnol. Oceanogr. **34:** 1305–1310.

———, D. Velinsky, R. Devereux, W. A. Price, and L. A. Cifuentes. 1990. Analyses of stable carbon isotopes of nucleic acids to trace sources of dissolved substrates used by estuarine bacteria. Appl. Environ. Microbiol. **56:** 2012–2020.

Cole, J. J., N. F. Caraco, and B. L. Peirels. 1992. Can phytoplankton maintain a positive carbon balance in a turbid, freshwater, tidal estuary? Limnol. Oceanogr. **37:** 1608–1617.

Day, J. W., Jr., C. J. Madden, F. Ley-Lou, R. L. Wetzel, and A. M. Navarro. 1988. Aquatic primary productivity in Terminos Lagoon, p. 221–236. *In:* A. Yanez-Arancibia and J. W. Day Jr. [eds.], Ecology of coastal ecosystems in the southern Gulf of Mexico: The Terminos Lagoon region. Universidad Nacional Autonoma de Mexico.

Deegan, L. A., J. W. Day, Jr., J. G. Gosselink, A. Yanez-Aranciba, G. Soberon Chavez, and P. Sanchez-Gil. 1986. Relationships among physical characteristics, vegetation distribution and fisheries yield in Gulf of Mexico Estuaries, p. 83–100. *In:* D. A. Wolfe [ed.], Estuarine variability. Academic Press.

Eadie, B. J., L. M. Jeffrey, and W. M. Sackett. 1978. Some observations on the stable carbon isotope composition of dissolved and particulate organic carbon in the marine environment. Geochim. Cosmochim. Acta **42:** 1265–1269.

———, B. A. McKee, M. B. Lansing, J. A. Robbins, S. Metz, and J. H. Trefrey. 1994. Records of nutrient-enhanced coastal ocean productivity in sediments from the Louisiana continental shelf. Estuaries **17:** 754–765.

Flores-Verduge, F. J., J. W. Day, Jr., L. Mee, and R. Briseno-Duenas. 1988. Phytoplankton production and seasonal biomass variation of seagrass, *Ruppa maritima L.,* in a tropical Mexican lagoon with an ephemeral inlet. Estuaries **11:** 51–56.

Fogel, M. L., and, L. A. Cifuentes. 1993. Isotopic fractionation during primary production, p. 73–98. *In:* M. Engel and S. A. Macko [eds.], Organic geochemistry. Plenum Press.

———, L. A. Cifuentes, D. J. Velinski, and J. H. Sharp. 1992. The relationship of carbon availability in estuarine phytoplankton to isotopic composition. Mar. Ecol. Prog. Series **82:** 291–300.

Fox, L. E. 1983. The removal of dissolved humic substances during estuarine mixing. Estuar. Coast. Mar. Sci. **16:** 431–440.

Fry, B., S. A. Macko, and J. C. Zieman. 1987. In review of stable isotopic investigations of food webs in seagrass meadows. Florida Marine Research Publications.

———, and E. B. Sherr. 1984. ^{13}C measurements as indicators of carbon flow in marine and freshwater ecosystems. Cont. Mar. Sci. **27:** 13–47.

———, and S. C. Wainright. 1991. Diatom sources of ^{13}C-rich carbon in marine food-webs. Mar. Ecol. Prog. Ser. **76:** 149–157.

Goericke, R., J. P. Montoya, and B. Fry. 1994. Physiology of isotopic fractionation in algae and cyanobacteria, p. 222–240. *In:* K. Lajtha and R. H. Michener [eds.], Stable isotopes in ecology. Blackwell Scientific.

Goni, M. A., K. C. Ruttenberg, and T. I. Eglinton. 1997. Source and contribution of terrigenous organic carbon to surface sediments in the Gulf of Mexico. Nature **389:** 275–278.

Grossman, E. O. 1984. Carbon isotopic fractionation in live benthic foraminifera—comparison with inorganic precipitate studies. Geochim. Cosmochim. Acta **48:** 1505–1512.

Happ, G., J. G. Gosselink, and J. W. Day, Jr. 1977. The seasonal distribution of organic carbon in a Louisiana estuary. Estuar. Coast. Mar. Sci. **5:** 695–705.

Hedges, J. I., W. A. Clark, and G. L. Cowie. 1988. Organic matter sources to the water column and surficial sediments of a marine bay. Limnol. Oceanogr. **33:** 1116–1136.

———, and D. C. Mann. 1979. The characterization of plant tissues by their lignin oxidation products. Geochim. Cosmochim. Acta **43:** 1803–1807.

———, and P. L. Parker. 1976. Land-derived organic matter in surface sediments of the Gulf of Mexico. Geochim. Cosmochim. Acta **40:** 1019–1029.

Henrichs, S. M. 1992. Early diagenesis of organic matter in sediments: Progress and perplexity. Mar. Chem. **39:** 1119–1149.

———. 1993. Early diagnosis of organic matter: The dynamics (rates) of cycling of organic compounds, p. 101–114. *In:* M. Engel and S. A. Macko [eds.], Organic geochemistry. Plenum Press.

Hobbie, J. E., R. J. Daly, and S. Jaspers. 1977. Use of Nucleopore filters for counting bacteria by fluorescence microscopy. Appl. Environ. Microbiol. **37:** 805–812.

Holligan, P. M., R. P. Harris, R. C. Newell, D. S. Harbour, R. N. Head, E. A. S. Linley, M. I. Lucas, P. R. G. Tranter, and C. M. Weekley. 1984. Vertical distribution and partitioning of organic carbon in mixed, frontal and stratified waters of the English Channel. Mar. Ecol. Prog. Ser. **14:** 111–127.

Hopkinson, C. S., Jr. 1985. Shallow-water benthic and pelagic metabolism: Evidence for heterotrophy in the nearshore Georgia Bight. Mar. Biol. **87:** 19–32.

Imberger, J., T. Berman, R. R. Christian, E. B. Sherr, D. E. Whitney, L. R. Pomeroy, R. G. Wiegert, and W. J. Wiebe. 1983. The influence of water motion on the distribution and transport of materials in a salt marsh estuary. Limnol. and Oceanogr. **28:** 201–214.

Ittekkot, V. 1989. Global trends in the nature of organic matter in river suspension. Nature **332:** 436–438.

Kaul, L. W., and P. N. Froelich. 1984. Modeling estuarine nutrient geochemistry in a simple system. Geochim. Cosmoschim. Acta **48:** 1417–1433.

Keil, R. G., and D. L. Kirchman. 1991. Contribution of dissolved free amino acids and ammonium to the nitrogen requirements of heterotrophic bacterioplankton. Mar. Eco. Prog. Ser. **73:** 1–10.

Kemp, W. M., and W. R. Boynton. 1984. Spatial and temporal coupling of nutrient inputs to estuarine primary production: The role of particulate transport and decomposition. Bull. Mar. Sci. **35:** 522–535.

Kenney, B. E., W. Litaker, C. S. Duke, and J. Ramus. 1988. Community oxygen metabolism in a shallow tidal estuary. Estuar. Coast. Shelf Sci. **27:** 33–43.

Kirchman, D. L., and M. P. Hoch. 1989. Bacterial production in the Delaware Bay estuary estimated from thymidine and leucine incorporation rates. Mar. Ecol. Prog. Ser. **45:** 169–178.

Kremer, J., and S. Nixon. 1978. A coastal marine ecosystem, simulation and analysis. Springer-Verlag.

Lee, S. H., and J. A. Fuhrman. 1987. Relationships between biovolume and biomass of naturally derived marine bacterioplankton. Appl. Environ. Microbiol. **53:** 1298–1303.

Lignell, R. 1990. Excretion of organic carbon by phytoplankton: Its relationship to algal biomass, primary productivity and bacterial secondary productivity in the Baltic Sea. Mar. Ecol. Prog. Ser. **68:** 85–99.

Lopez-Veneroni, D., and L. A. Cifuentes. 1994. Transport of dissolved organic nitrogen in Mississippi river plume and Texas-Louisiana continental shelf near-surface waters. Estuaries **17:** 796–808.

Macko, S. A. 1981. Stable nitrogen isotope ratios as tracers of organic geochemical processes. Ph.D. thesis, University of Texas at Austin.

Mallin, M. A., and H. W. Paerl. 1992. Effects of variable irradiance on phytoplankton productivity in shallow estuaries. Limnol. Oceanogr. **37:** 54–62.

Mariotti, A., C. Lancelot, and G. Billen. 1984. Natural isotopic composition of nitrogen as a tracer of origin for suspended organic matter in the Scheldt estuary. Geochim. Cosmochim. Acta **48:** 549–555.

McMillan, C., P. L. Parker, and B. Fry. 1980. $^{13}C/^{12}C$ ratios in seagrasses. Aquat. Bot. **9:** 237–249.

Meybeck, M. 1982. Carbon, nitrogen, and phosphorus transport by the world rivers. Am. J. Sci. **282:** 401–450.

Milliman, J. D., Q. Xie, and Z. Yang. 1984. Transfer of particulate organic carbon and nitrogen from the Yangtze River to the ocean. Am. J. Sci. **284:** 824–834.

Nixon, S. W. 1981. Freshwater inputs and estuarine productivity, p. 31–57. *In:* R. D. Cross and D. L. Williams [eds.], Proceedings of the National Symposium on Freshwater Inflow to Estuaries, U.S. Fish and Wildlife Service. Office of Biological Services, FWS/OBS-81/04.

Odum, H. T., and R. F. Wilson. 1961. Further studies on reaeration and metabolism of Texas bays, 1958–1960. U. Tex. Inst. Mar. Sci. Pub. **8:** 23–55.

Parsons, T. R., Y. Maita, and C. M. Lalli. 1985. A manual of chemical and biological methods for seawater analysis. Pergamon Press.

Pennock, J. R. 1985. Chlorophyll distributions in the Delaware estuary: Regulation by light-limitations. Estuar. Coast. Shelf Sci. **21:** 711–725.

———, W. W. Schroeder, J. L. W. Cowan, and R. P. Stumpf. Submitted. The loading, reactivity and distribution of carbon, nitrogen and phosphorus in a river-dominated estuary: Mobile Bay, Alabama (USA). Estuaries.

Qian, Y., M. C. Kennicutt II, J. Salvberg, S. A. Macko, R. R. Bidigare, and J. Walker. 1996. Suspended particulate organic matter (SPOM) in Gulf of Mexico estuaries: Compound-specific isotope analysis and plant pigment composition. Org. Geochem. **24:** 875–888.

Randall, J. M., and J. W. Day. 1987. Effects of river discharge and vertical circulation on aquatic primary production in a turbid Louisiana (USA) estuary. Neth. J. Sea. Res. **21:** 231–242.

Rice, D. L. 1982. The detritus nitrogen problem: New observations and perspectives from organic geochemistry. Mar. Ecol. Prog. Ser. **9:** 153–162.

Rivera-Monroy, V. H., J. W. Day, R. R. Twilley, F. Vera-Herrera, and C. Coronado-Molina. 1995. Flux of nitrogen and sediment in a fringe mangrove forest in Terminos Lagoon, Mexico. Estuar. Coast. Shelf Sci. **40:** 139–160.

Smith, S. V., and J. T. Hollibaugh. 1993. Coastal metabolism and the oceanic carbon balance. Rev. Geophysics **31:** 75–79.

———, J. T. Hollibaugh, S. J. Dollar, and S. Vink. 1991. Tomales Bay metabolism: C-N-P stoichiometry and ecosystem heterotrophy at the land sea interface. Estuar. Coast. Shelf Sci. **33:** 223–257.

———. 1997. Annual cycle and interannual variability of ecosystem metabolism in a temperate climate embayment. Ecological Monographs **67:** 509–533.

———, and F. T. MacKenzie. 1987. The ocean as a net heterotrophic system: Implications from the carbon biogeochemical cycle. Global Biogeochem. Cycles **3:** 187–198.

———, W. J. Wiebe, J. T. Hollibaugh, S. J. Doller, S. W. Hagar, B. E. Cole, G. W. Tribble, and P. A. Wheeler. 1987. Stoichiometry of C, N, P, and Si in a temperate-climate embayment. J. Mar. Res. **45:** 427–460.

Spiker, E. C., and Schemel, L. E. 1979. Distribution and stable-isotope composition of carbon in San Francisco Bay, p. 195–212. *In:* T. J. Conomos [ed.], San Francisco

Bay: The urbanized estuary. Pacific Division, American Association for the Advancement of Sciences.

Swenson, E. M. 1980. General hydrography of Lake Pontchartrain, Louisiana, volume 1. *In:* J. H. Stone [ed.], Environmental analysis of Lake Pontchartrain, Louisiana, its surrounding wetlands, and selected land uses. CEL, CWR, LSU, BR, LA 70803. Prepared for U.S. Army Engineer District, New Orleans. Contract No. DACW29-77-C-0253.

Tantichodok, P. 1989. Relative importance of phytoplankton and organic detritus as food sources for the suspension-feeding bivalve, *Mytilus edulis L.,* in Long Island Sound. Ph.D. dissertation, State University of New York at Stony Brook.

Teague, K. G., C. J. Madden, and J. W. Day, Jr. 1988. Sediment-water oxygen and nutrient fluxes in a river-dominated estuary. Estuaries **11:** 1–9.

Thurman, E. M. 1985. Organic geochemistry of natural waters. Martinus Nijhoff/Dr. W. Junk.

Trefry, J. H., S. Metz, T. A. Nelsen, R. P. Trocine, and B. J. Eadie. 1994. Transport of particulate organic carbon by the Mississippi River and its fate in the Gulf of Mexico. Estuaries **17:** 839–849.

Twilley, R. R. 1985. The exchange of organic carbon in basin mangrove forests in a southwest Florida estuary. Estuar. Coast. Shelf Sci. **20:** 543–557.

———. 1988. Coupling of mangroves to the productivity of estuarine and coastal waters. Lecture Notes on Coastal and Estuarine Studies **22:** 155–180.

———, A. E. Lugo, and C. Patterson-Zucca. 1986. Production, standing crop, and decomposition of litter in basin mangrove-forests in southwest Florida. Ecology **67:** 670–683.

Valiela, I. 1995. Marine ecological processes (2nd ed.). Springer-Verlag.

Velinsky, D. J., L. A. Cifuentes, J. R. Pennock, J. H. Sharp, and M. L. Fogel. 1989. Determination of the isotopic composition of ammonium-nitrogen from estuarine waters at the natural abundance level. Mar. Chem. **26:** 351–361.

Wetzel, R. G. 1983. Limnology (2nd ed.). Saunders.

———. R. G. 1984. Detrital dissolved and particulate organic carbon functions in aquatic ecosystems. Bull. Mar. Sci. **35:** 503–509.

Wheeler, P. A., and, D. L. Kirchman. 1986. Utilization of inorganic and organic nitrogen by bacteria in marine systems. Limnol. Oceanogr. **31:** 998–1009.

Chapter **9**

Dissolved Organic Matter in Estuaries of the Gulf of Mexico

Laodong Guo, Peter H. Santschi, and Thomas S. Bianchi

INTRODUCTION

Dissolved organic carbon (DOC) is one of the largest organic carbon pools on earth and plays a central role in the biogeochemistry of a variety of elements in estuarine, coastal, and oceanic environments (Farrington 1992). The riverine input of DOC is a significant contribution to the oceanic inventory of organic carbon. In addition, the transfer of carbon from land to sea by river systems is one of the important links in the global carbon cycle (Hedges 1992). A better understanding of the role of estuaries in the oceanic carbon cycle depends on detailed knowledge of the chemistry and fate of organic carbon in estuarine systems that regulate and modify, to some extent, the chemistry and flux of riverine dissolved organic matter (DOM).

Estuarine organic matter consists of inputs from riverine inflow, autochthonous primary production, and allochthonous contributions from adjacent coastal ecosystems (Burton and Liss 1976). Recent studies in estuaries and coastal waters have focused on sources of DOM (e.g., Sigleo 1996; Bianchi et al. 1997a; Guo and Santschi 1997a), and photochemical and bacterial con-

Biogeochemistry of Gulf of Mexico Estuaries, Edited by Thomas S. Bianchi, Jonathan R. Pennock, and Robert R. Twilley.
ISBN 0-471-16174-8   © 1999 John Wiley & Sons, Inc.

sumption of DOM (e.g. Wetzel et al. 1995; Amon and Benner 1996a, 1996b), as well as on the role of natural organic matter in controlling the transport and fate of trace metals and radionuclides (Baskaran and Santschi 1993; Benoit et al. 1994; Powell et al. 1996; Wen et al. in press). A significant fraction of the DOC in estuarine waters is composed of colloidal or macromolecular organic matter (e.g., Whitehouse et al. 1989; Martin et al. 1995; Guo and Santschi 1997a) which plays an important role in the carbon cycle, trace metal scavenging, and biogeochemical processes (Sigleo and Means 1990; Guo and Santschi 1997b; Santschi et al. 1997). However, colloidal organic carbon (COC) remains a relatively poorly quantified component of the organic carbon pool in estuarine environments (Sigleo and Means 1990; Buffle and Leppard 1995). Due to recent improvements of sampling and pre-concentration techniques, the abundance of colloids, DOC molecular weight distribution, and geochemical behavior of colloidal organic matter (COM) in estuaries can be effectively characterized (Benner et al. 1992a; Powell et al. 1996; Guo and Santschi 1997a).

The biogeochemistry of DOM in estuaries of the Gulf of Mexico has only recently received greater attention, with most of the available data, from estuaries along the U.S. coast. Some studies have also examined interactions between organic matter, phytoplankton, and mangrove forests in coastal lagoons and wetlands of the Gulf coast (e.g., Happ et al. 1977; Twilley 1985; Herrera-Silverira and Ramirez 1996). This chapter attempts to synthesize what is known about the cycling of DOM and COM in estuaries of the Gulf of Mexico from available literature. We also make comparisons between estuaries from different regions within the Gulf, as well as with other temperate estuaries. Major sections include (1) distributions and cycling of DOC; (2) characterization of estuarine COM; and (3) sources of DOM derived from characterization of elemental composition, biomarker, and isotopic tracers.

GULF OF MEXICO ESTUARIES

The Gulf of Mexico is a 1.5 million km^2 subtropical, semi-enclosed ocean basin that receives fresh water from almost half of the continental United States. Table 9-1 lists the major rivers and associated estuaries in the Gulf of Mexico region, where DOC measurements were reported. The Gulf of Mexico is divided into four separate regions for comparative purposes. Estuaries in region I are all in Texas and include Corpus Christi Bay, San Antonio Bay, Matagorda Bay, Galveston Bay, and Sabine/Neches estuary. The three largest estuaries of the Gulf are in Region II and include the Mississippi River plume, Lake Pontchartrain, and Mobile Bay, located in Louisiana and Alabama, respectively. Estuaries in Region III include the mixing zone between Florida rivers and the Gulf of Mexico. Region IV contains the estuaries in Mexico.

Gulf of Mexico estuaries have distinctive sub-tropical or tropical features compared with estuaries on the eastern or western coasts of the United States.

TABLE 9-1. Concentration of DOC and Its Mixing Behavior in Estuaries of the Gulf of Mexico

Region	Estuary	[DOC] (μM)[a]	Mixing Behavior of DOC	Reference
I (Texas)	Corpus Christi	560–630	Significant removal during mixing	Benoit et al. (1994)
	San Antonio Bay	330–480	Significant removal during mixing	Benoit et al. (1994)
	Lavaca Bay	NA	Significant removal during mixing	Benoit et al. (1994)
	Galveston Bay	460	Significant removal during mixing	Benoit et al. (1994)
	Sabine-Neches	530	Significant removal during mixing	Benoit et al. (1994)
		550	N.A.	Stordal et al. (1996)
		367–1350	N.A.	Bianchi et al. (1997a)
	Galveston Bay	420–480	Source input in low salinity region in July	Guo and Santschi (1997a)
II (Louisiana and Alabama)	Mississippi River plume	270–330	Conservative mixing during winter but with source input during summer	Benner et al. (1992a)
	Lake Pontchartrain	425–483	Well mixed	Argyrou et al. (1997)
	Mobile Bay	424 ± 105	Conservative	Pennock et al. (submitted)
III (Florida)	Ochlockonee	~1050	Non-conservative mixing with inputs for LMW fraction and removal for HMW fraction	Powell et al. (1996)
	Rookery Bay	580–1250	N.A.	Guentzel et al. (1996)
IV (Mexico)	Laguna Madre	312[b]	N.A.	Twilley (1985)
	Terminos Lagoon	60–330[c]	N.A.	Amon and Benner (1996a)
	Celestum Lagoon	[d]	N.A.	Rivera-Monroy et al. (1995)
				Herrera-Silverira and Ramirez-Ramirez (1996)

[a] Concentration of DOC from river end member or average in estuary.
[b] Data from samples with a salinity of ~22.
[c] Concentration of DOC was converted from that of DON and the Redfield ratio.
[d] Only tannic acid (a measure of natural phenolic material) concentrations were available (ranged from <1 to 18 mg liter^{-1}).

271

For example, the decomposition rates of fresh litter in the Gulf of Mexico coastal areas are among the highest in the continental United States (Fig. 9-1). As a result, most of these sub-tropical estuaries on the U.S. coast are characterized by high concentrations of DOC (e.g., Bianchi et al. 1996, 1997a; Powell et al. 1996). Therefore, the cycling of DOM is of great importance to water chemistry, complexation, and cycling of trace elements in this region.

DISTRIBUTIONS OF DOC

Estuarine Mixing Behavior of DOC

The mixing pattern of DOC along an estuarine salinity gradient can usually be used to examine the conservative or non-conservative behavior of DOC in estuarine regions (Burton and Liss 1976; Mantoura and Woodward 1983). Although concentrations of DOC generally decrease from fresh water to seawater through estuarine and coastal waters, conservative or non-conservative mixing behavior of DOC can depend on the types of rivers, the season, water residence times, nutrient inputs, primary productivity, and other controlling factors such as benthic processes taking place within estuaries (e.g., Mantoura and Woodward 1983; Thurman 1985). Concentrations and the conservative or non-conservative behavior of DOC in Gulf of Mexico estuaries are summarized in Table 9-1.

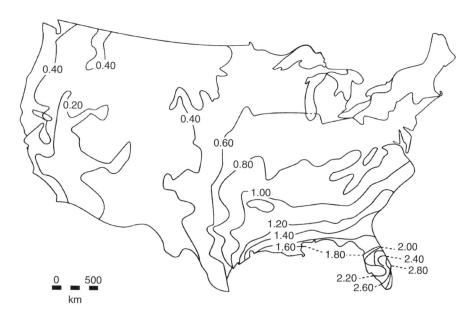

FIG. 9-1. *Fractional loss rate (see isopleth values) of mass from fresh litter during the first year of decay in the United States (from Meentemeyer 1978).*

The type of mixing behavior of DOC in Gulf of Mexico estuaries appears to be variable (Table 9-1). Benoit et al. (1994) investigated the distribution and behavior of DOC and trace metals as a consequence of estuarine mixing in the following Texas estuaries: the Sabine-Neches, Galveston Bay, Lavaca Bay, San Antonio Bay, Colorado River, and Corpus Christi Bay (Fig. 9-2).

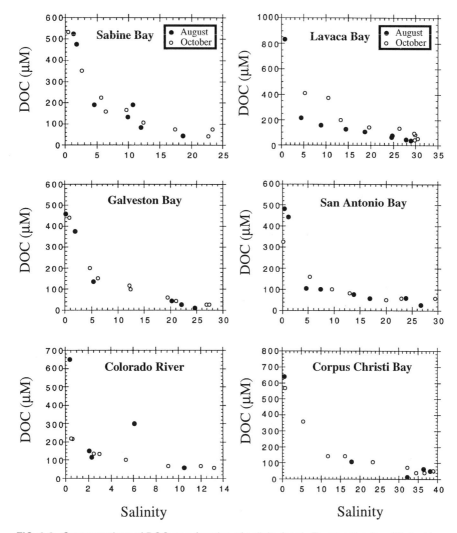

FIG. 9-2. Concentrations of DOC as a function of salinity for six Texas estuaries. (Plotted from data in G. Benoit et al., Partitioning of Cu, Pb, Ag, Zn, Fe, Al and Mn between filter-retained particles, colloids and solution in 6 Texas estuaries, Mar. Chem. **45**: 307–336, 1994 with kind permission from Elsevier Science-NL, Sara Burgerhartstraat 25, 1055 KV Amsterdam, The Netherlands.)

It becomes clear that concentrations of DOC in all six estuaries decreased non-conservatively with increasing salinity. The concave profiles of DOC in all estuaries investigated indicated net removal of DOC during the mixing between fresh water and seawater during that time (Benoit et al. 1994). Early laboratory experiments and field results showed the removal of fractions of DOC in estuaries by coagulation, flocculation, and other processes (e.g., Sholkovitz 1976; Fox 1983). The removal of DOC within six Texas estuaries (Fig. 9-2) seems to support the belief that there is flocculation of DOC with increasing ionic strength. If the removal of DOC in estuaries is primarily through the coagulation or flocculation of humic substances or other high molecular weight fractions (HMW DOC), then the removal of this HMW DOC, in most cases, is too small in magnitude to be observed on a DOC-salinity plot. This situation occurs because HMW DOC (e.g., >10 kDa) represents only a minor fraction (~10%) of the bulk DOC in estuaries (Guo and Santschi 1997a) even though the total COC (>1 kDa) comprises a major portion of the estuarine DOC (see the next sections).

While removal of DOC during estuarine mixing seems evident, an earlier detailed investigation of the distribution of DOC over 2.5 yr, in the Severn Estuary concluded that riverine DOC was conservative and that any biological and physical chemical processes did not affect the flux of DOC through the estuary (Mantoura and Woodward 1983). However, non-conservative DOC behavior with source inputs within estuaries has also been implicated. For example, non-conservative behavior of DOC during estuarine mixing has been observed in Galveston Bay (Fig. 9-3), with a significant input of DOC at the early stages of mixing in Trinity Bay (Guo and Santschi 1997a). The distribution pattern of DOC (Fig. 9-3) was consistent with that of particulate organic carbon (POC) from the same sampling period. This DOC input to Trinity Bay was supported by large benthic fluxes of nutrients and trace metals measured using benthic chambers and calculated from pore water profiles (Warnken 1998). Thus, the non-conservative mixing behavior of DOC in Galveston Bay could be largely attributed to the release of DOC from bottom sediments and/or to the enhanced primary productivity due to benthic nutrient inputs (Warnken 1998; Guo and Santschi 1997a). Interestingly, while a distinct source input of DOC was observed during July 1995 in Galveston Bay (Fig. 9-3), removal of DOC within the same estuary was found during August and October 1989 (Fig. 9-2). The difference in DOC behavior in this estuary could have resulted from differences in sampling seasons and thus hydrographic conditions (July 1995 versus August and October 1989).

Estuaries from Regions II and III in the Gulf of Mexico appear to have both conservative and non-conservative DOC mixing behavior (Fig. 9-3). For example, Benner et al. (1992a) found that the mixing behavior of DOC varied considerably over different seasons in the Mississippi River plume. Conservative mixing behavior of DOC was observed in winter, whereas non-conservative (input) behavior was found during the summer in the Mississippi River plume (Benner et al. 1992a). The substantial inputs of DOC at intermediate

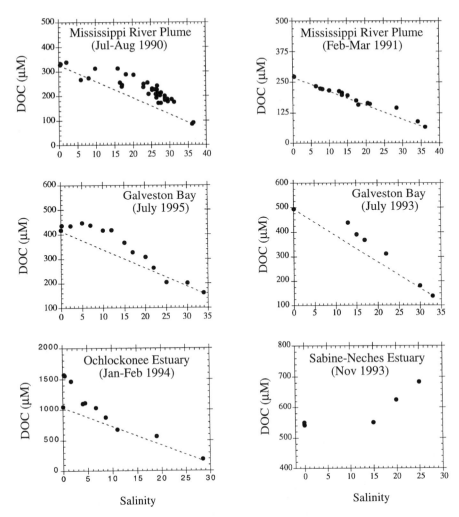

FIG. 9-3. Behavior of DOC during estuarine mixing from selected Gulf of Mexico estuaries (plotted from data in Stordal et al. (1996a) for the Sabine-Neches Estuary and from Powell et al. 1996 for the Ochlockonee estuary; modified from Benner et al. 1992a for the Mississippi River plume and from Guo and Santschi (1997a) for Galveston Bay).

salinities in the summer were attributed to increased primary productivity in that region of the Mississippi River plume (Benner et al. 1992a). This interpretation was further supported by stable nitrogen isotope (δ^{15}N) data in the HMW DOC (or colloidal) fraction (Benner et al. 1992a). In addition, a recent study in DOC distributions through 5 years of sampling at all seasons revealed seemingly conservative behavior of DOC in Mobile Bay, with variable DOC concentrations at the fresh water end member (Pennock et al. submitted). Moreover, non-conservative DOC behavior with large source inputs in low-salinity regions was also reported in the Ochlockonee estuary (Powell et al. 1996). However, considerable removal of HMW DOC was observed at the time of bulk DOC inputs (Powell et al. 1996). The removal of HMW DOC in the Ochlockonee Estuary was most efficient within a salinity of ~10, where bulk DOC concentrations ranged from 1000 to 15,000 μM, with 20–50% of bulk DOC in the HMW (>10 kDa) fraction. Thus, DOC input was largely in the low and medium molecular weight DOC fractions. If DOC input within this estuary was mostly from low molecular weight (LMW) fractions and if the removal of DOC resulted from the HMW fractions, then the coagulation of the small colloidal organic carbon fraction (e.g., between 1 kDa and 10 kDa) in the Ochlockonee estuary is also likely to be an important process. However, recent studies have shown that coagulation of COC may not be a quantitatively important removal process for DOC in estuarine (e.g., Stordal et al. 1996b) and oceanic (McCarthy et al. 1996) environments. It seems that the biogeochemical behavior of DOC in estuarine systems is more variable and complex than that in open ocean environments, where a significant correlation between DOC concentration and sigma-t exists, with little difference between study areas and seasons (Guo et al. 1995).

Seasonal Variations of DOC

Seasonal variation of DOC concentrations in the Sabine-Neches and Lake Pontchartrain estuaries (Region II) are very different in these fluvially dominated systems (Fig. 9-4). It has been reported that the Sabine-Neches estuary is one of the most organic-rich estuaries on the Gulf coast (Bianchi et al. 1996, 1997a). While high concentrations of DOC in the Sabine-Neches estuary were primarily derived from fresh-water wetlands in the Sabine-Neches Rivers, DOC concentrations did not significantly correlate with total river flow (Bianchi et al. 1997a). This correlation was attributed to substantial non-point source inputs from bordering wetlands. However, DOC concentrations were correlated with POC, not with chlorophyll-a, which indicated that much of this DOC was derived from allochthonous DOC sources, not from phytoplankton sources (Bianchi et al. 1997a). In the Sabine-Neches estuary, concentrations of DOC reported by Bianchi et al. (1997a) in most seasons were relatively high (from ~400 to 1500 μM), with little variation between the upper estuary and lower estuary except in August and October 1993 (Fig. 9-4). On the other hand, DOC concentrations in the Sabine-Neches estuary reported by Benoit et al. (1994) were considerably lower (42–533 μM) than those in Fig. 9-4, with significant decrease with increasing salinity in August and October (Fig. 9-2). A few data points of DOC concentrations in the Sabine-Neches estuary reported by Stordal et al. (1996a), however, showed the opposite trend; DOC concentrations increased with increasing salinity, from 542 μM at salinity 0% to 683 μM at salinity 25% (Fig. 9-3). Large differences in DOC concentration and its distribution patterns in the Sabine-Neches estuary between different studies seem beyond the natural variability. Further studies are needed.

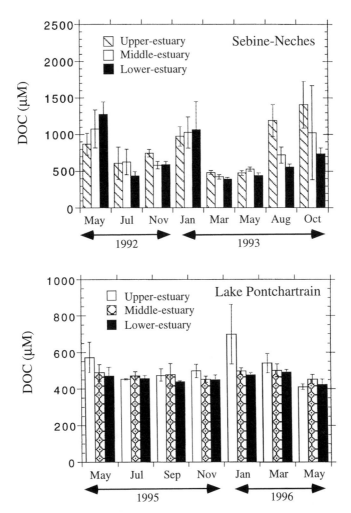

FIG. 9-4. *Seasonal variations of DOC in the Sabine-Neches and Lake Pontchartrain estuarine systems. (Modified from Bianchi et al. 1997a for the Sabine-Neches estuary; reprinted by permission of the Estuarine Research Federation. Modified from Kluwer Academic Publishers, Transport and fate of particulate and dissolved organic carbon in the Lake Pontchartrain estuary, Louisiana USA, Biogeochemistry 38, 1997, pp. 207–226, by M. E. Argyrou, T. S. Bianchi, and C. D. Lambert, Fig. 3, with kind permission from Kluwer Academic Publishers).*

While DOC concentrations in the Sabine-Neches estuary varied significantly (from ~500 to 1350 μM) with sampling seasons, seasonal variations in DOC concentrations in Lake Pontchartrain estuary were minimal (Fig. 9-4). The highest DOC concentration of ~700 μM in the upper estuary was observed during January 1996 (Argyrou et al. 1997). However, DOC concentrations were significantly different from year to year in the same sampling season, from 580 μM in May 1995 to ~415 μM in May 1996. These variations were apparently related to fresh-water inflow at different time periods (Argyrou et al. 1997). Moreover, DOC concentrations in the middle and lower estuary did not show significant seasonal variations (Fig. 9-4B). Although DOC in Lake Pontchartrain estuary appears to be largely derived from allochthonous sources, benthic fluxes of DOC to the overlying water column were found to be higher than those from riverine inputs to the Lake Pontchartrain estuary (Argyrou et al. 1997). Large DOC fluxes had led to a short residence time of 120 d calculated for the bulk DOC pool in this estuary. It was also suggested that heterotrophic processes were responsible for the high DOC turnover rate (Argyrou et al. 1997).

In another river-dominated estuary, Mobile Bay, fluctuation in DOC concentrations occurred mostly in the mid-bay and the lower bay (Pennock et al. submitted). In 5 years of observation at all seasons, concentrations of DOC were, on average, 380 μM and 323 μM at stations in the mid-bay and lower bay, respectively, whereas they were 300 μM at the near-coastal Gulf stations (Pennock et al. submitted). Seasonal variations in DOC concentrations were also observed in other estuaries of the Gulf of Mexico. Concentrations of DOC in Mississippi River waters decreased from summer (~330 μM) to winter (~275 μM) at the fresh-water station (Benner et al. 1992a). In Galveston Bay, the concentration of DOC (μM) in Trinity River waters was positively related to river flow (Q_W): DOC = $200Q_w + 343$, with Q_w in km^3 mo^{-1}, using data from three sampling trips (Guo 1995). It has been suggested that an inverse relationship between DOC concentration and river flow signifies that riverine DOC originates from populated catchments (Stumm and Morgan 1981). On the other hand, a positive linear relationship between DOC and river flow implies that riverine DOC has a natural origin or soil-derived source (Meybeck 1981). Thus, DOC in the Trinity River may mostly have a soil-derived input. The high rates of litter composition in soil around the Gulf are also consistent with the high DOC concentration in Gulf estuaries (Fig. 9-1).

Seasonal variation in DOC vertical distributions in the water column in the Gulf of Mexico (Guo and Santschi 1997c) was related to mixing and biological processes. While there was little difference in lower water-column DOC concentrations in the Gulf of Mexico (Guo et al. 1995), upper water-column DOC concentrations increased from winter to summer in the Gulf of Mexico (Guo and Santschi 1997c). Higher DOC concentrations in surface water in summer were related to the higher biological activities and higher riverine DOC inputs.

DOC Flux and Apparent Turnover Times

If the mixing behavior of a specific component in the water column is conservative, the annual flux (F_r) of that component from riverine inputs can be estimated by multiplying the concentration (C_r) in river water by the annual river flow (Q_w) as follows:

$$F_r = C_r \times Q_w \qquad (9\text{-}1)$$

For non-conservative components, the riverine flux (F_i) can be estimated by

$$F_i = \{C(s) - S(dC/dS)\}Q_w \qquad (9\text{-}2)$$

where S is salinity, $C(s)$ is concentration at salinity s, and dC/dS is the concentration gradient (e.g., Boyle et al. 1974). From Eq. 9-2, the flux difference between conservative ($d^2C / dS^2 = 0$) and non-conservative ($d^2C / dS^2 \neq 0$) behavior can be estimated by ($F_i - F_r$). Average annual inflow and DOC concentrations of rivers that flow into Gulf estuaries are listed in Table 9-2.

As a first approximation, the annual flux of DOC to estuaries by rivers was estimated to be $170\text{–}209 \times 10^{10}$ g C in the form of DOC in the Mississippi River plume. The second largest riverine DOC flux in Table 9-2 was that into Mobile Bay ($\sim 30 \times 10^{10}$ g C yr^{-1}). In Region I (Texas coast), river DOC fluxes decreased from the freshest estuary (Sabine-Neches) to the Trinity River, to rivers in the western Gulf coast in Texas (Table 9-2). A riverine DOC flux of $\sim 2.3 \times 10^{10}$ g C yr^{-1} to Lake Pontchartrain estuary was also

TABLE 9-2. Riverine DOC Flux to the Gulf of Mexico

River/Estuary	[DOC]$_{River}$ (μM)	Fresh Water Inflow (10^{10} m^3 yr^{-1})	Flux (10^{10} g C yr^{-1})	Reference
Region I				
Nueces/Corpus Christi	560–630	0.075	0.5–0.6	Benoit et al. (1994)
San Antonio-Guadalupa	330–480	0.028	0.1–0.2	Benoit et al. (1994)
Lavaca Bay	833	0.29	3	Benoit et al. (1994)
Colorado	650	0.24	2	Benoit et al. (1994)
Trinity/Galveston	420–480	1.24	6.7	Benoit et al. (1994)
Sabine-Neches	530	1.37	8.7	Benoit et al. (1994)
Region II				
Mississippi	270–330	52.9	170–209	Benner et al. (1992a)
Lake Pontchartrain	425–485	0.36	2.3	Argyrou et al. (1997)
Mobile Bay	424 ± 105	6.0	$\sim$31 ± 7	Pennock et al. (submitted)
Caminada/Barataria (Louisiana)	558	$\sim$0.02	4–28	Happ et al. (1977)

Note: Riverine water flow data are either from van der Leeden et al. (1990) or the corresponding reference.

reported by Argyrou et al. (1997). Using the removal or input flux, the turnover times (τ) of the bulk DOC can be estimated by dividing the total inventory by the flux, assuming steady-state conditions. For example, a turnover time of ~260 d and ~110 d was calculated for the bulk DOC and HMW DOC (>10 kDa), respectively, in estuarine waters of Galveston Bay (Guo 1995). Thus, the HMW COC (>10 kDa) seems to turn over on a shorter time scale than the bulk DOC pool in Galveston Bay. In addition to riverine flux, benthic DOC fluxes across the sediment-water interface could become significant (Warnken 1998). For example, a benthic DOC flux to the overlying water column in Lake Pontchartrain estuary was estimated to be about four times that of the riverine DOC flux (Argyrou et al. 1997). Using the sum of DOC fluxes, Argyrou et al. calculated a relatively short residence time of ~125 days for the bulk DOC in the Lake Pontchartrain Estuary (Argyrou et al. 1997).

COM IN ESTUARIES

Occurrence of COC

Organic carbon can be classified into at least three different physically defined pools: particulate (POC), colloidal (COC), and dissolved organic carbon (DOC). These fractions are operationally defined by the pore size of the membrane used to separate particles from colloids and colloids from the LMW dissolved fraction. Traditionally, DOC was defined as the fraction that passes through a glass fiber filter of 0.2–1 μm and thus includes colloidal fractions. Colloids are also operationally defined, usually between 0.2 or 0.4 μm as the upper limit and 1 or 10 kDa as the lower end of the size or molecular weight spectrum (e.g., Buffle and Leppard 1995). For example, COC_1 (1 kDa-0.2 μm) includes the colloidal subfractions COC_{10} (10 kDa and 0.2 μm) and COC_3 (3 kDa-0.2 μm).

Recently, the molecular weight distribution of DOC in estuarine waters has been examined using cross-flow ultrafiltration techniques (e.g., Guo and Santschi 1996). Results of DOC molecular distributions are available from a numbers of estuaries in the Gulf of Mexico, such as the Ochlockonee estuary (Powell et al. 1996), the Mississippi River plume (Benner et al. 1992a), Lake Pontchartrain estuary (Argyrou et al. 1997), Galveston Bay (Guo and Santschi 1997a), and some other Texas estuaries (Stordal et al. 1996a), as well as Gulf of Mexico waters (Guo et al. 1994, 1995).

Benner et al. (1992a) examined the distribution of COC in the Mississippi River plume and found that concentrations of COC_1 (1 kDa-0.2 μm) and COC_1/DOC fractions consistently decreased from fresh water to seawater. For example, the percentage of COC_1 in the bulk DOC pool was ~45% in riverine water and decreased to ~27% in open Gulf waters. There was little variation in the percentages of COC_1 isolated between summer and winter (Benner et al. 1992a). Guo and Santschi (1997a) also reported that the percentage of COC_1 in bulk DOC decreased from 66% in the Trinity River to 59–62% in estuarine waters of Galveston Bay (Fig. 9-5). For the HMW COC fraction (COC_{10}, 10 kDa–0.2 μm), the percentage of COC_{10} varied from 11% in the Trinity River to 6–8% in Galveston Bay waters (Fig. 9-5). In the Sabine-Neches estuary, the COC_1/DOC fraction changed from 37–42% at fresh-water stations to 14–27% at salinities between 15 and 25 (Stordal et al. 1996a). In Corpus Christi Bay, they ranged from 16 to 48% (Stordal et al. 1996a). Recently, Powell et al. (1996) reported for the Ochlockonee estuarine waters with salinities from 0.03 to 28.5 that, on average, ~48% of the bulk DOC was in a colloidal fraction (1 kDa–0.4 μm), while the COC_{10} (10 kDa–0.4 μm) fraction was ~23% of the bulk DOC (Fig. 9-6). In the low (<1) salinity regions, COC_{10} was up to 50% of the bulk DOC, with DOC concentrations as high

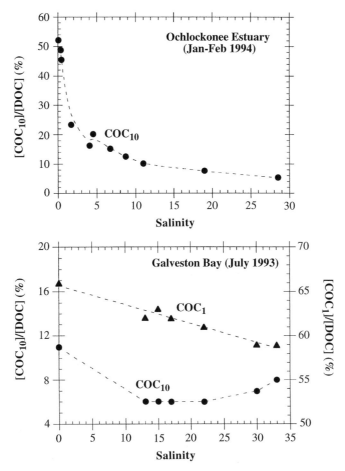

FIG. 9-5. *Removal of MMW DOC during estuarine mixing (upper panel: Ochlockonee estuary, data from Powell et al. 1996; lower panel: Galveston Bay, from Guo 1995).*

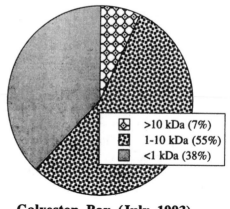

Galveston Bay (July 1993)

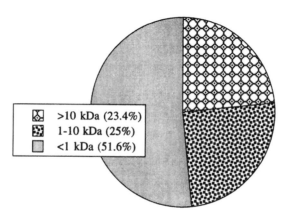

Ochlockonee Estuary (Jan-Feb 1994)

FIG. 9-6. *Molecular weight distribution of DOC in estuarine waters of Galveston Bay (data from Guo and Santschi 1997a) and Ochlockonee estuary (data from Powell et al. 1996).*

as 1550 μM (Powell et al. 1996). This value of COC_{10}/DOC is the highest percentage of COC_{10} reported in the Gulf. These high percentages are likely due to the high concentrations of humic substances in the Ochlockonee River and to the different membrane types or concentration factors used for ultrafiltration. However, as mentioned earlier, this region also has some of the highest litter decomposition rates, which is likely responsible for these high DOC concentration (Fig. 9-1). Using a 3-kDa ultrafilter, Argyrou et al. (1997) reported that only ~4 to 11% of the bulk DOC was in a colloidal fraction between 3 kDa and 0.2 μm (COC_3) in the Lake Pontchartrain estuary, with DOC concentrations of 425–700 μM. Therefore, much of the COC in Lake Pontchartrain estuary could be present in the medium molecular weight (e.g.,

TABLE 9-3. A Comparison of DOC Concentration and Colloidal Percentage between Rivers and Estuaries

Location	DOC (μM)	Ultrafilter Pore Size (kDa)	COC/DOC (%)	Reference
Sabine-Neches	550	1	37–42	Stordal et al. (1996a)
Corpus Christi	560–630	1	~36	[a]
Trinity River	495	1	66	Guo and Santschi (1997a)
Galveston Bay	145–469	1	59–63	Guo and Santschi (1997a)
Mississippi River	271–333	1	~45	Benner et al. (1992a)
Mississippi Delta	63–250	1	27–45	Benner et al. (1992a)
Ochlockonee estuary	189–1570	1	37–88	Powell et al. (1996)
Laguna Madre	312	1	~60	Amon and Benner (1996a)
Gulf of Mexico	70–86	1	35–42	Guo et al. (1995)
Lake Pontchartrain	425–483	3	4–11	Argyrou et al. (1997)
Gulf of Mexico	70	3	11	Guo et al. (1995)
Trinity River	495	10	11	Guo and Santschi (1997a)
Galveston Bay	145–469	10	6–8	Guo and Santschi (1997a)
Ochlockonee estuary	189–1570	10	5–52	Powell et al. (1996)
Gulf of Mexico	70–86	10	4–5	Guo et al. (1995)
Amazon	399	1	76	Benner and Hedges (1993)
Chesapeake Bay	118–215	1	52–65	Guo and Santschi (1997a)
Potomac/Patuxen	92–333	5	38–70	Sigleo (1996)
Mackenzie	50–300	10	~5–30	Whitehouse et al. (1989)
Rhone delta	82–148	10	8–30	Dai et al. (1995)
Chesapeake Bay	118–215	10	10–16	Guo and Santschi (1997a)

[a] Estimated from DOC in Benoit et al. (1994) and COC_1 in Stordal et al. (1996a).

1–3 kDa) colloidal fraction. In addition, concentrations of COC_3 decreased from the river to the mouth of the estuary (Argyrou et al. 1997).

Table 9-3 summarizes the COC concentrations and shows a comparison between rivers and estuaries in the Gulf of Mexico and other parts of the world. COC_1/DOC ratios in the Mississippi River (~45%) are relatively low compared with those in the Amazon River (~76%) and the Trinity River (~66%). Regardless of membrane molecular weight cutoffs, percentages of COC seem to be more variable in some estuaries than others (Table 9-3). Interestingly, while a large portion of bulk DOC in the Ochlockonee estuary was present in the COC_{10} fraction, the majority of the bulk DOC in Lake Pontchartrain estuary was present in the <3-kDa DOC fraction (Table 9-3). Large differences in COC fractions between estuaries could result from the nature of riverine DOM or the differences in ultrafiltration procedures—for example, concentration factors, membrane brands and models, and pre-filter pore size (Guo and Santschi 1996). Further studies are needed to investigate the importance of COC in estuarine systems.

Molecular Weight Distributions of DOC as Determined by Ultrafiltration

Analyses of molecular weight distributions of the bulk DOC pool in Galveston Bay waters (Region I) showed that, on average, ~7% of the total DOC was in the 10-kDa to 0.2-μm fraction and ~55% was in the 1- to 10-kDa fraction, leaving only 38% in the <1-kDa dissolved fraction (Fig. 9-6). Even though COC_{10} comprised only ~7% of the total DOC, COC_1 comprised, on average, 62% of the total DOC in Galveston Bay (Guo and Santschi 1997a). Indeed, COC outweighed the LMW or truly DOC fraction (<1 kDa) in this estuary. Therefore, molecular weight distributions of DOC in estuaries differ from oceanic environments, where the <1-kDa LMW component is the major fraction of the bulk DOC pool (Benner et al. 1992b; Guo et al. 1995). In the total organic carbon (TOC) pool, on average, POC comprised ~12% of the TOC, COC_{10} made up ~6%, COC between 1 and 10 kDa (COC_{1-10}) constituted ~47%, and the <1-kDa LMW organic carbon represented ~34% of the TOC in estuarine waters of Galveston Bay (Guo and Santschi 1997a). In the Sabine-Neches estuary, on average, COC_1 comprised ~25% of the DOC, and 75% of the bulk DOC was in the <1-kDa LMW fraction (Stordal et al. 1996a). Lower percentages of the COC_1 fraction are surprising since the Sabine-Neches system is one of the freshest estuaries on the Gulf coast, with high concentrations of DOC (Bianchi et al. 1996), and thus is expected to have high concentrations of HMW humic substances. Alternatively, differences in those ratios could have resulted from operational differences. However, lower percentages of HMW COC_3 (3 kDa–0.2 μm) were also measured in another estuary in Lake Pontchartrain estuary (Region II) with high DOC concentrations (Argyrou et al. 1997). Thus, the percentage of COC in estuarine waters may also be related to the composition of riverine DOC and to access of DOC to sunlight (Wetzel et al. 1995).

On average, ~40% of the DOC in the Mississippi River plume (Region II) was in the 1-kDa to 0–2 μm fraction, whereas ~60% of the DOC was in the <1-kDa LMW dissolved fraction (Benner et al. 1992a). Percentages of COC in Ochlockonee estuarine waters (Region III) were more variable. The percentage of HMW COC_{10} in the low-salinity region was up to ~50% of the bulk DOC, decreasing to 5% at a salinity of 28.5, while the percentage of the <1-kDa LMW dissolved organic carbon ranged from ~20% of the bulk DOC at a salinity of <1 to ~80% at a salinity of ~28.5 (Powell et al. 1996). An average molecular weight distribution of DOC in Ochlockonee estuarine waters was also shown to have 23% of the bulk DOC in >10-kDa fraction and 25% in the 1- to 10-kDa fraction, leaving ~45% in the <1-kDa LMW fraction (Fig. 9-6). In Galveston Bay waters, a major portion (55%) of bulk DOC was in the medium molecular weight (1–10 kDa) colloidal fraction, with no significant difference throughout the entire estuary during the sampling time (Guo and Santschi 1997a). However, in Ochlockonee estuarine waters, the <1-kDa LMW DOC was the largest component of the bulk DOC, with a significant portion of DOC in the COC_{10} fraction (Fig. 9-6), which decreased dramatically from low-salinity to high-salinity areas (Powell et al. 1996). In Laguna Madre of Region IV, ~60% of the DOC (~312 μM) was in the COC_1 colloidal fraction (Table 9-3). In general, estuarine waters in Regions I and III of the Gulf contain higher concentrations of DOC and COC (e.g., Ochlockonee Estuary, Sabine-Neches Estuary, Galveston Bay, and Corpus Christi Bay) than estuarine waters from Region II (e.g., Mississippi River plume, Lake Pontchartrain) (Table 9-3).

Conformation of COM Revealed by AFM

Atomic force microscopy (AFM) is a technique widely used to study biomolecules because of its high resolution at atomic and molecular scales (Namjesnik-Dejanovic and Maurice 1997). This technique has rarely been applied to the study of the physical composition of natural macromolecules. Recently, Santschi et al. (in press) analyzed a few COM_1 along with various blank samples using the AFM method. COM_1 samples isolated by ultrafiltration were collected from Galveston Bay and coastal waters in the Gulf of Mexico. Before being mounted on a freshly cleaved mica surface, colloidal samples were dissolved and equilibrated (Santschi et al. in press).

The preliminary results showed that a large fraction of COM from Galveston Bay and the Gulf of Mexico consists of fibrils material (Fig. 9-7). Abundant quasi-spherical particles and aggregates were also seen. These fibrils have a diameter of 1–3 nm and an length of 100–1000 nm and sometimes appear to aggregate sideways to each other. Fibrils have been described for marine and fresh waters by transmission electron microscopy (TEM) (e.g., Buffle and Leppard 1995; Leppard 1997). The fact that colloids contain fibrils with a typical size of 1–3 × 100 nm has the following consequences: (1) The molecular weight of COM derived from ultrafiltration using globular macromolecules as standard materials could be biased. For example, fibrils 1–2 × 100–1000 nm in size have molecular weights closer to 100,000 than to 10,000, i.e., the nominal cutoff at an average pore size of 3 nm. (2) Molecular diffusion coefficients of aquatic colloids are likely considerably smaller than those assumed for spherical macromolecules with the same diameter. These interesting results may have significant influences on the way we view the size spectrum of COC in aquatic systems.

CYCLING OF DOM

Elemental Composition of COM

In addition to the size distribution of DOM, chemical and biochemical characterizations of DOM are essential to our understanding of the sources and cycling of DOM within estuarine systems. For example, elemental (e.g., C,

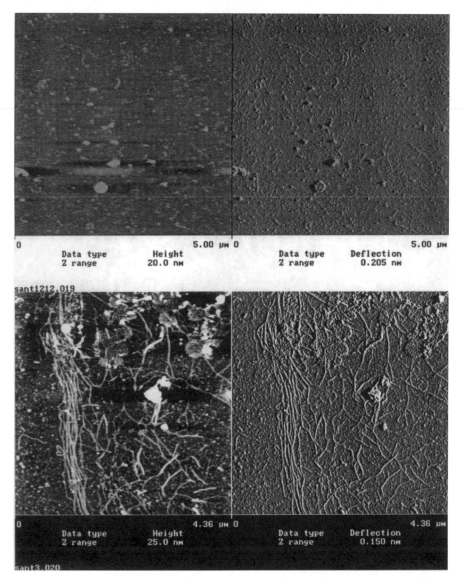

FIG. 9-7. Atomic force micrographs of COM collected from estuarine waters of Galveston Bay (upper panel) and coastal waters of the Gulf of Mexico (lower panel) (from Balnois and Wilkinson unpubl. results).

H, N, S, O), isotopic (^{13}C, ^{14}C, ^{15}N, etc.), and molecular (biomarker compounds, etc.) composition can be used to trace the origin, fate, and reactivity, as well as other biogeochemical processes of DOM within estuaries and coastal waters (Bianchi et al. 1995, 1997b; Canuel et al. 1995; Guo and Santschi 1997a).

Average elemental (C, N, and S) compositions of different COM samples are listed in Table 9-4. Carbon content in the isolated COM samples ranged from 23 to 37%, whereas nitrogen content varied from 0.9 to 4.9%. Using the highest organic carbon concentration (~37%) of COM_1 from Table 9-4, it seems that the ratio of COM to COC is about 2.7 for this >1-kDa COM material in estuarine environments. The COM:COC ratio is similar to the ratio of 2.5 derived from a simple formula of CH_2O, suggesting that major components in isolated COM are carbohydrates. Total sulfur concentration measured in Mississippi River plume COM_1 samples was 1.2–1.5% (Table 9-4). Organic sulfur concentration in the Trinity River COM sample was ~0.9% (Guo and Santschi 1997a) which is similar to those found in fulvic and humic acids samples (Table 9-4). According to results of elemental characterization, reported C:N ratios of estuarine COM ranged from 9 to 52 (Table 9-4). C:N ratios of HMW DOM were, in general, significantly higher than the Redfield ratio (~6–7). Elevated C:N ratios indicate that COM is a nitrogen-depleted organic matter pool and that the LMW DOM fraction must be highly nitrogen enriched relative to the bulk DOM.

Variations in C:N ratios of COM across the salinity gradient in estuaries have also been studied. In Galveston Bay waters, C:N ratios of COM, in general, decreased from fresh water to coastal waters (Fig. 9-8). This trend was observed not only for different COM fractions (both COM_1 and COM_{10}) but also in two separate sampling trips (July 1993 and July 1995). For example, C:N ratios of COM_1 in the Trinity River station were 24 to 26 compared to 13 to 17 at the mouth of Galveston Bay (Guo and Santschi 1997a). Similarly, C:N ratios of COM_{10} were ~25 in the Trinity River and decreased from ~18 at a salinity of ~15 to 16 at a salinity of ~30. The decrease of C:N ratios in COM_1 indicated that sources of the COM in this estuary changed from more terrestrial components in fresh-water and upper bay regions to more labile or freshly produced autochthonous DOM components in the coastal waters. This conclusion is further supported by the stable carbon isotope composition in COM (see Fig. 9-10 and discussion below). While the decrease of C:N ratios in COM from fresh water to coastal water was close to linear in Galveston Bay, the detailed variations indicated that the C:N ratio of COM first decreased rapidly upon entering the upper bay and then remained almost constant in salinity between 2 and 20, after which it dropped to the lowest value and then stayed nearly constant again in the lower part of Galveston Bay (Fig. 9-8). Therefore, carbon sources of COM seem to change discontinuously from the river to the upper estuary and to the lower part of the estuary (Guo and Santschi 1997a).

Interestingly, the C:N ratio in the higher molecular weight fraction (COM_{10}) in Galveston Bay was consistently lower than that of the COM_1 fraction, which contains mostly intermediate molecular weight (1–10 kDa)

TABLE 9-4. Average Elemental Composition (Organic C, N, and S) of COM and DOM From Aquatic Environments

River or Sample	Salinity	C (%)	N (%)	S (%)	C:N	Reference
Mississippi	0	36.5 ± 0.4	1.86 ± 0.05	1.5 ± 0.1	23	Guo and Santschi (unpubl. results)
Mississippi plume	15–28	35.4 ± 0.3	2.28 ± 0.19	1.2 ± 0.1	18	Guo and Santschi (unpubl. results)
Ochlockonee	0–28	—	—	—	20–90	Powell et al. (1996)
Trinity River	0	27	1.2	0.94	26	Guo and Santschi (1997a)
Galveston Bay	23	23	1.4	—	19	Guo and Santschi (1997a)
Amazon	0	24–37	0.86–1.4	—	27–52	Hedges et al. (1994)
Patuxent	0–8.4	24–37	2.8–4.9	—	9–12	Sigleo (1996)
Riverine FA	0	51.9	1.1	0.6	55	Thurman (1985)
Lake water FA	0	52.0	1.3	1.0	46	Thurman (1985)
Seawater FA	—	50.0	6.4	0.46	9	Thurman (1985)
Stream FA	0	54.56	0.87	0.74	73	Aiken (1985)
Stream HA	0	55.94	1.27	0.93	51	Aiken (1985)
Suwannee FA	0	53.6	0.87	0.66	72	Thorn et al. (1992)

FA = fulvic acids; HA = humeric acids.

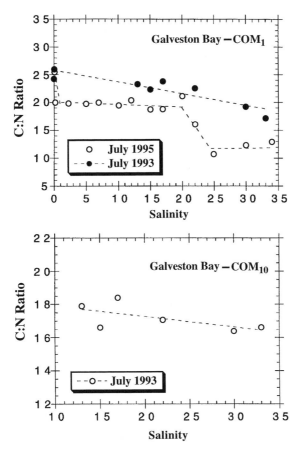

FIG. 9-8. *Variations of C : N ratios in macromolecular COM fractions (1 kDa < COM₁ > 0.2 μm and 10 kDa < COM₁₀ < 0.2 μm) with salinity in Galveston Bay waters. (Reprinted from L. Guo and P. H. Santschi, Isotopic and elemental characterization of colloidal organic matter from the Chesapeake Bay and Galveston Bay, Mar. Chem.* **59:** *1–15, 1997 with kind permission from Elsevier Science-NL, Sara Burgerhartstraat 25, 1055 KV Amsterdam, The Netherlands.)*

fractions and encompasses the COM_{10} fraction (Fig. 9-8). Lower C:N ratios in the COM_{10} imply that the higher molecular weight organic matter fraction in estuarine waters is derived mostly from freshly produced POM and may have higher reactivity than the lower molecular weight organic matter fractions (Amon and Benner 1994; Guo and Santschi 1997a). A comparison of the C:N ratios between organic matter of different sizes revealed that the C:N ratio increased from suspended particulate matter to COM_{10} to COM_1 in Galveston Bay (Fig. 9-9). The increase of C:N ratios with decreasing size suggests that the direction of organic matter degradation is from particulate to COM and then to LMW DOM (Guo and Santschi 1997a). As organic matter degrades from particulate to COM_{10} to COM_{1-10} and then to LMW DOM, nitrogen-

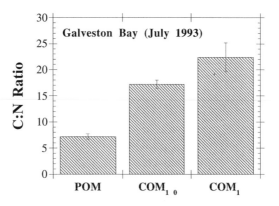

FIG. 9-9. Values of C : N ratios in suspended POM, HMW COM (COM$_{10}$), and COM$_1$. (Reprinted from L. Guo and P. H. Santschi, Isotopic and elemental characterization of colloidal organic matter from the Chesapeake Bay and Galveston Bay, Mar. Chem. **59**: 1–15, 1997 with kind permission from Elsevier Science-NL, Sara Burgerhartstraat 25, 1055 KV Amsterdam, The Netherlands.)

containing organic matter in larger-sized fractions is preferentially decomposed, leaving more carbon in smaller-sized organic matter fractions (higher C:N ratios) and at the same time increasing the nitrogen content (lower C:N ratio again) in the LMW DOM fraction (e.g., Sigleo 1996).

Direct measurements of organic C:N ratios in dissolved and colloidal phases were also reported for Ochlockonee estuarine waters; average values are listed in Table 9-4. C:N ratios of the bulk DOM pool ranged from 22 to 45, similar to those (14–55) in the LMW DOM fraction (Powell et al. 1996). On the other hand, C:N ratios in the HMW (>10-kDa) fraction were highly variable, ranging from 3 to 207, whereas C:N ratios in the 1- to 10-kDa COM fraction ranged from 20 to 90 (Powell et al. 1996). Higher C:N ratios in different DOM fractions and very high bulk DOC concentrations in low-salinity waters likely indicate that sources of DOM were largely terrestrial humic substances in the Ochlockonee estuary. However, C:N ratios in the colloidal fractions varied non-systematically along a salinity gradient in the Ochlockonee estuary. High C:N ratios were attributed to the low productivity in the estuary during the sampling season (Powell et al. 1996).

C:N ratios of COM could be an indicator of carbon sources and diagenetic processes of DOM in estuarine environments. A comparison of the elemental composition in different estuaries showed that C:N ratios of COM$_1$ were highly variable even though carbon contents were similar (Table 9-4). Significant difference in COM C:N ratios between rivers may point to the distinct source functions of rivers or possibly to procedural difference in sampling and characterizing COM. More elemental characterization of COM is needed.

Carbon Isotopic Signatures

Signatures of carbon isotopes have also been used to study source functions of DOM in estuaries of the Gulf of Mexico. Values of $\delta^{13}C$ in COM$_1$ increased with increasing salinity in the Mississippi River plume (Fig. 9-10). However, this trend is more scattered in Galveston Bay (Fig. 9-10). Lower values of $\delta^{13}C$ in COM_1 from low-salinity waters also indicates a terrestrial carbon source for the COM fraction. On the other hand, the enrichment in ^{13}C in coastal water COM_1 is likely related to autochthonous DOM sources, consistent with the source functions derived from C:N ratios (Fig. 9-8; see also the previous discussion). While measurements of stable carbon isotopes in DOM fractions are still limited, radiocarbon data are even fewer. Guo and Santschi (1997a) reported $\Delta^{14}C$ values of the COM_1 fraction from Galveston Bay and Chesapeake Bay. They found that $\Delta^{14}C$ values of the COM_1 fraction in estuarine waters of Galveston Bay were generally high (from 51 ± 6‰ to +97 ± 7‰), which corresponds to a ^{14}C age of 418 ± 52 y BP to contemporary, while

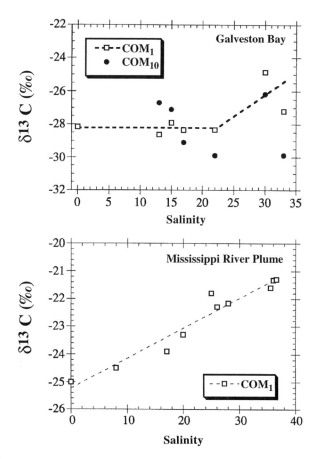

FIG. 9-10. *Stable carbon isotope composition of macromolecular COM fractions in the Mississippi delta (re-plotted from Benner et al., 1992a) and Galveston Bay waters. (Reprinted from L. Guo and P. H. Santschi, Isotopic and elemental characterization of colloidal organic matter from the Chesapeake Bay and Galveston Bay, Mar. Chem.* **59:** *1–15, 1997 with kind permission from Elsevier Science-NL, Sara Burgerhartstraat 25, 1055 KV Amsterdam, The Netherlands.)*

$\Delta^{14}C$ values of the COM_{10} fraction were consistently lower, ranging from $-189 \pm 5‰$ to $-74 \pm 6‰$ (Fig. 9-11). Lower $\Delta^{14}C$ values in the HMW fraction contradict the results observed in the upper water column in the Gulf of Mexico, where $\Delta^{14}C$ values of the COM_{10} fraction were generally higher than those of the COM_1 fraction (Santschi et al. 1995), but are consistent with the general pattern in COM samples from marine benthic nepheloid layers (Guo et al. 1996). If the COM_{10} fraction was mostly derived from phytoplankton sources, a higher value of $\Delta^{14}C$ would have been expected. On the other hand, if the COM_{10} fraction was produced or disaggregated from re-suspended old organic carbon, then lower $\Delta^{14}C$ values are also possible. Galveston Bay is a particle dynamic estuary (Benoit et al. 1994), with an average water depth of ~2 m. Sediment re-suspension prevails under even moderately windy conditions. Therefore, relatively low $\Delta^{14}C$ values suggest that sources of the COM_{10} fraction could be largely from older but more nitrogen-rich and labile organic matter preserved in sediments (Guo and Santschi 1997a).

Using the relationship between $\Delta^{14}C$ value and $C:N$ ratio, Guo and Santschi (1997a) demonstrated that values of $\Delta^{14}C$ in the COM_1 were significantly correlated with $C:N$ ratios in Galveston Bay (Fig. 9-12). In general, the COM from the fresh-water end member had a higher $C:N$ ratio and a higher $\Delta^{14}C$ value. As riverine water mixed with seawater within the estuary, the $\Delta^{14}C$ values and $C:N$ ratios in the COM_1 decreased toward the Gulf of Mexico (Fig. 9-12). Therefore, the carbon sources of the COM varied from terrestrially dominated (high $\Delta^{14}C$ and high $C:N$ ratio) in river water to older but more

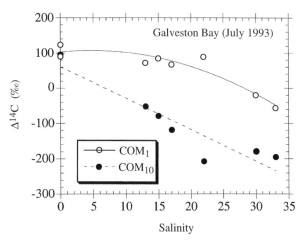

FIG. 9-11. *Values of $\Delta^{14}C$ (‰) in macromolecular COM fractions (1 kDa < COM_1 > 0.2 μm and 10 kDa < COM_{10} < 0.2 μm) in estuarine waters of Galveston Bay. (Reprinted from L. Guo and P. H. Santschi, Isotopic and elemental characterization of colloidal organic matter from the Chesapeake Bay and Galveston Bay, Mar. Chem.* **59:** *1–15, 1997 with kind permission from Elsevier Science-NL, Sara Burgerhartstraat 25, 1055 KV Amsterdam, The Netherlands.)*

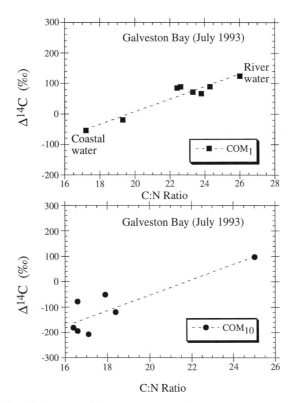

FIG. 9-12. *Relationship between $\Delta^{14}C$ values and $C:N$ ratios in macromolecular COM fractions in estuarine waters of Galveston Bay. (Reprinted from L. Guo and P. H. Santschi, Isotopic and elemental characterization of colloidal organic matter from the Chesapeake Bay and Galveston Bay, Mar. Chem.* **59:** *1–15, 1997 with kind permission from Elsevier Science-NL, Sara Burgerhartstraat 25, 1055 KV Amsterdam, The Netherlands.)*

recycled and labile sources (low $\Delta^{14}C$ and low $C:N$ ratio) in the lower part of the estuary.

THE ROLE OF DOM IN THE TRANSPORT AND FATE OF TRACE ELEMENTS

It has been suggested that the ultimate fate of chemical reactive trace contaminants, including organics, metals, and radionuclides, is strongly related to the natural organic matter in estuarine environments, largely through heterogeneous processes (e.g., Santschi et al. 1997 and references therein).

Measurements of mercury phase speciation revealed that a large fraction (57%, on average) of total dissolved Hg was actually associated with COM in three Texas estuaries (Stordal et al. 1996a), and 35–87% was associated with COM in Ochlockonee Estuary (Guentzel et al. 1996). They also reported a covariation of colloidal Hg with COC concentration, indicating that the cycling of Hg is controlled by the cycling of COM. In the Ochlockonee estuary, where DOC concentrations were as high as 1500 μM, colloidal Fe was significantly correlated with organic carbon in the COM fraction (Powell et al. 1996). A significant correlation of trace metals (e.g., Cu, Ni, Mn, Co, Zn, Cd, and Ag) with COM has also been found, either from direct measurements in colloidal solution phase (Wen et al. in press) or from measurements of isolated COM (Guo et al. submitted). Most metals measured were considerably enriched in COM compared to suspended particles but were similar in composition to marine organisms (Guo et al. submitted). It appears that the cycling of many transition metals in estuarine waters is controlled by the cycling of COM. Furthermore, high DOC concentrations in estuarine waters of the Sabine-Neches system have been held responsible for altering the distribution coefficients (K_d) of certain particle-reactive radionuclides (e.g., ^{7}Be and ^{210}Pb), the residence times of particles, and therefore the mobility of these nuclides (Baskaran et al. 1997). Thus, it is likely that COM plays a major role in regulating the transport of particle-reactive radionuclides, especially in high-DOC subtropical estuaries of the Gulf of Mexico.

SUMMARY

The behavior of DOC demonstrated varying features during estuarine mixing in different regions of Gulf of Mexico estuaries. Although removal of DOC was observed in most of the estuaries in Region I, non-conservative behavior with source inputs of DOC existed in estuaries of Regions II and III in the Gulf of Mexico. The Mississippi River had the largest riverine DOC flux into the Gulf of Mexico. Fluxes of riverine DOC seemed to decrease in either direction away from the Mississippi River.

A significant fraction of bulk DOC was found in a colloidal fraction with a molecular weight >1 kDa. Concentrations of COC >1 kDa appear to outweigh the <1-kDa DOC fraction in estuaries. However, features of molecular weight distributions of DOC in estuarine waters were different from estuary to estuary. More detailed studies are needed to find out if these variations are real or operator-dependent.

COM_{10} could be more nitrogen rich than its counterpart, COM_1, in estuaries. A general direction of organic matter degradation from particulate to COM to DOM was suggested by the increase in C:N ratio with decreasing size. Therefore, COM may have higher reactivities and thus is a more active component in the carbon cycle and other biogeochemical processes in estuarine environments.

Sources of HMW DOM can be constrained by its isotopic and elemental composition. In general, riverine water COM_1 showed higher Δ^{14}C values and higher C:N ratios. As river water mixed with seawater, values of $\Delta^{14}C$ and C:N ratios in COM_1 decreased from the upper estuary to the lower estuary. However, while C:N ratios decreased linearly, the decrease in $\Delta^{14}C$ values with salinity was highly non-linear. It thus appears that estuarine COM_1 is not a simple mixture of riverine and marine COM_1, but rather is cycled on short time scales and is strongly affected by internal sources.

Even though terrestrial organic matter was found in the main parts of estuaries, its composition is likely altered by selective removal of the HMW fraction; furthermore, internal cycling of DOC and COC fractions within estuaries sequentially alters the isotopic and elemental composition of COC fractions. Measurements of isotopic and elemental signatures on size-fractionated or chemically fractionated DOM components are more meaningful than those performed on the bulk DOM pool, as fractions of different organic sizes with distinct isotopic and elemental composition may turn over on different time scales and thus may have different geochemical pathways in aquatic systems.

ACKNOWLEDGMENTS

We thank Jon Pennock and Bill Landing for providing unpublished results, Robert Twilley for providing references, and Robert Wetzel for comments on our manuscript. The writing of this chapter was supported by the Department of Energy (Grant DE-FG05-92ER61421), the Office of Naval Research (Grant N00014-93-1-0877), a Texas Sea Grant, and the Texas Institute of Oceanography.

REFERENCES

Aiken, G. R. 1985. Humic substances in soil, sediment, and water. Wiley.

Amon, R. M. W., and R. Benner. 1994. Rapid cycling of high-molecular-weight dissolved organic matter in the ocean. Nature **369:** 549–552.

———. 1996a. Bacterial utilization of different size class of dissolved organic matter. Limnol. Oceanogr. **41:** 41–51.

———. 1996b. Photochemical and microbial consumption of dissolved organic carbon and dissolved oxygen in the Amazon River system. Geochim. Cosmochim. Acta **60:** 1783–1792.

Argyrou, M. E., T. S. Bianchi, and C. D. Lambert. 1997. Transport and fate of particulate and dissolved organic carbon in the Lake Pontchartrain estuary, Louisiana, U.S.A. Biogeochemistry **38:** 207–226.

Baskaran, M., M. Ravichandran, and T. S. Bianchi. 1997. Cycling of ^7Be and ^{210}Pb in a high DOC, shallow, turbid estuary of southeast Texas. Estuar. Coast. Shelf Sci. **45:** 165–176.

————, and P. H. Santschi. 1993. The role of particles and colloids in the transport of radionuclides in coastal environments of Texas. Mar. Chem. **43:** 95–114.

Benner, R., C. Chin-Leo, W. Gardner, B. Eadie, and J. Cotner. 1992a. The fates and effects of riverine and shelf-derived DOM on Mississippi River plume/Gulf shelf processes, p. 84–94. *In:* Nutrient Enhanced Coastal Ocean Productivity, NECOP Workshop Proceedings, October 1991, NOAA Coastal Ocean Program, TAMU-SG-92-109.

————, and J. I. Hedges. 1993. A test of the accuracy of fresh water DOC measurements by high-temperature catalytic oxidation and UV-promoted persulfate oxidation. Mar. Chem. **41:** 161–165.

————, J. D. Pakulski, M. McCarthy, J. I. Hedges, and P. G. Hatcher. 1992b. Bulk chemical characteristics of dissolved organic matter in the ocean. Science **255:** 1561–1564.

Benoit, G., S. Oktay, A. Cant, M. O. Hood, C. H. Coleman, O. Corapcioglu, and P. H. Santschi. 1994. Partitioning of Cu, Pb, Ag, Zn, Fe, Al and Mn between filter-retained particles, colloids and solution in 6 Texas estuaries. Mar. Chem. **45:** 307–336.

Bianchi, T. S., M. Baskaran, J. DeLord, and M. Ravichandran. 1997a. Carbon cycling in a turbid estuary of southeast Texas: The use of plant pigment biomarkers and water quality parameters. Estuaries **20:** 404–415.

————, M. E. Freer, and R. G. Wetzel. 1996. Temporal and spatial variability, and the role of dissolved organic carbon (DOC) in methane fluxes from the Sabine River floodplain (Southeast Texas, U.S.A.). Arch. Hydrobiol. **136:** 261–287.

————, C. Lambert, P. H. Santschi, M. Baskaran, and L. Guo. 1995. Plant pigments as biomarkers of high-molecular-weight dissolved organic carbon. Limnol. Oceanogr. **40:** 422–428.

————, C. Lambert, P. H. Santschi, and L. Guo. 1997b. Sources and transport of land-derived particulate and dissolved organic matter in the Gulf of Mexico (Texas shelf/slope). Org. Geochem. **27:** 65–78.

Boyle, E., R. A. Collier, A. T. Dengler, J. M. Edmond, A. C. Ng, and R. F. Stallard. 1974. On the chemical mass-balance in estuaries. Geochim. Cosmochim. Acta **38:** 1719–1728.

Buffle, J., and G. G. Leppard. 1995. Characterization of aquatic colloids and macromolecules. 1. Structure and behavior of colloidal material. Environ. Sci. Technol. **29:** 2169–2175.

Burton, J. D., and P. S. Liss. 1976. Estuarine chemistry. Academic Press.

Canuel, E. A., J. E. Cloern, D. B. Ringelberg, J. B. Guckert, and G. H. Rau. 1995. Molecular and isotopic tracers used to examine sources of organic matter and its incorporation into the food webs of San Francisco Bay. Limnol. Oceanogr. **40:** 67–81.

Dai, M.-H., J.-M. Martin, and G. Cauwet. 1995. The significant role of colloids in the transport and transformation of organic carbon and associated trace metals (Cd, Cu, and Ni) in the Rhone delta (France). Mar. Chem. **51:** 159–175.

Farrington, J. W. 1992. Marine organic geochemistry: Review and challenges for the future. Mar. Chem. **39:** 1–244.

Fox, L. E. 1983. The removal of dissolved humic acid during estuarine mixing. Estuar. Coast. Shelf Sci. **16:** 413–440.

Guentzel, J. L., R. T. Powell, W. M. Landing, and R. P. Mason. 1996. Mercury associated with colloidal material in an estuarine and an open ocean environment. Mar. Chem. **55:** 177–188.

Guo, L. 1995. Cycling of dissolved and colloidal organic matter in oceanic environments as revealed by carbon and thorium isotopes. Ph.D. dissertation, Texas A&M University.

———, C. H. Coleman, Jr., and P. H. Santschi. 1994. The distribution of colloidal and dissolved organic carbon in the Gulf of Mexico. Mar. Chem. **45:** 105–119.

———, and P. H. Santschi. 1996. A critical evaluation of the cross-flow ultrafiltration technique for sampling colloidal organic carbon in seawater. Mar. Chem. **55:** 113–127.

———. 1997a. Isotopic and elemental characterization of colloidal organic matter from the Chesapeake Bay and Galveston Bay. Mar. Chem. **59:** 1–15.

———. 1997b. Composition and cycling of colloids in marine environments. Rev. Geophys. **35:** 17–40.

———. 1997c. Measurements of dissolved organic carbon (DOC) in sea water by high temperature combustion method. Acta Oceanolog. Sin. **16:** 339–353.

———, P. H. Santschi, L. A. Cifuentes, S. Trumbore, and J. Southon. 1996. Cycling of high molecular weight dissolved organic matter in the Middle Atlantic Bight as revealed by carbon isotopic (^{13}C and ^{14}C) signatures. Limnol. Oceanogr. **41:** 1242–1252.

———, P. H. Santschi, and K. W. Warnken. 1995. Dynamics of dissolved organic carbon (DOC) in oceanic environments. Limnol. Oceanogr. **40:** 1392–1403.

———. Submitted. Trace metal composition of colloidal material from estuarine and marine environments. Limnol. Oceanogr.

Happ, G., J. G. Gosselink, and J. W. Day, Jr. 1977. The seasonal distribution of organic carbon in a Louisiana estuary. Estuar. Coast. Mar. Sci. **5:** 695–705.

Hedges, J. I. 1992. Global biogeochemical cycles: Progress and problems. Mar. Chem. **39:** 67–93.

———, G. L. Cowie, J. E. Richey, P. D. Quay, R. Benner, M. Strom, and B. R. Forsberg. 1994. Origins and processing of organic matter in the Amazon River as indicated by carbohydrates and amino acids. Limnol. Oceanogr. **39:** 743–761.

Herrera-Silverira, J. A., and J. Remirez-Remirez. 1996. Effects of natural phenolic material (tannin) on phytoplankton growth. Limnol. Oceanogr. **41:** 1018–1023.

Leppard, G. G. 1997. Colloidal organic fibrils of acid polysaccharides in surface waters: Electron-optical characteristics, activities and chemical estimates of abundance. Colloids and Surface A **120:** 1–15.

Mantoura, R. F. C., and E. M. S. Woodward. 1983. Conservative behavior of riverine dissolved organic carbon in the Severn Estuary: Chemical and geochemical implications. Geochim. Cosmochim. Acta **47:** 1293–1309.

Martin, J.-M., M.-H. Dai, and G. Cauwet. 1995. Significance of colloids in the biogeochemical cycling of organic carbon and trace metals in a coastal environment—example of the Venice Lagoon (Italy). Limnol. Oceanogr. **40:** 119–131.

McCarthy, M., J. Hedges, and R. Benner. 1996. Major biochemical composition of dissolved high molecular weight organic matter in seawater. Mar. Chem. **55:** 281–297.

Meentemeyer, V. 1978. Climate regulation of decomposition rates of organic matter in terrestrial ecosystems, p. 779–789, *In:* D. C. Adriand and I. L. Brisbin [eds.], Environmental chemistry and cycling processes. Conf. 760429. National Technical Information Service.

Meybeck, M. 1981. River transport of organic carbon to the ocean, p. 216–267. *In:* Flux of carbon by rivers to the oceans. Conf. 8009140, UC-11. U.S. Office of Energy Research.

Namjesnik-Dejanovic, K., and P. A. Maurice. 1997. Atomic force microscopy of soil and stream fulvic acids. Colloids and Surface A **120:** 77–86.

Pennock, J. R., W. W. Schroeder, J. L. W. Cowan, and R. P. Stumpf. Submitted. The loading, reactivity and distribution of carbon, nitrogen and phosphorus in a river-dominated estuary: Mobile Bay, Alabama (USA). Estuaries.

Powell, R. T., W. M. Landing, and J. E. Bauer. 1996. Colloidal trace metals, organic carbon and nitrogen in a Southeastern U.S. estuary. Mar. Chem. **55:** 161–176.

Rivera-Monroy, V. H., J. D. Day, R. R. Twilley, F. Vera-Herrera, and C. Coronado-Molina. 1995. Flux of nitrogen and sediment in a fringe mangrove forest in Terminos Lagoon, Mexico. Estuar. Coast. Shelf Sci. **40:** 139–160.

Santschi, P. H., L. Guo, E. Balnois, K. Wilkinson, and J. Buffle. (In press). Fibrillar polysaccharides in marine macromolecular organic matter as imaged by atomic force microscopy and TEM. Limnol. Oceanogr.

———, L. Guo, M. Baskaran, S. Trumbore, J. Southon, T. Bianchi, B. Honeyman, and L. Cifuentes. 1995. Isotopic evidence for the contemporary origin of high-molecular weight organic matter in oceanic environments. Geochim. Cosmochim. Acta **59:** 625–631.

———, J. J. Lenhart, and B. D. Honeyman. 1997. Heterogeneous process affecting trace contaminant distribution in estuaries: The role of natural organic matter. Mar. Chem. **58:** 99–125.

Sholkovitz, E. R. 1976. Flocculation of dissolved organic and inorganic matter during the mixing of river water and seawater. Geochim. Cosmochim. Acta **40:** 831–845.

Sigleo, A. C. 1996. Biochemical components in suspended particles and colloids: Carbo-hydrates in the Potomac and Patuxent estuaries. Organ. Geochem. **24:** 83–93.

———, and J. C. Means. 1990. Organic and inorganic components in estuarine colloids: Implications for sorption and transport of pollutants. Rev. Environ. Contam. Toxicol. **112:** 123–147.

Stordal, M. C., G. A. Gill, L.-S. Wen, and P. H. Santschi. 1996a. Mercury phase speciation in the surface waters of three Texas estuaries: Importance of colloidal forms. Limnol. Oceanogr. **41:** 52–61.

———, P. H. Santschi, and G. A. Gill. 1996b. Colloidal pumping: Evidence for the coagulation process using natural colloids tagged with ^{203}Hg. Environ. Sci. Technol. **30:** 3335–3340.

Stumm, W., and J. J. Morgan. 1981. Aquatic chemistry (2nd ed.). Wiley Interscience.

Thorn, K. A., J. B. Arterburn, and M. A. Mikita. 1992. ^{15}N and ^{13}C NMR investigation of hydroxylamine-derivatized humic substances. Environ. Sci. Technol. **26:** 107–116.

Thurman, E. M. 1985. Organic geochemistry of natural waters. Academic Press.

Twilley, R. R. 1985. The exchange of organic carbon in basin mangrove forests in a Southwest Florida estuary. Estuar. Coast. Shelf Sci. **20:** 543–557.

van der Leeden, F., F. L. Troise, and D. K. Todd. 1990. The water encyclopedia. Lewis.

Warnken, K. 1998. Benthic fluxes of trace metals (Fe, Mn, Cu, Pb, Zn, Cd) and nutrients (N, P, Si) in Galveston Bay. M.S. thesis, Texas A&M University.

Wen, L.-S., P. H. Santschi, G. A. Gill, and C. L. Paternostro. (In press). Estuarine trace metal distributions in Galveston Bay: Colloidal forms and dissolved phase speciation. Mar. Chem.

Wetzel, R. G., P. G. Hatcher, and T. S. Bianchi. 1995. Natural photolysis by ultraviolet irradiance of recalcitrant dissolved organic matter to simple substrates for rapid bacterial metabolism. Limnol. Oceanogr. **40:** 1369–1380.

Whitehouse, B. G., R. W. Macdonald, K. Iseki, M. B. Yunker, and F. A. McLaughlin. 1989. Organic carbon and colloids in the Mackenzie River and Beaufort Sea. Mar. Chem. **26:** 371–378.

Trace Element/ Organic Cycling

Chapter 10

Trace Element Behavior in Gulf of Mexico Estuaries

Liang-Saw Wen, Alan Shiller, Peter H. Santschi, and Gary Gill

INTRODUCTION

Estuaries link rivers with the ocean. Because of the dynamic nature of the chemical, physical, geological, and biological processes at work in estuaries, they can be vigorous geochemical reactors and thereby affect the flux of materials from the continent to the ocean. Thus, understanding the geochemistry of estuaries is an important facet of constructing oceanic mass balances. Furthermore, because the global physiography of estuaries probably changes with changing sea level, understanding the factors affecting estuarine chemistry is vital to understanding changes in oceanic paleochemistry. Estuaries are also a useful place to study environmental chemistry because the strong physical-chemical gradients present, as well as the high biological productivity, result in the magnification of certain geochemical processes.

The study of trace elements is a particularly important aspect of estuarine chemistry. Trace elements, despite their low concentrations, can be toxic agents or, in some cases, required nutrients. Indeed, many elements can be both, depending on their concentration. The availability and toxicity of trace

Biogeochemistry of Gulf of Mexico Estuaries, Edited by Thomas S. Bianchi, Jonathan R. Pennock, and Robert R. Twilley.
ISBN 0-471-16174-8 © 1999 John Wiley & Sons, Inc.

elements are more than just a simple function of concentration. Physicochemical characteristics of the medium (agents, ionic strength, adsorbing surfaces, redox state, pH, etc.) can affect an element's interaction with the biota. Estuaries, because of their strong chemical gradients, are therefore regions where the behavior and interactions of trace elements are more complex and dynamic than in other aquatic environments such as lakes or oceans. Because of the dynamic nature of estuaries and the changing chemical interactions within them, trace elements can also be indicators of various environmental processes occurring within estuaries such as sedimentary input, biological uptake, desorption, and flocculation. In principle, trace element distributions can be used to indicate the magnitudes or time scales of these processes relative to mixing. Thus, it is not surprising that considerable efforts in the United States have been devoted to studying trace elements in the coastal and estuarine environment. However, little is known about trace elements in Gulf of Mexico estuaries compared to U.S. east and west coast estuaries (Cutter 1990). This chapter provides the first overview of trace element research in Gulf coast estuaries, with emphasis on trace elements in the water column, sediments, suspended matter, biota and elemental fluxes, and their controlling factors.

The northern coastline of the Gulf of Mexico extends in a 2300-km roughly semicircular arc from the Florida Keys on the east to the mouth of the Rio Grande on the west. Across this coastline into the Gulf of Mexico flows about 51% of the riverine discharge of the coterminous United States, and approximately 73% of this is carried by the Mississippi River (Ward 1980). Along the Gulf shoreline lie numerous indentations, embayments, and inlets, which collectively constitute perhaps one of the most valuable estuarine resource of the United States (Boykin 1972). Most of these estuaries exhibit a surprising degree of similarity in their general hydrographic characteristics. In general, Gulf estuaries are lagoonal embayments, with the exceptions of the Mississippi River, which is a drainage estuary, and the Brazos and Suwannee Rivers, which are classic river-mouth estuaries. Because a significant part of the estuary of the Mississippi River is on the shelf and the rest of the estuaries are shallow, speciation and transport of trace metals and other trace contaminants are unique as well (Shiller 1997). With regard to sediment transport characteristics and salinity gradients, four types of estuaries can be distinguished on the Gulf Coast (Fig. 10-1). However, unlike many other estuaries studied in the past, most estuaries of the Gulf of Mexico are sub-tropical systems, are typically shallow, and contain high concentrations of suspended particulate and dissolved organic matter (e.g., Keeney-Kennicutt and Presley 1985; Shiller and Boyle 1991; Morse et al. 1993; Shiller 1993a; Benoit et al. 1994; Wen et al. 1996, 1997a,). These features make them different from the typical east and west coast estuaries, which are generally deeper. Furthermore, these estuaries are typically surrounded by salt marshes, which are among the most sensitive sites for pollutant effects on ecosystems. At the same time, they are also effective biogeochemical reactors for filtering large amounts of pollutants from the water.

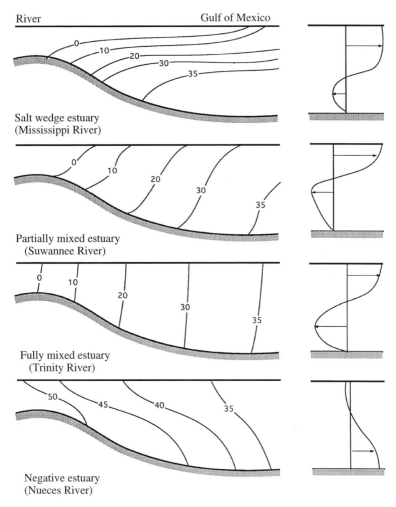

FIG. 10-1. *Schematics of types of estuaries found in the Gulf of Mexico.*

Estuarine Mixing Behavior

In an estuary, mixing occurs between natural waters of very different chemical composition and physicochemical properties, such as ionic strength, pH, and redox potential, which can have a dramatic impact on the transport, fate, and speciation of trace components in an estuary. For many minor and trace components, such as nutrients and many transition metals and metalloids, riverine concentrations are often considerably higher than those of the marine end member. Estuarine waters also receive large particulate inputs and can have significant suspended loads. Availability of reaction sites on suspended particles can undergo changes during estuarine mixing, and can markedly alter partitioning of trace elements between particulate and dissolved phases and hence the composition of estuarine sediments.

The reactivity (or estuarine mixing behavior) of a constituent is conventionally interpreted by plotting the concentration of that constituent against a conservative index of mixing, such as salinity (Officer 1979). In a *one-dimensional, two end-member, steady-state system,* a conservative constituent will plot linearly versus salinity (theoretical dilution line; Fig. 10-2). In contrast, for a non-conservative constituent, extrapolation of concentrations varying from high salinity to zero salinity yields an "effective" river concentration (C^*) that indicates the reactivity of a constituent within the estuary and can be used to calculate the total flux of the constituent to the ocean (Fig. 10-2). Specifically, when $C^* = C_0$ (the actual river concentration), the constituent is behaving conservatively; when $C^* < C_0$, there is removal of the constituent within the estuary; and when $C^* > C_0$, there is input of the constituent within the estuary. Fluxes to the ocean are estimated by multiplying the effective river concentration by the river water flux, R. The flux of a constituent entering the estuary from the river is $F_{riv} = RC_0$, the flux out is $F_{sea} = RC^*$, and the net internal input or removal flux is given by the difference: $F_{int} = R(C^* - C_0)$. This standard model of estuarine chemical distributions is a powerful tool when the underlying assumptions are satisfied.

The standard model has pitfalls when its assumptions are violated. Officer and Lynch (1981) demonstrated that variations in river concentrations on a

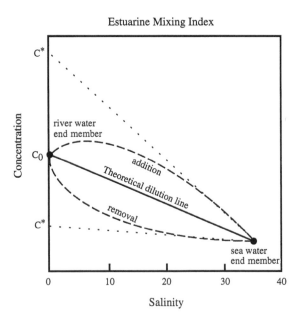

FIG. 10-2. Representation of the estuarine distribution and mixing behavior of a "dissolved" constituent under steady-state conditions.

time scale shorter than that of estuarine mixing can lead to curved mixing plots even when the constituent being plotted versus salinity is actually unreactive. Shiller and Boyle (1991), noting the temporal variability of certain trace elements in the Mississippi River (Shiller and Boyle 1987a), suggested that this might be a factor affecting the distribution of Cr in the Mississippi River outflow system. This particular estuarine system is especially prone to this type of interpretive problem because much of the mixing of fresh water with seawater occurs on the Louisiana Shelf, where brackish waters may reside for several months (Dinnel and Wiseman 1984). More recent work on the temporal variability of dissolved trace elements in the Mississippi River (Shiller 1997) provides an indication that the standard model must be applied cautiously to the interpretation of trace element distributions in this system.

Shiller (1996) discussed how recycling traps and upwelling are other processes that violate the assumptions of the standard model. In particular, it was pointed out that shelf-edge upwelling along the Louisiana-Texas Shelf could provide fluxes of elements such as Cd and Zn with magnitudes similar to or greater than those of their fluxes from the Mississippi and Atchafalaya Rivers. This adds a further potential complexity to the interpretation of trace element distributions in this estuarine system.

Processes in Estuaries That Regulate Trace Element Behavior

The dynamics of estuarine sediment and suspended particle cycling lead to particle fractionation processes through particle-selective re-suspension and settling, mudflat processes, and particle-particle interactions (i.e., coagulation by shear, Browning motion, and differential setting). These processes are driven by river flow, tidal energy, and storms, and therefore are spatially and temporally variable and tend to be accentuated in shallow environments (Bale et al. 1985, 1989; Dyer 1989; Eisma et al. 1990). The coupling of these particle fractionation processes to chemical transformations in the estuarine water column significantly impacts the distributions of most dissolved substances (Li 1981; Santschi 1988; Burton and Statham 1990; Wen et al. 1997b). Through these processes, a wide variety of trace metal ions become associated with particles in coastal marine environments. Trace metal uptake may result from (1) ion exchange or surface complexation; (2) co-precipitation with hydrous oxide coatings of iron and manganese; (3) incorporation into mineral lattices, cells of organisms, or fecal material; and (4) colloidal pumping, hydrophobic interactions of colloidally bound metals with the particle surface, and flocculation of colloidal organic and inorganic matter during mixing of river water and seawater. The transport, accumulation, and fate of chemically reactive trace metals in coastal marine environments are thus controlled by both particle and chemical dynamics (e.g., Price and Calvert 1973; Boyle et al. 1977; Buffle et al. 1990; Bilinski et al. 1991; Kornicker and Morse 1991; Santschi and Honeyman 1991; Wen et al. 1997b).

The complexation of metal ions with inorganic ligands has been evaluated theoretically in numerous studies on the basis of thermodynamic ion-association equilibrium models utilizing the stability constants for the major inorganic complexes (e.g., Turner et al. 1981; Millero 1985; Kester 1986; Kumar 1987; Altmann and Buffle 1988; Hering and Morel 1988, 1989; Bruno 1990). There has been some disagreement about the results of such studies, largely reflecting differences in the selection of complexes included in the models and uncertainties in the choice of stability constants.

These complexation calculations can give us, at best, only the inorganic metal speciation under equilibrium conditions. They do not take into account organic complexes and their slow exchange kinetics, which are generally affected by the dynamics of the physical and chemical processes found in estuaries. Complexation of trace elements with organic ligands often ameliorates trace metal behavior in the water column (Simkiss and Taylor 1989; Newman and Jagoe 1994). A number of studies have illustrated that many trace elements are strongly complexed by organic ligands (Huizenga and Kester 1983; Sunda and Ferguson 1983; Van den Berg et al. 1987; Coale and Bruland 1988), and many of these metal-organic complexes appear to be in the colloidal size range (Dai and Martin 1995; Guentzel et al. 1996; Powell et al. 1996; Stordal et al. 1996; Wen et al. 1996, 1997a, 1997b).

There is increasing awareness of the importance of colloids, which are traditionally defined as microparticles and macromolecules that reside in the 1-nm to 1-μm size range, in understanding the geochemistry of trace elements in river and estuarine waters (e.g., Benoit et al. 1994; Wen et al. 1994, 1996, 1997a, 1997b; Dai and Martin 1995; Martin et al. 1995; Guentzel et al. 1996; Powell et al. 1996; Stordal et al. 1996). Colloids have an important mediating effect on metal partitioning between particles and solution (e.g., Moran and Moore 1989; Sigleo and Means 1990; Whitehouse et al. 1990; Baskaran and Santschi 1993; Benoit et al. 1994; Stordal et al. 1996, 1997; Wen et al. 1997b). Furthermore, colloidally complexed metals have different physicochemical and biological properties and pathways than truly dissolved ionic forms (Sigleo and Means 1990; Buffle and Leppard 1995a, 1995b; Dai and Martin 1995; Martin et al. 1995; Stordal et al. 1996; Wen et al. 1996, 1997b). This topic is addressed further in Chapter 11. A schematic description of trace metal speciation in an estuarine environment is presented in Fig. 10-3 (Wen 1996; Santschi et al. 1997).

TRACE ELEMENTS IN THE WATER COLUMN

Estuarine Distributions—General Features

Prior to a discussion of trace element distributions in estuarine waters, it is important to specify what is meant by the terms "dissolved" and "particulate," because the phase and chemical speciation of many elements in aquatic sys-

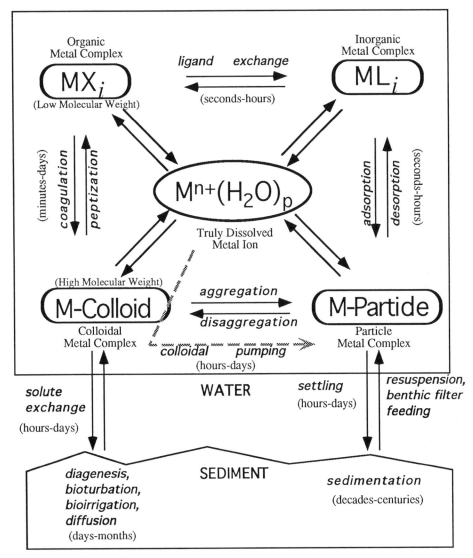

FIG. 10-3. *Schematic of trace element speciation and dynamics of cycling in estuarine environment (modified from Wen 1996 and Santschi et al. submitted).*

tems, such as estuaries, are not well understood. Components may occur in true solution (as single ions, ion pairs, and complexes) or as biopolymers having a wide range of molecular weights extending up to those that settle under gravity (Gustafsson and Gschwend 1997). Given the continuous size spectrum of environmental particles, the distinction between what is dissolved and what is particulate is not always clear. However, it is conventional to define dissolved material operationally as that fraction of the total material in the water that will pass through a membrane filter having a nominal pore size of 0.45 μm. Colloidal-sized particles are, therefore, included in the dissolved fraction, although ultrafilters of nominal pore size or molecular weight cutoffs are available for their isolation. The size distribution of typical components in an estuarine or aquatic system is shown in Fig. 10-4.

Dissolved trace element concentrations as a function of salinity have been reported for a variety of northern Gulf of Mexico estuaries and have generally been interpreted within the context of the standard mixing model discussed previously. The method has been widely used to interpret trace element distributions in a variety of Gulf of Mexico estuaries including the Mississippi (Hannor and Chan 1977—Ba; Shiller and Boyle 1991—Cd, Cu, Cr, Fe, Mo, Ni, V, Zn), Ochlockonee (Froelich et al. 1985—Ge; Byrd and Andreae 1986—Sn; Guentzel et al. 1996—Hg, MeHg; Powell et al. 1996—Cd, Cu, Fe, Mn, Ni), Suwannee (Byrd and Andreae, 1986—Sn), Brazos (Keeney-Kennicutt and Presley, 1985—Cu, Fe, Mn, Pb), and Trinity (Stordal, 1996—As, Hg, Sb, Sn; Wen et al. 1996, 1997a, 1997b—Ag, Cu, Cd, Co, Fe, Ni, Pb, Zn) Rivers. Representative behaviors of dissolved trace elements in estuaries inferred from constituent-salinity diagrams are presented in Table 10-1.

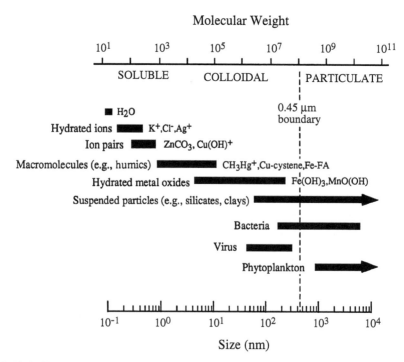

FIG. 10-4. *Size spectrum of soluble, colloidal, and particulate species found in estuarine environments.*

TABLE 10-1. Field Observations of "Dissolved" Trace Elements in Estuaries of the Gulf of Mexico

Estuary	Element	Concentration	Behavior	References
Colorado	Ag	1~278 pM	A	Benoit et al. (1994)
	Cu	9~21 nM	A&R	Benoit et al. (1994)
	Pb	77~608	R	Benoit et al. (1994)
	Zn	6~29 nM	A	Benoit et al. (1994)
Lavaca	Ag	16~153 pM	A	Benoit et al. (1994)
	Cu	6~27 nM	R	Benoit et al. (1994)
	Pb	48~651 pM	R	Benoit et al. (1994)
	Zn	6~72 nM	R&R	Benoit et al. (1994)
Mississippi	Cd	8~302 pM	A	Shiller (1993a,b)
	Cr	0.95~2.61 nM	C	Shiller and Boyle (1987a,b)
	Cu	1.19~26.8 nM	C	Shiller and Boyle (1987a,b)
	Ge	5~70 pM	A	Hannor and Chan (1977a,b)
	M	9~104 nM	C	Shiller and Boyle (1987a,b)
	Ni	2~30 pM	C	Shiller and Boyle (1987a,b)
	V	1.0~38.4 nM	A	Shiller (1991)
Nueces	Ag	1~77 pM	A	Benoit et al. (1994)
	As	8~80 nM	A	Stordal (1996)
	Cu	4~31 nM	A	Benoit et al. (1994)
	Hg	1.2~6.9 pM	C	Stordal (1995)
	Pb	154~960 pM	C	Benoit et al. (1994)
	Sb	2~4 nM	A	Stordal (1996)
	Zn	10~89 nM	A	Benoit et al. (1994)
Ochlokonee	Cd	37~54 pM	A	Powell et al. (1996)
	Cu	2.0~5.4 nM	A	Powell et al. (1996)
	Fe	0.4~7 μM	R	Powell et al. (1996)
	Ge	2~92 pM	A	Froelich et al. (1985)
	Hg	2~30 pM	R	Guentzel et al. (1996)
	Ni	2.6~5.5 pM	A	Powell et al. (1996)
	Sn	9~16 pM	C	Byrd and Andreae (1986)
Peace	As	14~19 nM	R	Froelich et al. (1985)
	Ba	30~206 nM	A	Froelich et al. (1985)
	Ge	6~92 pM	R	Froelich et al. (1985)
	Sn	8~115 pM	A	Froelich et al. (1985)
Sabine	Ag	22~134 pM	A&R	Benoit et al. (1994)
	Cu	6~33 nM	A	Benoit et al. (1994)
	Hg	2.4~3.6 pM	A	Stordal et al. (1995)
	Pb	106~1472 pM	R	Benoit et al. (1994)
	Zn	6~32 nM	A	Benoit et al. (1994)
San Antonio	Ag	1~76 pM	A	Benoit et al. (1994)
	Cu	7~50 nM	A	Benoit et al. (1994)
	Pb	96~1206 pM	A&R	Benoit et al. (1994)
	Zn	7~275 nM	C&R	Benoit et al. (1994)

TABLE 10-1. (*Continued*)

Estuary	Element	Concentration	Behavior	References
Suwanne	Sn	5~10 pM	C	Byrd and Andreae (1996)
Trinity	Ag	3~115 pM	A	Wen et al. (in press)
	As	7~30 nM	A	Stordal (1996)
	Cd	31~166 pM	A	Wen et al. (in press)
	Co	1~10 nM	A	Wen et al. (in press)
	Cu	5~30 nM	C&A	Wen et al. (in press)
	Fe	0.05~1 μM	R	Wen et al. (in press)
	Hg	1.0~6.8 pM	A	Stordal et al. (in press)
	Ni	3~30 nM	C&A	Wen et al. (in press)
	Pb	50~770 pM	R	Wen et al. (in press)
	Sb	0.8~6.2 nM	A	Stordal (1996)
	Se	0.2~2.6 nM	A	Stordal (1996)
	Zn	4.6~68.8 nM	A	Wen et al. (in press)

C = conservative, R = non-conservative removal, A = non-conservative addition.

The information given in Table 10-1 is largely based on information from limited surveys in which the temporal intra-estuarine variability was not always apparent. Intra-estuarine variations in trace metal end-member concentrations and axial distributions may arise through the seasonal cycling of trace elements, mediated by temperature and biological activity. The latter is of particular significance to the distributions of methylated Hg and Sn species generated by estuarine organisms (Byrd and Andreae 1986; Guentzel et al. 1996). Short-term fluctuations result mainly from tide- and wind-induced processes such as infusion from pore waters. In a hydrological sense, increasing runoff acts to dilute a steady source. Additional chemical consequences are less predictable because the reaction-controlling variables of pH and suspended particulate matter (SPM) concentration vary non-linearly with discharge (e.g., Shiller and Boyle 1987a). Moreover, the discharge-dependent flushing characteristics may impose time constraints on metal-ligand interactions in fresh waters, and the subsequent sorptive behavior of metals in the turbid mixing zone can become kinetically controlled (e.g., Nyffeler et al. 1984, 1986).

Biogeochemical and hydrodynamic processes can vary substantially among estuaries; thus, it is useful to examine trace element distributions estuary by estuary while comparing intra-estuary behavior and processes. Axial distributions of a given trace metal exhibit considerable inter-estuarine variability, indicating sensitivity to the unique biogeochemical and hydrodynamical conditions of a system. In the absence of complementary laboratory studies, mechanistic interpretations of trace metal behavior rely on knowledge of these conditions. This may be exemplified by contrasting the chemical behavior of trace elements (Cu, Ni, Cd) in two well-characterized but distinctly dissimilar estuaries: the Trinity and the Mississippi (Fig. 10-5) (Shiller and Boyle 1991; Wen et al. in press).

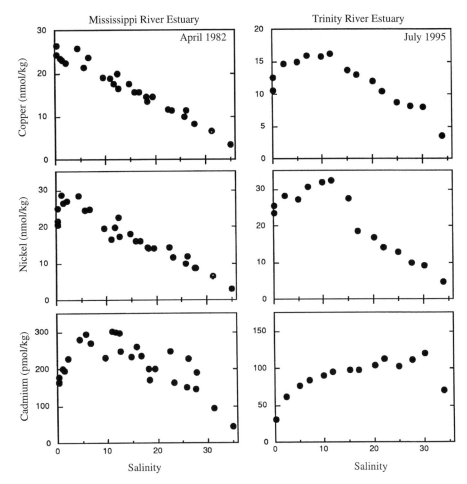

FIG. 10-5. *Dissolved trace element (Cu, Ni, Cd) distributions in the Mississippi River (data from Shiller 1991) and Trinity River estuaries (data from Wen et al. in review).*

Mississippi and Atchafalaya Estuaries. The largest estuarine system in the Gulf of Mexico is that of the Mississippi River. Approximately 70% of the discharge of this river enters the Gulf through the tributaries of the birdfoot delta, mixing with Gulf waters near the shelf break. In contrast, the other 30% of the water flows to the Gulf through the Atchafalaya River, which enters the Gulf in a shallow bay and a broad, shallow shelf. The distribution of flow is regulated by the U.S. Army Corps of Engineers. Shiller and Boyle (1991) described the distributions of eight trace elements in the delta outflow during a period of high discharge. Mixing experiments were used to help resolve controlling processes. Dissolved Cu, Ni, and Mo behaved conservatively. The Cr-salinity relationship suggested removal of this element during mixing; however, the lack of Cr reactivity in mixing experiments, as well as its large temporal variability in the river (Shiller and Boyle 1987a), suggested non-linear, conservative behavior (Loder and Reichard 1981; Officer and Lynch 1981). Iron and Zn were also observed to be generally unreactive in this estuary, probably because of the alkaline nature of the Mississippi River, which acts to lower Fe concentrations, combined with the high suspended load, which may buffer dissolved Fe concentrations. Alternatively, both elements might be complexed with organic matter, thus rendering them less reactive. The V distribution showed substantial removal, even to the point of the estuary's being a net sink for V. Shiller and Boyle (1987b, 1991) attributed this to biological uptake of V. However, recent work (Shiller et al. in press) shows a correlation between dissolved V and O_2 in bottom waters of this region, suggesting that V depletion may result from uptake into reducing shelf sediments. The estuarine Cd distribution showed evidence of desorption from the suspended load, as has been observed for this element in many other estuaries. Hannor and Chan (1977) observed Ba to undergo a similar desorptive release in this region.

Shiller (1993b) compared Cu, Ni, and Cd distributions from the Mississippi River delta outflow with distributions from the Atchafalaya outflow. Despite the chemical, biological, and climatic similarities of these two parts of the Mississippi River outflow region, the physiographic differences between the two environments (shelf break versus bay and broad, shallow shelf) resulted in observable differences in the estuarine chemical distributions between the two sub-regions. In particular, significant removal of Ni and Cd was observed in the Atchafalaya outflow but not in the delta outflow. The fate of the fluvially suspended load, rates of mixing, and extent of nutrient recycling were identified as factors connecting the physiographic differences between the sub-regions with the differences in chemical distribution.

Ochlockonee River Estuary. The estuary of the Ochlockonee River in the Florida panhandle has been utilized as a site for the study of the behavior of several trace elements in a comparatively pristine coastal plain environment. Froelich et al. (1985) compared the distributions of dissolved Ge and Si in this system. Their study, composed of 11 monthly transects through the estuary, is one of the few studies of seasonal variability of a trace element in an estuary. Germanium and Si transects were very similar, with net removal by diatoms in the upper bay during winter and spring and net input of these elements from bottom opal dissolution during summer and fall.

Byrd and Andreae (1986) found the Sn distribution in the Ochlockonee estuary to be conservative, within the scatter of the data. They also found a conservative Sn distribution in the nearby Suwannee River estuary, another comparatively pristine coastal plain estuary in the Florida panhandle. In general, their work suggests that dissolved Sn is largely unreactive in uncontaminated estuaries.

Guentzel et al. (1996) found that Hg displayed non-conservative net removal behavior during estuarine mixing, with concentrations decreasing as salinity increased. Colloidal Hg (>1 kDa) represented 35% to ~87% of the total dissolved Hg within the estuary. Their equilibrium Hg speciation modeling supports the hypothesis that colloidal Hg is bound by thiol-type functional groups associated with colloidal organic matter.

Powell et al. (1996) found that total dissolved Fe and Mn behaved nonconservatively in the Ochlockonee estuary. Removal from the water column resulted principally from loss of the high molecular weight (HMW, >10 kDa) colloidal fraction, with Mn removed at a lower salinity than Fe. For Ni, Cu, and Cd, elements that had relatively conservative estuarine behavior, the HMW fraction is very important in the river, but these elements are quickly converted from HMW to low molecular weight (LMW, <1 kDa) species at increasing salinity.

Charlotte Harbor/Peace River Estuary. Froelich et al. (1985) examined trace element distributions in the Charlotte Harbor (Peace River) estuary in south Florida. Drainage from phosphate deposits, mining activities, and agriculture results in the Peace River's having ~90 μM PO_4 and ~50 μM NO_3. As a consequence, large phytoplankton blooms typically occur near the river mouth. Inorganic Ge was removed in the bloom with diatom productivity, whereas methylated Ge forms behaved conservatively. Inorganic As was slightly depleted in the bloom, with concomitant production of monomethyl As. The bloom appeared to affect the Sn distribution, showing a low-salinity input possibly associated with an input of dissolved organic matter. However, above 15 salinity, the Sn concentration decreased almost linearly to the seawater end member. Barium showed a large input in the mid-salinity range. In other estuaries, Ba showed desorptive release at low salinity (e.g., Hannor and Chan 1977), and the higher salinity and more diffuse nature of the Ba input in Charlotte Harbor were suggested to result from slow release of Ba from clays deposited in the bay by catastrophic phosphate slime spills into the Peace River.

Brazos River Estuary. Keeney-Kennicutt and Presley (1985) examined the distributions of dissolved Cu, Fe, Mn, and Pb in the Brazos River Estuary in Texas. This river has a high dissolved salt content due to outcrops of halite, gypsum, and limestone in the watershed; it also has a high suspended load. Estuarine distributions were examined during winter, spring, and fall transects through the salt wedge. In general, trace element distributions were observed to have maxima in the mid-salinity range. These apparent metal inputs were attributed to diagenetic re-mobilization from reducing sediments within the estuary.

Trinity River Estuary. In the Trinity River Estuary, Wen et al. (1997a, in press) found both conservative and non-conservative estuarine mixing behavior for several trace elements investigated during different study periods. During July 1995, non-conservative estuarine mixing behavior was observed for Ag, Cd, Co, Cu, Ni, and Zn, resulting from a large mid-salinity source. At the same time, Fe and Pb showed net estuarine removal behavior. The Cd distribution also showed evidence of desorption from the suspended load, as has been observed for Cd in many other estuaries. The mid-estuarine sources resulted either from desorption from resuspended particles or benthic fluxes of pore water constituents aided by benthic macrofauna and wind-driven turbulence (Santschi, 1995). A mid-estuarine source was also observed by Benoit et al. (1994) for Cu and Pb but not for Ag and Zn.

Wen et al. (1997a, in press) found significant colloidal (>1 kDa) fractions for Ag (43 ± 15%), Cd (45 ± 9%), Cu (55 ± 4%), Co (19 ± 6%), Ni (36 ± 6%), Pb (64 ± 9%), Zn (91 ± 5%), and Fe (79 ± 11%). With the occasional exception of Fe and Pb, colloidal element concentrations correlated significantly with colloidal organic carbon concentrations (>1 kDa), suggesting that colloidal elements resulted from metal-organic complexation. Iron was preferentially associated with HMW (>10 kDa) colloids, while Cu, Ni, and Pb were associated mostly with LMW (1–10 kDa) colloidal material. Molecular weight or size distribution studies of organic carbon and trace elements revealed evidence of two competing processes: (1) coagulation and colloidal pumping and (2) production of LMW species, possibly mediated through photochemical or microbiological reactions.

Stordal et al. (1996) examined the estuarine mixing behavior of filter-passing (0.45 μm) Hg in the Trinity River estuary during different seasons and found it conservative, within the scatter of the data. The colloidal Hg fraction ranged from 12 to 93% of the total dissolved pool. They also found that colloidal Hg covaried with colloidal organic carbon concentration, indicating that a major portion of the Hg was associated with submicron-sized particles and organic macromolecules. In a study of metalloid elements (Stordal 1996), concentrations of filter-passing (<0.45 μm) and colloidal As, Sb, and Se were measured in surface waters from two Texas estuaries, Galveston Bay and Corpus Christi Bay. Filter-passing concentrations of As, Sb, and Se ranged from 7.3 to 65 nM, 0.82 to 6.2 nM, and 0.18 to 2.6 nM, respectively. For all three elements, non-conservative mixing was observed within both estuaries. The colloidal As and Sb were generally less than 5% of the filter-passing concentration and were independent of salinity, indicating that the majority of these species are truly dissolved (<1 kDa). In contrast, colloidal Se was approximately 35% of the filter-passing Se for most samples, indicating that a significant portion of the dissolved Se was associated with colloids.

Suspended Particulate Trace Elements

Knowledge of the mixing ratios of marine to fluvial material in SPM is essential for an understanding of the transport processes, as well as for constructing a flux balance of sediments and particulate metals in the estuarine environment.

Axial distributions of trace metals in surface sediments and SPM generally reflect the mixing of fluvial material of high metal content with marine material of low metal content and a high autochthonous organic component. Although particulate matter is instrumental for biogeochemical phase transformations, these interactions are often thought to be inconsequential for the bulk solid composition because of the magnitude of the sedimentary metal reservoir (e.g., Santschi et al. 1984, 1987, 1990; Li et al. 1989). However, the large-scale phase redistributions of Fe and Mn can alter the composition of non-detrital components (Wen and Santschi, submitted), as exemplified by mid-estuarine maxima of acid-leachable particulate Fe and Mn in the Trinity River Estuary caused by precipitation of the respective oxidized species.

Using a selective leaching method (Landing and Lewis 1991) on suspended particles, Wen and Santschi (submitted) found that particle-reactive (strongly hydrolyzed) elements (Mn, Cd, Pb) in Galveston Bay were mostly found in the surface-adsorbed fraction, while "nutrient-type" elements (Cu, Ni, Zn) were found equally associated with surface-adsorbed and Fe-Mn carrier phases. The majority of Al and Fe and a significant fraction of other elements (Cu, Ni, Pb, Zn) were in the refractory phase, indicating association with silicate phases or iron sulfides (Fig. 10-6). Strong correlations between Fe and several elements (Al, Cu, Ni, Pb, Zn) in suspended particles were in agreement with the view that these elements were closely associated with the formation of Fe oxyhydroxide/sulfide phases. It is clear that during these redistributions of Fe and Mn phases, trace elements such as Ag, Cd, Ni, Pb, and Zn also undergo a similar change in partitioning patterns (Wen et al. 1997b; Wen and Santschi, submitted).

Warnken (1998) found elemental distribution patterns in surface sediments similar to those observed for trace elements in suspended particles by Wen and Santschi (submitted) in the Trinity River Estuary. This suggests that the estuarine distributions of suspended particulate trace elements can, at times, be controlled solely by re-suspension events. Trefry et al. (1986) also found identical trace element concentrations in SPM of the Mississippi River and in the delta's sediments.

New insights into colloidal pumping, that is, the movement of colloidally bound metals into particulate phases, as well as into other interactions during estuarine mixing, have been obtained recently from studies using radioactive tracers. Wen et al. (1997b) and Stordal et al. (1997) found that suspended particles sequester colloidally bound trace metals and aggregate from colloids into larger suspended particles in a matter of hours to days in estuarine waters. These experimental results suggest that interactions between surface-reactive fractions of the colloidal material and particles can play a crucial role in the solid-solution partitioning of many trace elements.

Correlations of Trace Metals with Nutrients and Organic Matter

It is well known that complexation by organics, as well as uptake by biota, can exert important influences on trace element distributions in natural waters.

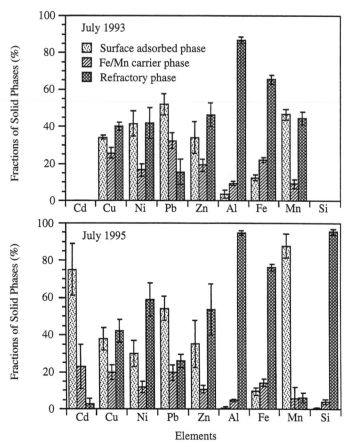

FIG. 10-6. *Particulate trace element fractionation in suspended particles in the Trinity River estuary (data from Wen and Santschi, 1997a).*

Such relationships have been suggested in estuarine environments based on the co-variance of a trace element distribution with nutrients or dissolved organic carbon (DOC) distributions. For example, Cd, Zn, Fe, Cu, and Ni are intimately linked with nutrient cycles in the ocean, and many of these elements also exist predominantly as organic complexes (e.g. Martin and Gordon 1988; Nguyen et al. 1988; Bruland 1989, 1992; Coale and Bruland 1988; Bruland et al. 1991; Rue and Bruland 1995).

Significant correlations between Cu and Ni and nutrient concentrations (phosphate and silicate) were found in studies of Mississippi River and Trinity River Estuaries (Fig. 10-7) (Shiller and Boyle 1991; Shiller 1993a; Wen et al. in press). These relationships may have resulted from biological uptake of trace elements and/or from strong macromolecular ligand complexes related by biota, thus producing organic forms of colloidal Cu and Ni. However, this

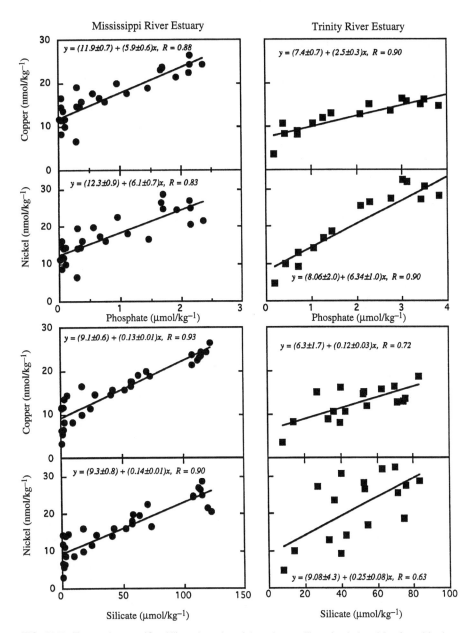

FIG. 10-7. Trace element (Cu, Ni) and nutrient (phosphate, silicate) relationships found in the Mississippi River and Trinity River estuaries (data from Shiller 1991; Wen et al. in press).

process does not appear to have had a major influence on the estuarine mixing behavior of trace metals in the Trinity River estuary, but rather provided additional colloidal material from planktonic sources.

In the delta outflow plume of the Mississippi River, correspondence of the V distribution with that of phosphate was taken as evidence of biological uptake of V (Shiller and Boyle 1987b, 1991). Shiller and Boyle (1991) also used phosphate removal in the Mississippi plume along with oceanic metal/phosphate removal ratios to predict the removal of various trace elements from plume waters. They predicted that Cd, Zn, and Fe should be completely removed by biological uptake in these waters. That such removal of these elements was not observed was taken as evidence of "buffering" of trace element concentrations by the suspended load.

In an examination of chemical distributions in the Mississippi/Atchafalaya outflow system, Shiller (1993a) used the contrasting nutrient distributions of the delta and shelf regions to help explain differences in trace element behaviors. Specifically, it was concluded that the apparently greater recycling of nutrients in the shelf region, combined with differing nutrient and metal recycling efficiencies, resulted in more significant trace element removal in this region than in the delta outflow. Furthermore, it was suggested that lower particle loads in shelf waters than in delta waters would lessen the buffering effect of adsorbed metals.

Despite the utility of these apparent nutrient-trace element relationships, such comparisons should be made cautiously. The dynamics of estuaries can result in correlations of chemical parameters that merely reflect mixing processes rather than biogeochemistry. For instance, although Shiller and Boyle (1987b, 1991) reported that V was removed biologically based on the similarity of the V and phosphate distributions in the Mississippi River outflow, newer data suggest another possible explanation. Shiller et al. (in press) show a positive correlation between dissolved V and O_2 in Louisiana Shelf bottom waters, indicating that V is being removed into reducing bottom sediments. Because salty shelf bottom waters mix with the fresher shelf surface waters, the apparent surface water V removal might simply reflect mixing with the V-depleted bottom waters, whereas surface water phosphate removal could still be biological. This type of multi-end member upwelling artifact is one of the pitfalls of viewing estuarine mixing one-dimensionally (Shiller 1996).

Because colloids offer a much greater number of surface sites per unit mass than SPM, and because they consist mostly of organic macromolecules (Santschi et al. 1995), it is conceivable that colloidal organic matter (COM) is more important than SPM in the transport and biogeochemical cycling of trace elements in estuarine and other aquatic systems (Honeyman and Santschi 1988, 1989; Buffle and Leppard 1995a, 1995b; Santschi et al. 1997). As those studies showed, colloidal organic carbon can represent a large fraction of the total DOC in the water column. It is therefore not surprising to find that there is a very good correlation between colloidal trace metals (Hg, Ag, Cd, Co, Fe, Zn, Cu, Ni, Pb) and colloidal organic carbon (Guentzel et al. 1996; Powell et al. 1996; Stordal et al. 1996; Wen et al. 1997a, in press). Hence, COM appears to play a major role in the transfer of the truly dissolved fraction to particles and in further transport of some trace metals in Gulf coast estuaries (see Chapter 11 this volume).

Particle-Water Distribution Coefficients

Solid-solution interactions are often interpreted and quantified in terms of a conditional particle-water partition coefficient, K_d (e.g., O'Connor and Connolly 1980; Olsen et al. 1982; Li et al. 1984a, 1984b; Nyffeler et al. 1984; Buchholtz et al. 1986; Morris 1986; Honeyman and Santschi 1989), defined as

$$K_d = M_p/M_s$$

where M_p (w/w) and M_s (w/v) are the metal concentrations in the particulate and solution phases, respectively. Such ratios have proven to be conceptually useful in elucidating elemental removal mechanisms and residence times in the coastal and open ocean (e.g., Cherry et al. 1978; Whitfield and Turner 1979, 1982; Fisher 1986; Santschi et al. 1987; Honeyman and Santschi 1989; Li et al. 1989). However, determination of K_d values in estuarine and coastal waters (e.g., Valenta et al. 1986; Santschi et al. 1987; Balls 1989; Li et al. 1989; Turner et al. 1993; Benoit et al. 1994) has been criticized for failing to discriminate between different solid phase associations, particularly the solid component available for exchange processes between the carrier phase and solution (Rapin 1986; Bourg 1987; Martin et al. 1987; Ciceri et al. 1988). Recent work on the solid phase speciations of suspended marine particles that employed ultra-clean techniques has successfully elucidated the reactivity of trace metals on the suspended solid phases (Landing and Lewis 1991; Sholkovitz et al. 1994; Wen et al. 1997a, 1997b).

Benoit et al. (1994) found that SPM concentration was the only variable related to systematic variations in partitioning of trace metals and Fe and Al concentrations between filter-passing (<0.45 μm) and filter-retained (>0.45 μm) fractions across the salinity gradient in six Texas estuaries. Inverse relationships were observed for several metals between K_d and SPM concentrations. This inverse dependence can be explained by the "particle concentration effect," which is caused by the presence of colloidal matter in the filtrate fraction. Wen et al. (in press) found that partition coefficients between colloids and true solutions in the Trinity River Estuary were considerably higher than those between particles and true solution for all trace metals except Fe.

TRACE ELEMENTS IN SEDIMENTS

Trace elements enter estuaries primarily through riverine, atmospheric, or anthropogenic inputs (De Groot et al. 1976; Kennish 1992; Windom 1992;

Long et al. 1996). Historically, the riverine transport of trace elements originating from the weathering of natural rocks and mineralized deposits has been the major source of trace elements. Anthropogenic inputs of trace elements often exceed those of the natural inputs (Summers et al. 1996). Trace element concentrations in marine and estuarine sediments are not only determined by elemental inputs but are also affected by reactions at particle surfaces that influence the quantity of element adsorbed, reduction/oxidation reactions, and differences in solution chemistry. Moreover, estuaries contain a greater density of living organisms than any other part of the sea, and their chemistries are correspondingly more influenced by these organisms than in other marine zones. It is well known that a substantial portion of products of plant growth enters estuarine sediments, where they concentrate in fine-grained sediments, undergo degradation, and release dissolved inorganic and organic molecules. The fate of elements, and hence the capacity of an estuary to receive them, is usually related to the abundance of fine-grained, organic-rich sediments that may be concentrated, re-suspended, or distributed by physical processes.

Studies of trace metals in sediments of the Gulf Coast are far more abundant than for any other environmental reservoir. Perhaps the most comprehensive study of metals in sediments from all along the Gulf coast is that of Summers et al. (1996). They examined sediment metal concentrations from 497 sites within 140 estuaries over a 3-yr period. Metal concentrations were normalized to Al to help distinguish anthropogenic from natural metal concentrations. They found strong relationships between Al and Cr, Cu, Pb, Ni, and Zn; moderate relationships between Al and As and Ag; and weak but significant relationships between Al and Hg and Cd. They concluded that about 39% of the sites located near population centers, industrial discharge sites, or military bases were contaminated by at least one metal. At other sites, contamination was believed to result from non-point sources.

Considerable effort has been put forth in recent years to reduce or eliminate point discharges of pollutants such as trace metals and petroleum hydrocarbons into Texas and other Gulf coast estuaries. This effort has resulted in considerable improvements in the water quality in Galveston Bay estuary (The Galveston Bay Plan 1995). Despite this effort, considerable harmful impacts on water quality and toxicity potentially exist from historical discharges that are stored in estuarine sediments (Valette-Silver et al. 1993). Episodic sediment re-suspension events (e.g., storms, shrimp trawling, or dredging) re-introduce sediments and associated pore waters from depth into the water column. Hence, sediments containing previous discharge releases potentially represent an infinite source of contamination. An excellent example of this is given by the recent work of Santschi et al. (submitted). Based on radiochemical assessments of recovery rates of Hg in surface sediments at a historically contaminated site in Lavaca Bay, Texas, they concluded that sediment burial is fast enough to limit Hg re-mobilization from contaminated sediments.

Distributions as a Function of Depth

Warnken (1998) determined trace element distributions and their partitioning into different carrier phases in the sediments from the Trinity River estuary (Fig. 10-8). All elements showed decreasing trends with depth, with no visible

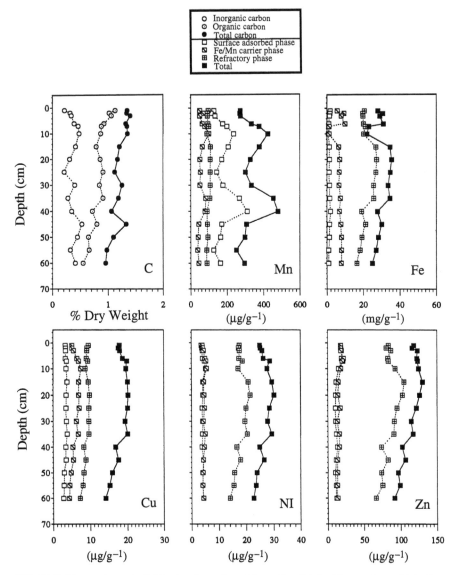

FIG. 10-8. *Trace element distribution in sediment from the Trinity River estuary (data from Warnken, 1998).*

sub-surface maxima (recent anthropogenic inputs) of trace elements, and most of the trace elements resided in the refractory phase, except for Mn. In a study of Calcasieu River/Lake (Louisiana), Mueller et al. (1989) found that Hg in the sediment reached a maxima at ~10–15 cm (between 1960 and 1976), after which there was a steady decrease. In a similar study from Lavaca Bay, Santschi et al. (submitted) found sub-surface Hg peaks in sediments at 5–60 cm. This anthropogenic signal was attributed to point sources from the local chloralkali industry, which is known to discharge Hg into the environment.

Estuarine sediments can pose a potential threat to the surrounding ecosystem from re-mobilization of historically deposited and now buried contaminants. For example, radiochemical assessments of recovery rates of Hg in surface sediments at a historically contaminated site in Lavaca Bay, Texas (e.g., Santschi et al. submitted) indicate that vertical mixing rates are low compared to burial rates, suggesting relatively fast recovery rates. However, lateral movement of surface sediments is sufficiently fast to have caused elevated Hg concentrations in open bay area and in neighboring wetlands, the most sensitive parts of the estuarine ecosystem, the breeding and feeding grounds of many of the most sensitive organisms.

Normalization of Trace Element Concentrations

Estuarine sediments are generally poorly sorted mixtures of sand, silt, and clay, and exhibit major variations in mean grain sizes, total organic carbon (TOC), and elemental contents. Variation in these parameters can result in concentrations in estuarine sediments varying by a factor of ~3. To understand trace element concentrations in sediments as a function of variable composition factors, trace element concentrations are often "normalized" to a carrier phase, such as Al, Fe, Li, TOC, or grain size. The choice of an appropriate sediment constituent to be used for normalization, however, is not straightforward.

Because trace elements tend to concentrate both within and on the surface of finer-grained sediments (Horowitz 1991), coarser-grained materials and carbonates often act only as diluents of trace element concentrations in bulk sediments. Such effects can be reduced by applying grain size corrections. One commonly used method is to analyze only the grain size fraction of <63 μm (e.g., Morse et al. 1993). However, this method requires additional quantitative separation and careful interpretation because concentrations in the finer fraction may not reflect the concentrations in the total sediments. In areas where there is significant down-core and lateral variation in grain size distribution, this approach can result in erroneous values (Salomons and Forstner 1984; Ravichandran et al. 1995a, 1995b).

Elemental ratios have been used to study sediment transport processes and to deduce sources of particles in river, estuarine, and ocean water. Aluminum has often been used to normalize trace element concentrations because of its high natural concentration and minimal anthropogenic contamination and because it is a structural element of clays (Duinker 1983; Windom et al. 1988, 1989). Thus, the metal/Al ratio can be used to indicate the importance of pollution sources (Windom et al. 1988, 1989; Hanson et al. 1993; Daskalakis and O'Connor 1995; Summers et al. 1996) and the relative importance of aluminum-silicates compared to other minerals and organic matter as surfaces for metal complexation (Duinker 1983).

In two comprehensive studies, Windom et al. (1989) and Summers et al. (1996) analyzed a wide range of sediment samples from estuarine and coastal marine Gulf areas. Although these sediments were compositionally diverse, trace elements (As, Co, Cr, Cu, Fe, Pb, Mn, Ni, and Zn) co-varied significantly with Al, suggesting that natural aluminum-silicates minerals are the dominant natural metal-bearing phases. Using this normalization method, Alexander et al. (1993) generated historical profiles of metal accumulation for the St. Johns River (Florida) and Hillsborough Bay (Florida) Estuaries. They found that Pb enrichment has decreased since the mid-1970s because of reduced use of leaded gasoline. Trefry et al. (1986) found similar results in Mississippi Delta sediments. Windom et al. (1989) also found a correlation of trace elements with TOC, a weaker correlation than with Al for all trace elements except Cd. Use of TOC for normalizing trace elements is based on the assumption that it is a matrix on particle surfaces for complex formation. One weakness of TOC as a normalizing parameter, however, is that TOC is subject to considerable augmentation by anthropogenic inputs.

In a study of Galveston Bay and Baffin Bay (Texas) sediments, Morse et al. (1993) normalized trace element concentrations (1 M HCl extractable + pyrite metal) to total reactive iron. Normalization to total reactive Fe concentrations was done to determine if anomalous high total reactive trace metal concentrations were present in any of the sediments. Such anomalies may be indicative of contamination. However, Fe is not a matrix element, like Al, but is often present as surface coatings. Its suitability as a normalizing parameter derives from the fact that it directly associates with trace elements, its natural concentrations in sediments are more uniform than those of Al (Goldberg et al. 1979; Trefry et al. 1986), and it is not anthropogenically impacted (Paulson et al. 1989).

Trace Element Accumulation Rates

Radiochemical approaches are widely used to determine sedimentation and particle re-working rates in sediments. When coupled to numerical transport models, this approach provides a powerful tool for assessing the potential for present and future particle-related vertical and horizontal re-distribution of estuarine sediment contaminants such as Hg (Santschi et al. submitted). The most commonly used methods of dating and estimating sedimentation and accumulation rates of recent near-shore sediments are based on the concentration gradients of natural radionuclides such as ^{210}Pb (<150 yr) and on the sediment depth of bomb fallout radioisotopes such as ^{137}Cs and ^{239,240}Pu (Santschi et al. 1984). Using ^{210}Pb in shallow environments is complicated by the post-depositional mixing of particles by physical and biological processes. Hence, a second particle-reactive tracer such as ^{239,240}Pu is often used in conjunction with ^{210}Pb to deconvolute mixing from sedimentation. Plutonium is an excellent analogue for studies of trace element accumulation rate in estuarine environments because it is particle reactive, is effectively retained in sediments (^{137}Cs is more soluble), and has a well-defined input function (Santschi et al. 1980, 1983). Using radiochemical approaches, Ravichandran et al. (1995a, 1995b) determined sediment accumulation rates of 0.3 to ~0.9 g cm^{-2} yr^{-1} for several locations in the Sabine-Neches (Texas) estuary, which is typical for Gulf coast estuaries, as found by other studies (Mueller et al. 1989; Beck et al. 1990; Baskaran et al. 1993; Santschi et al. submitted), except for the Mississippi River/Delta system, where it can be as high as 3.4 g cm^{-2} yr^{-1} (e.g., Mckee et al. submitted).

Present-day trace element accumulation fluxes (F) can be estimated using sedimentation rate and trace element concentration profiles, as shown in the following equation:

$$F = \frac{C_0}{\Delta_t} \Sigma(1 - \phi_i)\rho\Delta_z \tag{10-1}$$

where C_0 is the trace element concentration at the sediment-water interface, Δ_t is the time between peak fallout of Pu (or ^{137}Cs) and the period of sampling, ϕ and ρ are the porosity and density of the sediments, respectively, and Δ_z is the thickness of the sediment section above the Pu (or ^{137}Cs) peak. Using this approach, the recent accumulation rates of trace elements in surface sediments of selected estuaries of the Gulf of Mexico are summarized in Table 10-2.

Post-depositional Diagenesis

Whether sediments are important sources or sinks for trace elements depends on a variety of post-depositional diagenetic reactions. These processes can have important implications for the geochemical cycling and environmental fate of many trace elements. For example, many trace elements are sequestered and become much less bioavailable and/or toxic when co-precipitated with iron sulfides. The primary diagenetic reactions involve the oxidation of organic matter by a succession of electron acceptors [O_2, NO_3, SO_4, HCO_3, and to some extent the solid metal oxide phases of Fe and Mn (Richards 1965; Froelich et al. 1979; Kersten and Forstner 1986, 1987)] in the sediments and pore waters. In anoxic marine systems, the sulfate ion becomes the predominant oxidizing agent (sulfate reduction), releasing CO_2, NH_4, PO_4, and S into the sediment pore waters. Subsequently, the released sulfide is available to react with trace elements.

TABLE 10-2. Trace Element Concentrations in Sediments and Selected Estimated Particulate Trace Element Fluxes to Surface Sediments of Gulf Coast Estuaries

Estuary	Sedimentation Rate (cm yr^{-1})	Element	Mean Surface Concentration (μg g^{-1})	Estimated Flux (μg cm^{-2} yr^{-1})
Sabine-Neches	0.30~0.89	Co	8.29	2.40~7.21
		Cr	40.8	13.0~33.1
		Cu	11.3	3.88~7.73
		Ni	13.9	4.14~12.3
		Pb	52.7	5.77~14.2
		Zn	76.7	25.5~59.0
Calcasieu	0.74	Ag	0.067	0.05
		As	0.72	0.53
		Cd	0.98	0.73
		Cr	19.1	14
		Cu	6.91	5.1
		Pb	9.9	7.3
		Zn	35.6	26
Hillsborough	1.1	As	7.6	8.4
		Cd	2.6	2.9
		Cr	156	172
		Cu	54.5	60
		Pb	119	131
		Ni	26.6	29
		Zn	223	245
Lavaca	0.3~2.0	Hg	0.2	0.04~0.4
Galveston	0.5	Ag	0.15	0.08
		As	11	6.3
		Cd	0.16	0.09
		Co	68	36
		Cu	14	11
		Hg	0.08	0.07
		Pb	27	14
		Sn	2	1
		Zn	107	93
		Ba	800	444
Mississippi Delta	0.4	Ag	0.16	0.06
		As	12	4.6
		Cd	0.18	0.07
		Cr	72	27
		Cu	21	8
		Hg	0.14	0.05
		Pb	27	10
		Sn	1.9	0.7
		Zn	144	54
		Ba	870	327

Source: Data from Beck et al. (1990), Ravichandran et al. (1995), Basharan et al. (submitted), and Santschi et al. (submitted).

Recent studies indicate that pyritization (Huerta-Diaz and Morse 1990, 1992; Morse et al. 1993; Morse 1994) or co-precipitation with FeS (DiToro et al. 1992) are important processes of early diagenesis for many metals. For sediments and trace metals tested to date, element activity (bioavailability and toxicity) in the sediment-interstitial water system is above a threshold value where the total released metal concentration is higher than the available sulfides to bind them, positively correlated with the relative abundance of acid volatile sulfides to trace elements that are extracted by cold hydrochloric acid (Cornwell and Morse 1987; DiToro et al. 1992; Huerta-Diaz et al. 1993; Morse 1994). This sulfide fraction (mostly FeS and MeS, with Me = As, Cd, Co, Cr, Cu, Hg, Mn, Mo, Ni, Pb, and Zn) is conventionally referred to as "acid volatile sulfide (AVS)." Trace elements associated with AVS can potentially be released into the aquatic environment when reduced sediments are transported to oxic environments through re-suspension, bioturbation, or dredging activities.

The incorporation of trace elements into the pyrite phase apparently is influenced by the presence of other competing mineral phases, such as iron and manganese oxides found in the surface reactive fraction. The degree of pyritization (DOP; Berner 1970), which can be regarded as a measure of the extent to which nonsilicate iron (i.e., reactive iron) has been transformed to pyrite, is defined as

$$DOP = \frac{(Pyrite - Fe)}{(Pyrite - Fe) + (Reactive - Fe)} \qquad (10\text{-}2)$$

Oakley et al. (1980) stated that adsorbing surfaces (e.g., clays, iron, and manganese oxyhydroxides) can increase significantly the quantities of metals present outside of a sulfide phase even under conditions where sulfide precipitation appears to be favorable. In order to account for this factor, Huerta-Diaz and Morse (1990) introduce the concept of "degree of trace element pyritization (DTMP)," which is equivalent to DOP for iron. This term can be expressed as

$$DTMP = \frac{(Pyrite - Me)}{(Pyrite - Me) + (Reactive - Me)} \qquad (10\text{-}3)$$

in which qualitative comparison of the pyritization of a given metal to that of Fe can be made. Pyritization reactions of reactive trace elements in different estuarine sediments from the Gulf of Mexico have been investigated (Huerta-Diaz and Morse 1992; Morse et al. 1993; Cooper and Morse 1996). Correlations between DTMP and DOP of Cu, Cr, and Mo in sediments of the Trinity River estuary are shown in Fig. 10-9 (Morse et al. 1993). In these studies of Texas estuaries, Morse and co-workers showed that a general pattern could be observed in which some metalloid elements (e.g., Hg, As) underwent greater pyritization than Fe. Transition metals underwent pyritization similar

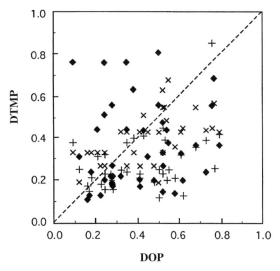

FIG. 10-9. *The relationships between trace element (DTMP) and iron (DOP) pyritization in Galveston Bay, Texas (data from Morse et al. 1993) (+ = Cu; × = Mo; ♦ = Cr).*

to that of Fe, and class B metals (Cd, Pb, Zn) underwent less pyritization than Fe. These studies also indicated that a variety of metals can be strongly adsorbed and/or co-precipitated with a variety of iron sulfide minerals, such as pyrite (Kornicker and Morse 1991; Morse et al. 1993) or mackinawite (FeS), a common metastable iron sulfide mineral (Arakaki and Morse 1993; Morse and Arakaki 1993).

TRACE ELEMENT FLUXES

Riverine Sources

Few reliable data on dissolved trace elements are available for rivers, principally because of problems during sampling and analysis (Shiller and Boyle 1987a; Flegal and Coale 1989; Windom et al. 1991; Taylor and Shiller 1995). One of the few rivers entering the Gulf of Mexico for which reliable data do exist, including data on temporal variability, is the Mississippi River. Shiller (1997) described a 27-month time series of dissolved trace elements in the lower Mississippi River and compared the results with data obtained a decade earlier using similar clean methods (Shiller and Boyle 1987a). Dissolved concentrations of a number of elements had significant seasonal variability, though long-term trends appeared to be absent. Manganese and Fe showed the greatest seasonal variability, with concentrations for both ranging from <10 to >70 nM. The seasonal trend showed concentrations of these two elements increasing rapidly in the fall and then decreasing in the spring. Zinc and Pb follow a similar seasonal trend, though the variability in their concentrations was only half that of Mn and Fe. Vanadium, Mo, and U followed an opposite seasonal trend. Redox processes were suggested to play an important role in determining seasonal dissolved concentration variability. Other elements, including Ba, Cd, Cu, Ni, and Rb, were observed to have more minor temporal variability and less distinct seasonal trends. Means and standard deviations of the dissolved concentrations are shown in Table 10-3. An implication of these results is that in estuaries with fresh-water transit times of a month or more, end-member variability may contribute to any non-linearity observed in estuarine property-salinity diagrams (Loder and Reichard 1981; Officer and Lynch 1981).

Estuarine Sources

Wen et al. (in press) estimated fluxes associated with internal sources and sinks of trace elements in the Trinity River estuary using a steady-state non-conservative estuarine mixing model. A summary of the modeling results for several trace elements is given in Table 10-4. This modeling effort predicts significant net internal inputs of dissolved and colloidal Cd, Co, Cu, Ni, and Zn. Even though there was an indication of large benthic inputs of Fe and Pb in the Trinity Bay, overall, dissolved and colloidal Fe and Pb were rapidly scavenged and settled out within the estuarine regions of Galveston Bay. Extensive inputs of colloidal metals were indicated by highly non-conservative distributions. Positive net internal fluxes of dissolved Co, Zn, Cd, Ni, and Cu increased the export of these elements from the estuary (relative to riverwater inputs) by 71%, 128%, 142%, 130%, and 520%, respectively. A significant

TABLE 10-3. Flow-Weighted Mean Dissolved Concentrations (and Standard Deviations) in the Lower Mississippi River Near Baton Rouge, Louisiana, October 1991–December 1993

Element	Mean	SD	
Ba	446	71	nM
Cd	137	34	pM
Cu	23	4	nM
Fe	44	26	nM
Mn	24	20	nM
Mo	23	11	nM
Ni	24	4	nM
Pb	45	20	pM
Rb	13	2	nM
U	5.8	1.9	nM
V	19	9	nM
Zn	4.3	1.4	nM

Source: Adapted from Schiller (in press).

TABLE 10-4. Estimates of Trace Element Fluxes in the Water Column of the Trinity River Estuary during July 1995

Element	C_s (nM)	C_0 (nM)	C^* (nM)	F_{riv} (mol d^{-1})	F_{est} (mol d^{-1})	F_{int} (mol d^{-1})
Dissolved ($<0.45\ \mu$m)						
Cd	0.071	0.031	0.075	0.6	1.5	0.9
Co	0.37	1.36	2.32	26.6	45.4	18.8
Cu	3.52	10.55	21.8	68.9	426	359
Fe	4.7	121	22	2366	425	−1941
Ni	4.66	14.1	32.5	276	635	359
Pb	0.15	0.24	0.16	4.6	3.1	−1.5
Zn	1.49	2.09	4.77	41	93	52
Colloidal (1 kDa–0.45 μm)						
Cd	0.008	0.018	0.102	0.4	2.0	1.6
Co	0.09	0.22	0.53	4.3	10.4	6.1
Cu	1.31	1.72	6.93	34	136	101
Fe	2.7	116	16.63	2269	325	−1943
Ni	0.83	9.44	15.18	185	297	112
Pd	0.078	0.200	0.103	3.9	2.0	−1.9
Zn	1.14	2.01	4.71	39	92	52

Note: The Trinity River discharge (R) was 1.96×10^{16} liters d^{-1} during the sampling period. C_s is the concentration of the sea-water end member. C_0 and C^* are the concentrations of the constituent in the river water end member for the actual and hypothetical cases, respectively. F_{riv} and F_{est} represent the flux into and out of the estuary, respectively. F_{int} is the net result of all source and removal fluxes occurring internally within the estuary.

fraction of this flux enhancement was due to the presence of colloidal forms of these trace metals.

Atmospheric Sources

There is a paucity of information on the atmospheric deposition fluxes of trace metals to the Gulf coast in the peer-reviewed literature. Information on atmospheric concentrations and deposition of Hg and several trace metals has been obtained for nine sites in the State of Florida as part of the Florida Atmospheric Mercury Study (FAMS) (Gill et al. 1995; Guentzel et al. 1995, in press; Landing et al. 1995, in press; Pollman et al. 1995; Perry et al. 1996). Average annual volume-weighted Hg concentrations in rain ranged from 15 to 17 ng liter^{-1} and depositional fluxes from 22 to 25 μg m^{-2} yr^{-1} throughout south Florida. A strong seasonal trend was observed in both the concentration and the flux of Hg; Hg concentrations increase by a factor of 2–4 during summer months and deposition is enhanced by a factor of 5 to 8, due in part to increased rainfall during summer months. The authors hypothesize that the seasonal trends in Hg concentration and flux result from seasonal trade wind transport of reactive gaseous Hg species coupled with strong convective thunderstorm activity during the summertime.

Beryllium-7 and ^{210}Pb are two radionuclides that have been widely used as tracers and geochronometers in aquatic and atmospheric systems. Even though the sources of these two nuclides to the atmosphere are distinctly different, both are highly particle reactive and thus become attached to aerosols in the atmosphere soon after production. Because these two radionuclides have different sources, they are useful in elucidating mechanisms and rates of removal of aerosols and trace element depositional fluxes. The bulk depositional fluxes of ^{210}Pb and ^{7}Be to Galveston Bay were determined between 1989 and 1991 (Baskaran et al. 1993). The authors found that the annual depositional fluxes of ^{7}Be and ^{210}Pb can vary by a factor of about 2.5, between 8.9 and 23.2 disintegrations per minute (dpm) cm^{-2} y^{-1}, with a mean of 14.7 for ^{7}Be, and 0.67 and 1.71, with a mean of 1.03, for ^{210}Pb, respectively. Thus, when using only ^{210}Pb and ^{7}Be as sediment, geochronology that assumes that the annual flux of these radionuclides to the sediment remains constant is questionable at best.

Sediment-Water Exchange

Because Gulf coast estuaries are typically very shallow and broad, the surface sediment area to water-column volume ratio is high. This feature tends to accentuate the importance of sediment-water exchange processes in Gulf coast estuaries compared to estuaries in other regions. Preliminary evidence indicating the importance of estuarine sediment-water exchange processes in controlling or mediating the concentration of trace elements and nutrients can be obtained from estuarine distributional transects, such as those shown in Fig. 10-5. If sediments are an important source, this is often reflected in elevated concentrations relative to conservative end-member mixing. The mid-estuary elevations for Cu and Ni observed in Galveston Bay do not correspond to any point source input, but rather depict broad-scale inputs from sediment-water exchange and sedimentary diagenetic processes. In contrast, the Mississippi River transects do not show evidence of a strong sediment source (Fig. 10-5).

Two approaches are usually used to obtain information on the sediment-water exchange of interstitial pore fluids with the overlying water column. An indirect assessment can be made from measurements of the vertical distribution of the constituent of interest in interstitial pore fluids using diffusion modeling approaches (Boudreau 1996). Sampling methods such as whole core squeezers are convenient tools to extract pore water to obtain information on pore water trace element concentration gradients (Presley et al. 1967, 1980; Presley and Trefry 1980; Jahnke 1988). A more direct approach is to enclose a portion of the water column in contact with bottom sediments using a chamber and monitor the change in concentration of the constituent of interest over time (Rowe et al. 1990, 1992). The gradient approach provides only steady-state flux information, while the chamber method has the potential to provide non-steady-state information on the temporal scales for which sampling is conducted. In a study of the Trinity River estuary (Warnken 1998), pore water concentration profiles in sediments and the concentration change in benthic chambers for NH_4, PO_4, $Si(OH)_4$, Mn, Fe, Ni, and Zn are shown in Fig. 10-10 and Fig. 10-11.

The diffusion of ions or molecules in interstitial pore fluids, in the absence of biological irrigation, is usually modeled based on a modification of Fick's First Law (Berner 1980). For diffusion in a sediment-pore water mixture, a generally accepted form of Fick's First Law is

$$F = -\left(\frac{\varphi D_w}{\theta^2}\right)\frac{\partial C}{\partial x} \qquad (10\text{-}4)$$

where, F is the flux of a solute with concentration C at depth x, θ is the tortuosity (dimensionless), φ is the sediment porosity, and D_w is the diffusion

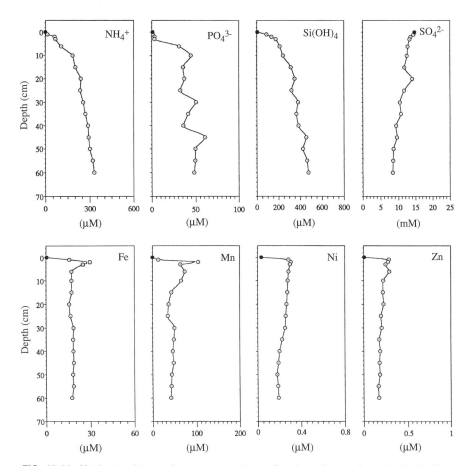

FIG. 10-10. Nutrient and trace element pore water profiles in sediments from the Trinity River estuary (data from Warnken 1998).

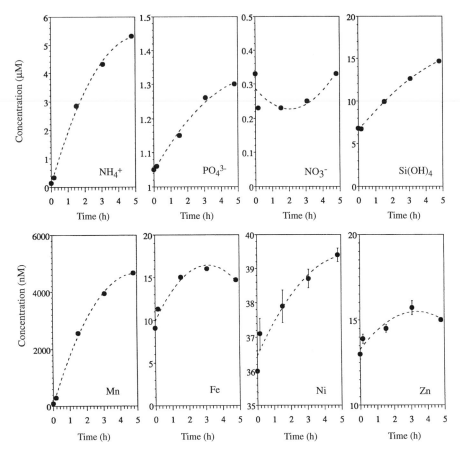

FIG. 10-11. Concentration of nutrient and trace element in benthic chamber water from the Trinity River estuary (data from Warnken 1998).

coefficient of the solute of interest in water without the presence of the sediment matrix. Tortuosity is a parameter that is not readily measured, but fortunately it has been shown that it is related to porosity, a parameter that is readily determined. Numerous studies have been conducted to relate tortuosity to sediment porosity. Boudreau (1996) recently proposed that the relationship between tortuosity and porosity is given by

$$\theta^2 = 1 - \ln(\varphi^2) \qquad (10\text{-}5)$$

Limitations in this type of flux estimation come from averaging over the large sediment collection intervals and also in the choice of diffusion constants. This model does not explicitly include the effects of processes such as bioturbation or bioirrigation, and also ignores horizontal gradients and pore water advection. Further difficulties arise when measuring the concentration gradients occurring at the sediment-water interface due to chemical co-precipitation reactions leading to barriers to diffusion across the sediment-water interface. However, despite these difficulties, this method provides a first-order view of the magnitude of the benthic flux.

The magnitude of diffusive benthic fluxes of several trace elements along the salinity gradient in the Trinity River estuary was determined using whole core squeezing techniques and by measuring changes in overlying water column concentrations with benthic chambers (Warnken et al. 1994; Warnken 1998) (Figs. 10-10 and 10-11). The benthic flux of trace metals calculated from diffusion alone showed considerable variability arising from both temporal and spatial changes. Manganese fluxes ranged from 20 to 367 μmoles m^{-2} d, with an average value of 193 $\pm$ 115 μmoles m^{-2} d^{-1}, while Fe fluxes were in the range of 7 to 94 μmoles m^{-2} d^{-1}, with an average value of 54 $\pm$ 24 μmoles m^{-2} d^{-1}. Nickel fluxes ranged from 120 to 910 nmoles m^{-2} d^{-1}, with an average value of 510 $\pm$ 320 nmoles m^{-2} d^{-1}, while Zn fluxes ranged from 180 to 1180 nmoles m^{-2} d^{-1}, with an average value of 560 $\pm$ 320 nmoles m^{-2} d^{-1}. All fluxes were out of the sediment into the water column, and the fluxes decreased in the order Mn > Fe > Zn > Ni. Manganese and Fe reduction zones were shallow, occurred at approximately 2 cm, and did not vary seasonally.

To determine the extent of biological activity on the flux, incubation studies using benthic flux chambers were conducted. Results showed that the overall benthic fluxes determined by this method were considerably larger. Only approximately 5–40% of the net flux from the sediments to the water column could be attributed to diffusion (Warnken 1998).

In a study at Lavaca Bay (Gill et al. submitted), pore water fluxes of Hg and monomethyl Hg (MMHg) were measured using a benthic flux chamber and also for interstitial porewater gradients. MMHg fluxes determined from pore water Hg gradients at 15 sites varied from 0.21 to 1500 ng m^{-2} d^{-1}. At one site, the MMHg flux observed using a benthic flux chamber during the dark period (35 ng m^{-2} hr^{-1}) was about five times greater than the estimated diffusional flux, suggesting that biological and/or chemical processes near the sediment-water interface are strongly mediating the sediment-water exchange of MMHg in Lavaca Bay.

SUMMARY AND CONCLUSIONS

To begin to understand trace element behavior in Gulf coast estuaries, one first needs to consider how basic biogeochemical and physical mixing processes are controlling trace element distributions. Biogeochemical processes include sorption/desorption, photosynthetic uptake, and re-mineralization reactions in the water and sediments. Of particular importance are different organic phases that can exhibit strong affinity to particular trace metals, thus affecting and often ameliorating their bioavailability and toxicity. For example, in a number of Gulf coast estuaries, many dissolved trace elements were found to be significantly associated with colloidal fractions (>1 kDa) composed primarily of organic macromolecules. Often, estuarine colloidal trace element concentrations correlated significantly with colloidal organic carbon concentrations (>1 kDa), suggesting metal-organic complexation and behavior dictated by processes influencing the cyclic of organic colloids in estuaries. Iron was preferentially associated with HMW (>10 kDa) colloids, while Cu, Hg, Ni and Pb were associated mostly with LMW (1 to ~10 kDa) colloidal material.

Trace element concentrations in sediments from selected Gulf coast estuaries have been used to reveal contamination histories, contaminant sources, and temporal and spatial variation. Results from some of these studies demonstrate that fluxes of Pb from leaded gasoline in the environment significantly declined in recent decades. Despite demonstrated improvements in trace metal releases into Gulf coast estuaries, much more work needs to be carried out to better understand the coupling of various transport processes across the sediment-water interface, including rates of sedimentation, particle reworking, sediment re-suspension, particle sorption and desorption, pore water diffusion and advection, and benthic photosynthesis and respiration reactions. Furthermore, amounts of AVS, DOP, and DTMP are important measurements that need to be taken into account more precisely when assessing the bioavailability and toxicity of trace elements in these sediments.

Despite continually improving analytical techniques for the determination of trace element distributions in estuarine environments, we do not yet have a clear mechanistic understanding of the important processes influencing the concentration and distribution of trace elements in Gulf coast estuaries. Controlled experiments are sorely needed to define the mechanisms and extent of complexation, bioavailability, and kinetics. In particular, information on the extent of trace metal complexation and the identity of specific functional groups responsible for metal complexation are needed. In sediments, information on the specific carrier phases is lacking. Such information is necessary to assess reliably the bioavailability of contaminated sediments. Furthermore, better information on the role of sediment transport processes in controlling the burial of trace contaminants is required to address long-term trends on the impact of polluted sediments on the estuarine ecosystem. This is particularly important in salt marshes, the most sensitive part of the estuarine ecosystem.

Furthermore, there are very limited data on the temporal and spatial variability of trace element distributions, and on many key sources, such as atmospheric deposition, industrial and municipal discharges, and non-point source inputs. This paucity of key information precludes modeling efforts, which in turn hinders our ability to predict current and future status of trace element transport, cycling, and toxicity in Gulf coast estuaries.

ACKNOWLEDGMENTS

We would like to thank William Landing for the critical review of this manuscript. This work was supported, in part, by the Office of Naval Research

(Grant N00014-93-1-0877), the Texas Seagrant (Grant 424025), and the Texas Institute of Oceanography.

REFERENCES

Alexander, C. R., F. D. Calder, and H. L. Windom. 1993. The historical record of metal enrichment in two Florida estuaries. Estuaries **16:** 627–637.

Altman, R. S., and J. Buffle. 1988. The use of differential equilibrium functions for interpretation of metal binding in complex ligand systems: Its relation to site occupation and site affinity distribution. Geochim. Cosmochim. Acta **52:** 1505–1520.

Arakaki, T., and J. Morse. 1993. Coprecipitation and adsorption of Mn^{2+} with mackinawite (FeS) under conditions similar to those found in anoxic sediments. Geochim. Cosmochim. Acta **57:** 1–15.

Bale, A., C. Barrett, J. West, and K. Oduyemi. 1989. Use of in situ laser diffraction particle sizing in estuaries, p. 133–138. *In:* J. McManus and M. Elliott [eds.], Developments in estuarine and coastal study techniques. Olsen and Olsen.

———, A. Morris, and R. Howland. 1985. Seasonal sediment movement in the Tamar Estuary. Oceanol. Acta **8:** 1–6.

Balls, P. W. 1989. Trend monitoring of dissolved trace metals in coastal sea water—a waste of effort? Mar. Poll. Bull. **20:** 546–548.

Baskaran, M., C. Coleman, and P. Santschi. 1993. Atmospheric depositional fluxes of ^7Be and ^{210}Pb at Galveston and College Station, Texas. J. Geophys. Res. **98:** 555–576.

———, B. J. Presley, S. Asbill, P. H. Santschi, and R. Taylor. (Submitted). Reconstruction of historical contamination of trace metals in Mississippi River Delta, Tampa Bay and Galveston Bay sediments. Environ. Sci. Technol.

———, and P. Santschi. 1993. The role of particles and colloids in the transport of radionuclides in coastal environments of Texas. Mar. Chem. **43:** 95–114.

Beck, J. N., G. J. Ramelow, R. S. Thompson, C. S. Mueller, C. L. Webre, J. C. Young, and M. P. Langley. 1990. Heavy metal content of sediments in the Calcasieu River/ Lake complex, Louisiana. Hydrobiology **192:** 149–165.

Benoit, G., S. D. Oktay-Marshall, A. Cantu II, E. M. Hood, C. H. Coleman, M, O, Corapcioglu, and P. H. Santschi. 1994. Partitioning of Cu, Pb, Ag, Zn, Fe, Al, and Mn between filter-retained particles, colloids, and solution in six Texas estuaries. Mar. Chem. **45:** 307–336.

Berner, R. A. 1970. Sedimentary pyrite formation. Amer. J. Sci. **268:** 1–23.

———. 1980. Early diagenesis: a theoretical approach. Princeton University Press.

Bilinski, H., S. Kozar, M. Palvasic, Z. Kwokal, and M. Branica. 1991. Trace metal adsorption on inorganic solid phases under estuarine conditions. Mar. Chem. **32:** 225–233.

Boudreau, B. P., 1996. The diffusive tortuosity of fine-grained unlithified sediments. Geochim. Cosmochim. Acta **60:** 3139–3142.

Bourg, A. 1987. Trace metal adsorption molding and particle-water interactions in estuarine environments. Cont. Shelf Res. **7:** 1319–1332.

Boykin, R. 1972. Texas and the Gulf of Mexico. Texas A&M University Press.

Boyle, E. A., J. M. Edmond, and E. R. Sholkovitz. 1977. The mechanism of iron removal in estuaries. Geochim. Cosmochim. Acta **41:** 1313–1324.

Bruland, K. W. 1989. Complexation of zinc by natural organic ligands in the central North Pacific. Limnol. Oceanogr. **34:** 269–285.

———. 1992. Complexation of cadmium by natural organic ligands in the central North Pacific. Limnol. Oceanogr. **37:** 1008–1017.

———, J. Donat, and D. Hutchins. 1991. Interactive influences of bioactive trace metals on biological production in oceanic waters. Limnol. Oceanogr. **36:** 1555–1577.

Bruno, J. 1990. The influence of dissolved carbon dioxide on trace metal speciation in seawater. Mar. Chem. **30:** 231–240.

Buchholtz, M., P. H. Santschi, and W. S. Broecker. 1986. Comparison of K_d-values by batch equilibration with in-situ determinations in the deep-sea using the MANOP Lander: The importance of geochemical mechanisms in controlling ion uptake and migration, p. 192–205. *In:* T. Sibley and C. Myttenaire [eds.], Application of distribution coefficients to radiological assessment models. Elsevier Applied Science.

Buffle, J. 1990. Complexation reactions in aquatic systems: An Analytical approach. Ellis Horwood.

———, and G. G. Leppard. 1995a. Characterization of aquatic colloids and macromolecules, 1. Structure and behavior of colloidal material. Environ. Sci. Technol. **29:** 2169–2175.

———. 1995b. Characterization of aquatic colloids and macromolecules, 2. Key role of physical structures on analytical results. Environ. Sci. Technol. **29:** 2169–2175.

Buffle, J., R. Altmann, M. Fitella, and A. Tessier. 1990. Complexation by natural heterogenous compounds: Site occupation distribution functions, a normalized description of metal complexation. Geochim. Cosmochim. Acta. **54:** 1535–1553.

Burton, J. D., and P. J. Stantham. 1990. Trace metals in seawater, p. 5–25. *In:* R. W. Francis and P. S. Rainbow [eds.], Heavy metals in the marine environment. CRC Press.

Byrd, J. T., and M. O. Andreae. 1986. Geochemistry of tin in rivers and estuaries. Geochim. Cosochim. Acta **50:** 835–845.

Chan, L. H., and J. S. Hanor. 1982. Dissolved barium in some Louisiana offshore waters: Problems in establishing baseline values. Contrib. Mar. Sci. **25:** 149–159.

Cherry, R. D., J. Higgo, and S. Fowler. 1978. Zooplankton fecal pellets and element residence time in the ocean. Nature **274:** 246–248.

Ciceri, G., A. Traversi, W. Martinotti, and G. Queirazza. 1988. Radionuclide partitioning between water and suspended matter: Comparison of different methodologies, p. 353–375. *In:* Pawlowski, L., E. Mentasti, W. Lacy and C. Sarzanini [eds.], Chemistry for the protection of the marine environment 1987. Studies in environmental science. Elsevier.

Coale, K., and K. Bruland. 1988. Copper complexation in the Northeast Pacific. Limnol. Oceanogr. **33:** 1084–1101.

Cooper, D. C., and J. W. Morse. 1996. The chemistry of Offatts Bayou, Texas: A seasonally highly sulfidic basin. Estuaries **19:** 595–611.

Cornwell, J. C., and J. W. Morse. 1987. The characterization of iron sulfide minerals in marine sediments. Mar. Chem. **22:** 193–206.

Cutter, G. A. 1990. Trace elements in estuarine and coastal waters—U.S. studies from 1986–1990. Rev. Geophys. **42:** 639–644.

Dai, M.-H., and J. M. Martin. 1995. First data on trace metal level and behavior in two major Arctic river-estuarine systems (Ob and Yenisey) and in the adjacent Kara Sea, Russia. Earth Planet. Sci. Lett. **131:** 127–141.

Daskalakis, K. D., and T. P. O'Connor. 1995. Normalization and elemental sediment contamination in the coastal united states. Environ. Sci. Technol. **29:** 470–477.

De Groot, A. J., W. Salomons, and E. Allersma. 1976. Processes affecting heavy metals in estuarine sediment, p. 131–157. *In:* D. Burton and P. Liss [eds.], Estuarine chemistry. Academic Press.

Di Toro, D. M., J. D. Mahony, D. J. Hansen, K. S. Scott, A. R. Carlson, and G. T. Ankley. 1992. Acid volatile sulfide predicts the acute toxicity of cadmium and nickel in sediments. Environ. Sci. Technol. **26:** 96–101.

Dinnel, S. P., and W. J. Wiseman. 1984. Fresh water on the Louisiana and Texas shelf. Cont. Shelf Res. **6:** 765–784.

Duinker, J. C. 1983. Effects of particle size and density on the transport of metals in the oceans, p. 34–45. *In:* C. S. Wong, K. England, E. Boyle, J. Burton, and E. Goldberg [eds.], Trace metals in seawater. Plenum.

Dyer, K. 1989. Sediment processes in estuaries: Future research requirements. J. Geophys. Res. **94:** 14327–14339.

Eisma, D., T. Schumacher, H. Boekel, J. Van Heerwaarden, H. Franken, M. Laan, A. Vaars, F. Eugenraam, and J. Kalf. 1990. A camera and image-analysis system for in situ observation of flocs in natural waters. Netherlands J. Sea Res. **27:** 43–56.

Fisher, N. 1986. On the reactivity of metals for marine phytoplankton. Limnol. Oceanogr. **31:** 443–450.

Flegal, A. R., and K. H. Coale. 1989. Discussion of "trends in lead concentrations in major U.S. rivers and their relation to historical changes in gasoline-lead consumption" by R. A. Alexander and R. A. Smith. Water Res. Bull. **25:** 1275–1277.

Froelich, P. N., L. W. Kaul, J. T. Byrd, M. O. Andreae, and K. K. Roe. 1985. Arsenic, barium, germanium, tin, dimethylsulfide and nutrient biogeochemistry in Charlotte Harbor, Florida, a phosphate-enriched estuary. Estuar. Coast. Mar. Sci. **20:** 239–264.

―――, G. Klinkhammer, M. Bender, N. Luedtke, G. Heath, D. Cullen, P. Dauphin, D. Hammond, B. Hartman, and V. Maynard. 1979. Early oxidation of organic matter in pelagic sediments of the eastern equatorial Atlantic: Suboxic diagenesis. Geochim. Cosmochim. Acta **43:** 1075–1090.

Gill, G., N. S. Bloom, S. Cappellino, C. T. Driscoll, C. Dobbs, L. McShea, R. Mason, and J. Rudd. (Submitted). Sediment-water fluxes of mercury in Lavaca Bay, Texas. Enviro. Sci. Technol.

―――, J. J. Guentzel, W. M. Landing, and C. D. Pollman. 1995. Total gaseous mercury measurements in Florida: The FAMS Project (1992–1994). Water, Air, and Soil Pollution **80:** 235–244.

Goldberg, E. D., J. J. Griffin, V. Hodge, M. Koide, and H. Windom. 1979. Pollution history of the Savannah River estuary. Environ. Sci. Technol. **13:** 588–594.

Guentzel, J. L., W. M. Landing, G. A. Gill, and C. D. Pollman. (In press). Mercury and major ions in rainfall, throughfall and foliage from the Florida Everglades. Atmospheric Environ.

————. 1995. Atmospheric deposition of mercury in Florida: The FAMS Project (1992–1994). Water, Air, and Soil Pollution **80:** 393–402.

————, T. R. Powell, W. M. Landing, and R. P. Mason. 1996. Mercury associated with colloidal material in an estuarine and an open-ocean environment. Mar. Chem. **55:** 177–188.

Gustafsson, O., and P. Gschwend. 1997. Aquatic colloids: Concepts, definitions, and current challenges. Limnol. Oceanogr. **42:** 519–528.

Hannor, J. S., and L.-H. Chan. 1977. Non-conservative behavior of barium during mixing of Mississippi River and Gulf of Mexico waters. Earth Planet. Sci. Lett. **37:** 242–250.

Hanson, P. J., D. W. Evans, D. R. Colby, and V. S. Zdanowicz. 1993. Assessment of elemental contamination in estuarine and coastal environments based on geochemical and statistical modeling of sediments. Mar. Environ. Res. **36:** 237–266.

Hering, J., and M. Morel. 1988. Kinetics of trace metal complexation: Role of alkaline-earth metals. Environ. Sci. Technol. **22:** 1469–1478.

————. 1989. Slow coordination reactions in seawater. Geochim. Cosmochim. Acta **53:** 611–618.

Honeyman, B. D., and P. H. Santschi. 1988. Critical review: Metals in aquatic systems. Predicting their scavenging residence times from laboratory data remains a challenge. Environ. Sci. Technol. **22:** 862–871.

————. 1989. A Brownian-pumping model for trace metal scavenging: Evidence from Th isotopes. J. Mar. Res. **47:** 950–995.

Horowitz, A. J. 1991. A primer on sediment-trace element chemistry. Lewis.

Huerta-Diaz, M. A., R. Carignan, and A. Tessier. 1993. Measurement of trace metals associated with acid volatile sulfides and pyrite in organic freshwater sediments. Environ. Sci. Technol. **27:** 2367–2372.

————, and J. Morse. 1990. A quantitative method for determination of trace metal concentration in sedimentary pyrite. Mar. Chem. **29:** 119–144.

————. 1992. Pyritization of trace in anoxic marine sediments. Geochim. Cosmochim. Acta **56:** 2681–2702.

Hutzenga, D. L., and D. R. Kester. 1983. The distribution of total and electrochemically available copper in the northwestern Atlantic Ocean. Mar. Chem. **13:** 281–291.

Jahnke, R. A. 1988. A simple, reliable, and inexpensive porewater sampler. Limnol. Oceanogr. **33:** 483–487.

Kennish, M. J. 1992. Ecology of estuaries: Anthropogenic effects. CRC Press.

Keeney-Kennicutt, W., and B. J. Presley. 1985. The geochemistry of trace metals in the Brazos River estuary. Estuar. Coast. Shelf Sci. **22:** 459–477.

Kersten, M., and U. Forstner. 1986. Chemical fractionation of heavy metals in anoxic estuarine and coastal sediments. Water Sci. Technol. **18:** 121–130.

————. 1987. Effect of sample pretreatment on the reliability of solid speciation data on heavy metals—implications for the study of early diagenetic processes. Mar. Chem. **22:** 299–302.

Kester, D. 1986. Equilibrium models in seawater: Applications and limitations, p. 337–363. *In:* M. Berbhard et al. [eds.]: The importance of chemical "speciation" in environmental processes. Springer-Verlag.

Kornicker, W., and J. Morse. 1991. The interactions of divalent cations with the surface of pyrite. Geochim. Cosmochim. Acta **55:** 2159–2172.

Kumar, M. 1987. Cation hydrolysis and the regulation of trace metal composition in seawater. Geochim. Cosmochim. Acta **51:** 2137–2145.

Landing, W., J. L. Guentzel, J. J. Perry, Jr., G. A. Gill, and C. D. Pollman. (In press). Methods for measuring mercury and other trace species in rainfall and aerosols in Florida. Atmospheric Environ.

———, and B. Lewis. 1991. Analysis of marine particulate and colloidal material for transition metals, p. 263–272. *In:* D. C. Hurd and D. W. Spencer [eds.], Marine particles: Analysis and characterization. American Geophysical Union.

———, J. J. Perry, Jr., J. L. Guentzel, G. A. Gill, and C. D. Pollman. 1995. Relationships between the atmospheric deposition of trace elements, major ions, and mercury in Florida: The FAMS Project (1992–1994). Water, Air, and Soil Pollution **80:** 343–352.

Li, Y.-H. 1981. Ultimate removal mechanisms of elements from the ocean. Geochim. Cosmochim. Acta **45:** 1659–1664.

———, L. Burkhardt, M. Buchholtz, P. O'Hara, and P. Santschi. 1984a. Partition of radiotracers between suspended particles and seawater. Geochim. Cosmochim. Acta **48:** 2011–2019.

———, L. Burkhardt, and H. Teraoka. 1984b. Desorption and coagulation of trace elements during estuarine mixing. Geochim. Cosmochim. Acta **48:** 1659–1664.

———, P. H. Santschi, P. O'Hara, M. Amdurer, and P. Doering. 1989. The importance of a benthic ecosystem to the removal of radioactive trace elements from coastal waters. Environ. Technol. Lett. **10:** 57–70.

Liss, P. S. 1976. Conservative and non-conservative behavior of dissolve constituents during estuarine mixing. *In:* J. D. Burton and P. S. Liss [eds.], Estuarine chemistry. Academic Press.

Loder, T. C., and R. P. Reichard. 1981. The dynamics of conservative mixing in estuaries. Estuaries **4:** 64–69.

Long, E. R., A. Robertson, D. A. Wolfe, J. Hameedi, and G. M. Sloane. 1996. Estimates of the spatial extent of sediment in major u.s. estuaries. Environ. Sci. Technol. **30:** 3585–3592.

Lord, C. J., and T. M. Church. 1983. The geochemistry of salt marshes: Sedimentary ion diffusion, sulfate reduction, and pyritization. Geochim. Cosmochim. Acta **47:** 1381–1391.

Martin, J. H., and R. M. Gordon. 1988. Northeast Pacific iron distributions in relation to phytoplankton productivity. Deep-Sea Res. **35:** 177–196.

Martin, J. M., M. H. Dai, and G. Cauwet. 1995. Significance of colloids in the biogeo-chemical cycling of organic carbon and trace metals in the Venice Lagoon (Italy). Limnol. Oceanogr. **40:** 119–131.

———, P. Nirel, and A. J. Thomas. 1987. Sequential extraction techniques: Promises and problems. Mar. Chem. **22:** 313–320.

McKee, B. A., J. G. Booth, and P. W. Swarzenski. (Submitted). Sediment deposition redistribution and accumulation in the Mississippi River Bight. Cont. Shelf Res.

Millero, F. 1985. The effect of ionic interactions in the oxidation of metals in natural waters. Geochim. Cosmochim. Acta **49:** 547–553.

Moffett, J. W., R. G. Zika, and L. E. Brand. 1990. Distribution and potential sources and sinks of copper chelators in the Sargasso Sea. Deep-Sea Res. **37:** 27–36.

Moran, B., and R. Moore. 1989. The distribution of colloidal aluminum and organic carbon in coastal and open ocean waters off Nova Scotia. Geochim. Cosmochim. Acta **53:** 2519–2527.

Morris, A. W. 1986. Removal of trace metals in the very low salinity region of the Tamar Estuary, England. Sci. Total Environ. **49:** 297–304.

Morse, J. 1994. Interactions of trace metals with authigenic sulfide minerals: Implications for their bioavailability. Mar. Chem. **46:** 1–6.

Morse, J., and T. Arakaki. 1993. Adsorption and coprecipitation of divalent metals with mackinawite (FeS). Geochim. Cosmochim. Acta **57:** 3635–3640.

———, B. J. Presley, R. J. Taylor, G. Benoit, and P. Santschi. 1993. Trace metal chemistry of Galveston Bay: Water, sediments and biota. Mar. Environ. Res. **36:** 1–37.

Mueller, C. S., G. J. Ramelow, and J. N. Beck. 1989. Mercury in the Calcasieu river/lake complex, Louisiana. Bull. Environ. Contam. Toxicol. **42:** 71–80.

Newman, M., and C. Jagoe. 1994. Ligands and the bioavailability of metals in aquatic environments, p. 39–71. *In:* J. Hamelink et al. [eds.] Bioavailability, physical, chemical, and biological interactions. Lewis.

Nguyen, B., S. Belviso, N. Mihalopoulos, J. Gostan, and P. Nival. 1988. Dimethyl sulfide production during natural phytoplanktonic blooms. Mar. Chem. **24:** 133–141.

Nyffeler, U. P., Y.-H. Li, and P. H. Santschi. 1984. A kinetic approach to describe trace-element distribution between particles and solution in natural aquatic systems. Geochim. Cosmochim. Acta **48:** 1513–1522.

———, P. H. Santschi, and Y.-H. Li. 1986. The relevance of scavenging kinetics to modelling of sediment-water interactions in natural waters. Limnol. Oceanogr. **31:** 277–292.

Oakley, S. M., C. E. Delphey, K. J. Williamson, and P. O. Nelson. 1980. Kinetics of trace metal partitioning in model anoxic marine sediments. Water Res. **14:** 1067–1072.

O'Connor, D. J., and J. P. Connolly. 1980. The effect of concentration of adsorbing solids on the partition coefficient. Water Res. **14:** 1517–1523.

Officer, C. B. 1979. Discussion of the behavior of nonconservative dissolved constituents in estuarine. Estuar. Coast. Mar. Sci. **9:** 91–94.

———, and D. R. Lynch. 1981. Dynamics of mixing in estuaries. Estuar. Coast. Shelf Sci. **12:** 525–534.

Olsen, C. R., N. Cutshall, and I. Larsen. 1982. Pollutant-particle associations and dynamics in coastal marine environments: A review. Mar. Chem. **11:** 501–533.

Paulson, A. J., R. A. Feely, H. C. Curl, and D. A. Tenn. 1989. Estuarine transport of trace metals in a buoyant riverine plume. Estuar. Coast. Shelf Sci. **26:** 231–248.

Perry, J. J., W. M. Landing, J. L. Guentzel, G. A. Gill, and C. D. Pollman. 1996. The relationships between mercury, trace metals, and major ions in precipitation collected in Florida: The Florida Atmospheric Mercury Study (FAMS). Paper presented at the 4th International Conference on Mercury as a Global Pollutant, August 4–8, Hamburg, Germany.

Pollman, C. D., G. A. Gill, W. M. Landing, D. A. Bare, J. L. Guentzel, D. Porcella, E. Zillioux, and T. Atekson. 1995. Overview of the Florida Atmospheric Mercury Study (FAMS). Water Air Soil Pollution **80:** 285–290.

Powell, R. T., W. M. Landing, and J. E. Bauer. 1996. Colloidal trace metals, organic carbon, and nitrogen in a southeastern U.S. estuary. Mar. Chem. **55:** 165–175.

Presley, B. J., R. R. Brooks, and H. M. Kapper. 1967. A simple squeezer for removal of interstitial waters from ocean sediments. J. Mar. Res. **25:** 355–357.

——, J. H. Refry, and R. F. Shokes. 1980. Heavy metal inputs to Mississippi Delta sediments. Water Air Soil Pollution **13:** 481–494.

——, R. J. Taylor, and P. N. Boothe. 1992. Trace metal concentrations in sediments of the eastern Mississippi bight. Mar. Environ. Res. **33:** 267–282.

——, and J. H. Trefry. 1980. Sediment-water interactions and the geochemistry of interstitial waters, p. 187–222. *In:* E. Olausson and I. Cato [eds.], Chemistry and biogeochemistry of estuaries. Wiley.

Price, N. B., and S. E. Calvert. 1973. A study of the geochemistry of suspended particulate matter in coastal waters. Geochim. Cosmochim. Acta **1:** 169–189.

Rapin, F. 1986. Potential artifacts in the determination of metal partitioning in sediments by a sequential extraction procedure. Environ. Sci. Technol. **20:** 836–840.

Ravichandran, M., M. Baskaran, P. H. Santschi, and T. S. Bianchi. 1995a. Geochronology of sediments of Sabine-Neches Estuary, Texas. Chem. Geol. **125:** 291–306.

——. 1995b. History of trace pollution in Sabine-Neches Estuary, Beaumont, Texas. Environ. Sci. Technol. **29:** 1495–1503.

Richard, F. A. 1965. Anoxic basins and fjords, p. 611–645. *In:* J. Riley and G. Skirrow [eds.], Chemical oceanography. Plenum.

Rowe, G. T., G. S. Boland, W. C. Phoel, R. F. Anderson, and P. E. Biscaye. 1992. Deep sea-floor respiration as an indication of lateral input of biogenic detritus from continental margins. Cont. Shelf Res. **24:** 132–139.

——, and A. P. McNichol. 1990. Carbon cycling in coastal sediments: Estimating remineralization in Buzzards Bay, Massachusetts. Geochem. Cosmochim. Acta **55:** 2989–2991.

Rue, E. L., and K. W. Bruland. 1995. Complexation of iron (III) by natural organic ligands in the Central North Pacific as determined by a new competitive ligand equilibration/adsorptive cathodic stripping voltammetric method. Mar. Chem. **50:** 117–138.

Salomons, M., and U. Forstner. 1984. Analytical aspects of environmental chemistry. Wiley.

Santschi, P. 1995. Seasonality of nutrient concentration in Galveston Bay. Mar. Env. Res. **40:** 337–362.

Santschi, P., M. Allison, S. Asbill, A. Eek, S. Cappellino, C. Dobbs, and L. McShea. (Submitted). Sediment transport and Hg recovery in Lavaca Bay, as evaluated from radionuclide and Hg distributions. Environ. Sci. Technol.

——, L. Guo, M. Baskaran, S. Trumbore, J. Southon, T. Bianchi, B. Honeyman, and L. Cifuentes. 1995. Isotopic evidence for the contemporary origin of high-molecular-weight organic matter in oceanic environments. Geochim. Cosmochim. Acta **59:** 625–631.

——, P. Hoehener, G. Benoit, and M. Buchholtz-Ten Brink. 1990. Chemical processes at the sediment-water interface. Mar. Chem. **30:** 269–315.

——, and B. Honeyman. 1991. Are thorium scavenging and particle fluxes in the ocean regulated by coagulation?, p. 107–115. *In:* P. Kershaw and D. Woodhead [eds.], Radionuclides in the study of marine processes. Elsevier Applied Science.

————, J. Lenhart, and B. Honeyman. 1997. Heterogeneous processes affecting trace contaminant distribution in estuaries: The role of natural organic matter. Mar. Chem. **58:** 99–125.

————, Y.-H. Li, J. Bell, D. Adler, M. Amdurer, and U. P. Nyffeler. 1983. The relative mobility of natural (Th, Pb, Po) and fallout (Pu, Cs, Am) radionuclides in the coastal marine environment: Results from model ecosystems (MERL) and Narragansett Bay studies. Geochim. Cosmochim. Acta **47:** 201–310.

————. 1984. Particle flux and trace metal residence times in natural waters. Limnol. Oceanogr. **29:** 1100–1108.

————, Y.-H. Li, J. Bell, R. M. Trier, and K. Kawtaluk. 1980. Plutonium in the coastal marine environment. Earth Planet. Sci. Lett. **51:** 248–265.

————, Y.-H. Li, P. O'Hara, M. Amdurer, D. Adler, and P. Doering. 1987. The relative mobility of radioactive trace elements across the sediment-water interface of the MERL model ecosystems of Narragansett Bay. J. Mar. Res. **45:** 1007–1048.

————. 1988. Factors controlling the biogeochemical cycles of trace elements in fresh and coastal marine waters as revealed by artificial radioisotopes. Limnol. Oceanogr. **33:** 848–866.

————, S. Nixon, M. Pilson, and C. Hunt. 1984. Accumulation of sediments, trace metals (Pb, Cu) and hydrocarbons in Narragansett Bay, Rhode Island. Est. Coast. Shelf Sci. **19:** 427–450.

Shiller, A. M. 1993a. Comparison of nutrient and trace element distributions in the delta and shelf outflow regions of the Mississippi/Atchafalaya River. Estuaries **16:** 541–546.

————. 1993b. A mixing rate approach to understanding nutrient distributions in the plume of Mississippi river. Mar. Chem. **43:** 211–216.

————. 1996. The effect of recycling traps and upwelling on estuarine chemical flux estimates. Geochimi. Cosmochim. Acta **60:** 3177–3185.

————. 1997a. Dissolved trace elements in the Mississippi River: Seasonal, interannual, and decadal variability. Geochim. Cosmochim. Acta **61:** 4321–4330.

————. (In press). An overview of the marine chemistry of the Gulf of Mexico. *In:* H. Kumpf et al. [eds.], The Gulf of Mexico—A large marine ecosystem.

————, and E. A. Boyle. 1987a. Variability of dissolved trace metals in the Mississippi River. Geochim. Cosmochim. Acta **51:** 3273–3277.

————. 1987b. Dissolved vanadium in rivers and estuaries. Earth Planet. Sci. Lett. **86:** 214–224.

————. 1991. Trace elements in the Mississippi River Delta outflow region: Behavior at high discharge. Geochim. Cosmochim. Acta **55:** 3241–3251.

————, L. Mao, and J. Cramer. (In press). Determination of dissolved vanadium in natural waters by flow injection analysis with colorimetric detection. Limnol. Oceanogr.

Sholkovitz, E., W. Landing, and B. Lewis. 1994. Ocean particle chemistry: The fractionation of rare earth elements between suspended particles and seawater. Geochim. Cosmochim. Acta **56:** 1567–1580.

Sigleo, A. C., and J. C. Means. 1990. Organic and inorganic components in estuarine colloids: Implications for sorption and transport of pollutants. p. 123–147. *In:* Reviews of environmental contamination and toxicology. Springer-Verlag.

Simkiss, K., and M. Taylor. 1989. Metal fluxes across the membrane of aquatic organisms. Rev. Aquat. Sci. **1:** 173–188.

Stordahl, M. C. 1996. The biogeochemistry of As, Se, Hg, and Sb in Texas estuaries. Ph.D. dissertation, Texas A&M University.

Stordal, M. C., G. A. Gill, L.-S. Wen, and P. H. Santschi. 1996. Mercury phase speciation in the surface waters of three texas estuaries: importance of colloidal forms. Limnol. Oceanogr. **41:** 52–61.

———, P. H. Santschi, and G. Gill. 1997. Colloidal pumping: Evidence for the coagulation process using natural colloids tagged with [203]Hg. Environ. Sci. Technol. **30:** 3335–3340.

Summers, J. K., T. L. Wade, and V. D. Engle. 1996. Normalization of metal concentrations in estuarine sediments from the Gulf of Mexico. Estuaries **19:** 581–594.

Sunda, W. G., and R. L. Ferguson. 1983. Sensitivity of natural bacterial communities to additions of copper and to cupric ion activity: A bioassay of copper complexation in seawater, p. 871–891. *In:* C. Wong et al. [eds.], Trace metals in sea water. Plenum.

Taylor, H. E., and A. M. Shiller. 1995. The Mississippi River methods comparison study: Implications for water quality monitoring of dissolved trace elements. Environ. Sci. Technol. **29:** 1313–1317.

The Galveston Bay Plan. 1995. The Comprehensive Conservation and Management Plan for the Galveston Bay Ecosystem. A project of the Galveston Bay National Estuary Program.

Trefry, J. H., T. A. Nelsen, R. P. Trocine, S. Metz, and T. W. Vetter. 1986. Trace metal fluxes through the mississippi river delta system. Cons. Int. Explor. Mer. **186:** 277–288.

———, and B. J. Presley. 1976. Heavy metal transport from the Mississippi River to the Gulf of Mexico, p. 39–76. *In:* H. Windom and R. Duce [eds.], Marine pollution transfer. Heath & Co.

Turner, A., G. E. Millward, A. J. Bale, and A. W. Morris. 1993. Application of the KD concept to the study of trace metal removal and desorption during estuarine mixing. Estuar. Coast. Shelf Sci. **36:** 1–13.

Turner, D., M. Whitefield, and A. Dickson. 1981. The equilibrium speciation of dissolved components in freshwater and seawater at 25°C and 1 atm pressure. Geochim. Cosmochim. Acta **45:** 855–871.

Valenta, P., E. K. Duursma, A. Merks, H. Rutzel, and H. Nurnberg. 1986. Distribution of Cd, Pb and Cu between the dissolved and particulate phase in the eastern Scheldt and western Scheldt Estuary. Sci. Total Environ. **53:** 41–76.

Valette-Silver, N. J., S. B. Bricker, and W. Salomons. 1993. Estuaries, **16:** 28–42.

Van Den Berg, C. M. G., A. G. Merks, and E. Duursma. 1987. Organic complexation and its control of the dissolved concentrations of copper and zinc in the Scheldt estuary. Estuar. Coast. Shelf Sci. **24:** 785–797.

Ward, G. H. 1980. Hydrography and circulation processes of Gulf estuaries, p. *In:* P. Hamilton and K. MacDonald [eds.], Estuarine and wetland processes, Plenum.

Warnken, K. L. 1998. Sediment-water exchange of trace metals and nutrients in Galveston Bay, Texas. Master's thesis, Texas A&M University.

———, Griffin, and G. Gill. 1994. Benthic exchange of Nutrients and trace Metals in Galveston Bay, Texas. EOS, **75:** 331.

————, and P. Santschi. (Submitted). Estuarine trace metal distributions in Galveston Bay II: Solid phase speciation and heterogeneous processes. Geochim. Cosmochim. Acta.

Wen, L.-S., P. Santschi, G. Gill, R. Lehman, and C. Paternostro. 1997a. Colloidal and particulate silver in river and estuarine waters of Texas. Environ. Sci. Technol. **31:** 723–731.

————, P. Santschi, and C. Paternostro. 1994. Trace metal speciation in the waters of the Trinity River Estuary. EOS **75:** 122.

————, P. Satschi, G. Gill, and C. Paternostro. (In review). Estuarine trace metal distributions in Galveston Bay I: Colloidal forms and dissolved phase speciation. Mar. Chem.

————, P. Satschi, and D. Tang. 1997b. Interaction between radioactively labeled colloids and natural particles: Evidence for colloidal pumping. Geochim. Cosmochim. Acta **61:** 2867–2878.

————, M. C. Stordal, D. Tang, G. A. Gill, and P. H. Santschi. 1996. An ultraclean cross-flow ultrafiltration technique for the study of trace metal phase speciation in seawater. Mar. Chem. **55:** 129–152.

Whitfield, M., and D. R. Turner. 1979. Water-rock partition coefficients and the composition of seawater and river water. Nature **278:** 132–137.

————. 1982. Chemical periodical and the speciation and cycling of the elements. *In:* C. S. Wong, et al. [eds.], Trace metals in seawater, Plenum.

Whitehouse, B., P. Yeats, and P. Strain. 1990. Cross-flow filtration of colloids from aquatic environments. Limnol. Oceanogr. **35:** 1368–1375.

Windom, H. I. 1992. Contamination of the marine environment from land-based sources. Mar. Pollut. Bull. **25:** 32–36.

————, J. T. Byrd, R. G. Smith, Jr., and F. Huan. 1991. Inadequacy of nasqan data for assessing metal trends in the nation's rivers. Environ. Sci. Technol. **25:** 1137–1142.

————, R. Smith, and C. Rawlinson. 1989. Particulate trace metal composition and flux across the southeastern U.S. continental shelf. Mar. Chem. **27:** 283–297.

————, R. Smith, C. Rawlinson, M. Hungspreugs, S. Dharmvanij, and J. Wattayakorn. 1988. Trace metal transport in a tropical estuary. Mar. Chem. **24:** 293–305.

————, S. L. Schropp, F. D. Calder, J. D. Ryan, R. G. Smith, Jr., L. C. Burney, F. G. Lewis, and C. H. Rawlinson. 1989. Natural trace metal concentrations in estuarine and coastal marine sediments of the southeastern united states. Environ. Sci. Technol. **23:** 314–320.

Chapter 11

Natural Organic Matter Binding of Trace Metals and Trace Organic Contaminants in Estuaries

Peter H. Santschi, Laodong Guo, Jay C. Means, and M. Ravichandran

INTRODUCTION

Natural organic matter (NOM) can interact strongly, both chemically and physically, with trace metals and trace organics. Therefore, NOM can profoundly affect speciation of trace substances and their fate and transport in the aquatic environment. In this chapter, we will first provide general concepts of organic complexation with trace substances, followed by the role of NOM in affecting trace metal and trace organics cycling in estuarine and coastal waters of the Gulf of Mexico.

Characteristics of NOM

The concentration of NOM is analytically expressed in units of μM or mg of carbon per liter of water. It is operationally defined as particulate (POC),

Biogeochemistry of Gulf of Mexico Estuaries, Edited by Thomas S. Bianchi, Jonathan R. Pennock, and Robert R. Twilley.
ISBN 0-471-16174-8 © 1999 John Wiley & Sons, Inc.

dissolved (DOC), and colloidal (COC) organic carbon. Concentrations of NOM in natural waters are typically in the mg liter^{-1} range. This material is generally composed of hundreds of compounds such as proteins, polysaccharides, lipids, and humic or fulvic acids. NOM is heterogeneous in composition, size, or molecular weight, and radiocarbon age (e.g., Sharp 1973; Mantoura and Woodward 1983; Guo et al. 1996) and is polymeric, polyelectrolytic, polyfunctional, polydisperse, internally porous, and hydrated to a variable degree (Buffle 1990); furthermore, its conformation is affected by pH and ionic strength. One of the most important properties of macromolecular NOM is its amphilic character, that is, the fact that each macromolecule contains both hydrophilic and hydrophobic parts, which can strongly affect its sorption and complexation behavior to charged and uncharged trace molecules and inorganic ions. The interactions between NOM and trace metal ions and hydrophobic organic contaminants (HOC) are depicted in Fig. 11-1. Because of the polydispersivity and organization (e.g., formation of aggregates or micelles) of NOM, average chemical (e.g., complexation, acid-base, or redox) and physical (e.g., age, residence time, exchange time) properties are often not very meaningful parameters, and distribution spectra of these properties are used instead (Buffle 1990).

Characteristics of Trace Elements and Trace Organics

NOM in aquatic environments is found in all size and molecular weight ranges. Typically, it is associated with mineral particle surfaces, where photochemical and microbiological processes are facilitated (Fig. 11-1B). Trace elements and trace organics occur in natural waters at very low concentrations, usually one part per billion (i.e., μg liter^{-1}) or less. Since NOM concentrations in estuarine and coastal waters are typically 100–500 μM C (Guo and Santschi 1997a), with functional group concentrations of up to μM levels, there usually is adequate complexation capacity for trace metals. For example, Santschi et al. (1995) reported a concentration of 1.4 meq/g of proton reactive sites in macromolecular organic matter (>1 kDa) from Galveston Bay and the Gulf of Mexico, resulting in a concentration of 1–10 μM of proton reactive sites. These concentrations are more than sufficient to potentially complex most trace metals in estuarine and coastal waters, provided that a significant number of the proton reactive sites are strong binding sites for certain trace metals.

Since NOM in aquatic systems is amphilic, with significant hydrophobic parts at pH = 8, trace organics can sorb to NOM both through van der Waals interactions to hydrophobic parts and through electrostatic interactions to more hydrophilic parts. A significant fraction of the organic matter in freshwater systems is composed of humic substances. However, natural humic substances vary widely in their capacity to sorb hydrophobic trace organic (HTO) compounds (Chiou et al. 1987).

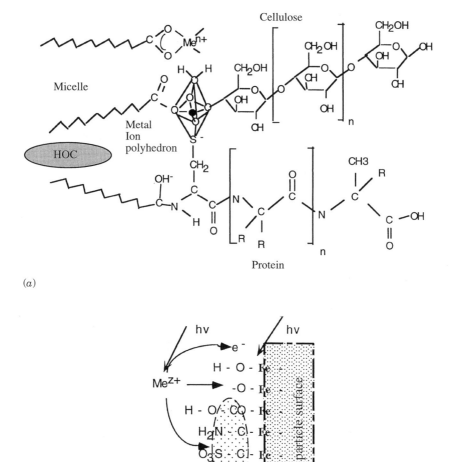

FIG. 11-1. Schematic of DOM complexation of HOC (A) and trace metal ions (B).

COMPLEX FORMATION BETWEEN METAL IONS AND ORGANIC LIGANDS

Trace metal complexation in the simplest case can be expressed as a ligand exchange

$$M_1L_1 + M_2L_2 = M_1L_2 + M_2L_1 \qquad (11\text{-}1)$$

where

M_1 = trace metal
M_2 = major metal or proton
L_1 = water or inorganic ligand
L_2 = organic ligand

Ligand exchange between trace metals and natural organic compounds can, however, be slow to occur (Hering and Morel 1989). The specific functional groups of NOM, such as -COOH, -OH, -NR$_2$, and -SR$_2$, with R as -CH$_2$- or -H, can strongly complex (or chelate) trace metals in natural waters. Strong functional groups for trace metal complexation can be found in

- Siderophores and other phytochelatins
- Biopolymers (e.g. proteins and lipopolysaccharides)
- Humic substances

While there is equivocal evidence that humic substances are significant chelators for trace metals other than Hg (Mantoura et al. 1978), recent evidence suggests the importance of NOM ligands in the form of siderophores and other phytochelatins (Grill et al. 1985, 1987; Donat and Bruland 1995) as strong trace metal chelators in the organic matter of fresh and sea-water environments. Moreover, in estuarine and coastal waters, complexation by dissolved organic matter was suggested to be important for trace metals such as Cu and Zn (Van den Berg et al. 1987; review in Donat and Bruland 1995).

The binding constants (K) of relevant functional groups for transition metals follow the Irving-Williams Series, as they are a function of ionic radius, atomic number, and valence. Traditional models of trace metal binding to natural organic macromolecules include discrete-site and multi-site types. Discrete-site models characterize the NOM as individual ligands of one or two types (e.g., carboxyl and phenol groups) with specific K values (Perdue and Lytle 1983; Dzombak et al. 1986). The multi-site approach expands on the discrete-site approach by allowing for a range of K values for each type of site. The site distribution model allows for the inherent complexity of NOM by using a continuous distribution of non-identical functional groups (Perdue and Lytle 1983; Dzombak et al. 1986). However, most marine electrochemists who study trace metal complexation to NOM still use discrete-site models (e.g., Donat and Bruland 1995 and references therein).

An alternative approach to the two traditional models has been offered by Tipping and Hurley (1992) and Tipping (1993). In this model, binding is assumed to occur at eight distinct functional groups distinguished by individual intrinsic K terms. Like many theoretically based approaches, this approach incorporates terms that account for the molecular weight, size, and electrostatic behavior of NOM molecules. This model differs from other approaches by incorporating an empirical term for electrostatics and by including terms for metal binding in the diffuse layer as counter ions. Since this approach does not require elaborate numerical solution techniques, it is more easily incorporated into transport and geochemical equilibrium codes (Tipping 1994).

There is no reason to assume that NOM and NOM-trace metal complexing in the Gulf of Mexico is much different from that in other marine environments; therefore, many results from work carried out in other estuarine and coastal waters should be directly applicable to Gulf coast estuaries. At the very least, they should be useful for comparative purposes.

There are several experimental approaches for studying trace metal complexation to NOM. The chemical approach involves the direct determination of concentrations of electrochemically labile trace metals and ligands and their specific binding constants (Buffle 1990; Donat and Bruland 1995). For example, results from potentiometric titrations of COC_1 (COC >1 kDa) indicate apparent acidity constants, pK_{app} of about 4.9, 7.1, and 9.5 (Fig. 11-2; Honeyman et al. unpublished results). Alternative approaches use phase speciation and chemical competition techniques to isolate metal-organic compounds, either by specific resin adsorption, membrane separation, or organic solvent extraction techniques (Buffle 1990). For the Gulf of Mexico estuaries and coastal

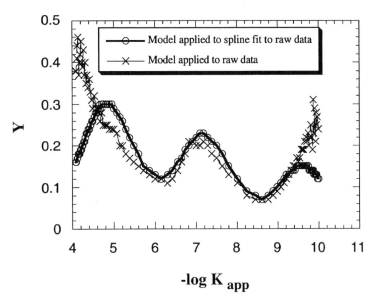

Apparent CA affinity distributions of COC obtained from charge-pH curves

FIG. 11-2. Results from complexometric titrations of COM. (Honeyman et al. unpublished results).

waters, a limited number of trace metal speciation studies have been carried out, which will be described in later sections.

TRACE ORGANIC SORPTION BY NOM

Trace organic sorption by NOM of neutral hydrophobic compounds and ioniz-able charged molecules proceeds by different mechanisms. Therefore, the two processes will be considered separately.

Hydrophobic (Neutral) Trace Organic Sorption on NOM

Among the various interactions that influence the transport and fate of organic contaminants in environmental systems, sorptive interactions with various particulate materials are considered to be of prime importance in determining the ultimate fate and, to some degree, the toxicity of organic contaminants to target species. Sorption may also play a role in determining the volatilization and bioavailability of the contaminants of concern.

Sorption results when a component in the gas phase or in solution is concentrated at an interface (Bailey and White 1970). For sediment-water systems, the interface of primary concern is the liquid-solid interface. The sorbing species is referred to as the "solute" when it is in solution and as the "sorbate" when it is in contact with the solid (sorbent) surface. Sorption occurs when the forces of attraction between the sorbate and the solid (sor-bent) surface are greater than the sum of the repulsive forces between the solid and the sorbate and the forces of attraction between the solute and the solvent (Adamson 1976).

For neutral hydrophobic organic chemicals (NHOC) or solutes having low solubilities in water, sorption occurs due to a weak solute-solvent interaction rather than a large specific sorbate-sorbent interaction. Here, even very weak sorbate-sorbent attraction (e.g. van der Walls forces; Horvath and Melander 1978) will overcome the weak solute-solvent interaction and result in the sorption of the compound from solution (Mackay 1979; Mackay and Patterson 1981). A weak interaction between solute and sorbent is the result of a large decrease in the entropy of the system upon solvation (Karickhoff et al. 1979; Means et al. 1980; Karickhoff 1984) and the absence of significant hydrophilic character in the molecule. This process is termed "hydrophobic sorption" because of the existence of the weak solute-solvent interaction.

In research related to the fate and transport of HOC in sediment-aquatic systems, the role of sediment organic matter in controlling the sorption process of HOC was established, and predictive equations for the estimation of K_{oc} (partition coefficient between particulate organic and water) from compound properties such as aqueous solubility or K_{ow} (octanol/water partition coeffi-cient) were developed (Karickhoff et al. 1979; Means et al. 1980, 1982). In subsequent research, the roles of pH and ionic strength in modifying aqueous sorption properties were also established (e.g., Means 1995). These predictive equations are now commonly used in both the geochemical and environmental regulatory processes.

$$\log K_{oc} = 1.00 \log K_{ow} - 0.317 \quad (r^2 = 0.980) \tag{11-2}$$

$$\log K_{oc} = -0.686 \log S + 4.273 \quad (r^2 = 0.933) \tag{11-3}$$

where K_{ow} is the octanol-water partition coefficient and S is the aqueous solubility expressed in terms of μg ml^{-1}. Other molecular parameters, such as molecular topology, have been proposed as predictors of the nature and strength of sorption to sediment organic matter (Sabljic 1984). Other forms of the same equations are currently used in literature as well (e.g., Kenega and Goring 1980; Curtis et al. 1986; Henry et al. 1989), and differences reflect the nature of the sediments tested.

The limits of the predictability of solute sorption using these models stem from two sources. The first is the breakdown of some of the assumptions required to use these models when dealing with some real-world aquatic systems. In many systems of environmental concern, the assumptions of homogeneity and rapid, reversible equilibration between the liquid and solid phases do not hold due to kinetic or steric restrictions on equilibration. The presence of colloidal organic matter may also significantly alter some of the K_d constants used. Secondly, for compounds that are highly sorbed, the sorption capacity of the sorbent may be exceeded, particularly in low organic carbon solids, causing the assumption of non-saturation of phases to be exceeded. As the sorption capacity of the sediment is approached, non-linearity in the sorption isotherm may result in either self-adsorption on the solid phase or micelle formation in the solution phase. Evidence from batch sorption experiments suggests that non-linearity in the sorption constant (K_d) for such compounds may be observed experimentally at values exceeding 40–60% of solute solubility (Karickhoff 1984). Multiple mechanisms of interaction between a contaminant and sediments may also lead to deviations from predicted sorption behavior (Means et al. 1982). The reversibility of sorption processes onto organic matter has been the source of some disagreement in the literature. DiToro and Horzempa (1982) reported that sorption of certain polychlorinated biphenyls (PCBs) to sediment organic matter was irreversible, while other investigators reported the complete reversibility of sorption equilibria. Freeman and Cheung (1981) proposed a gel matrix model of sediment organic matter that included a three-dimensional network of highly branched polymeric material that could swell and contract in response to solution properties. Gschwend and Wu (1985) proposed a three-dimensional diffusion-based model of the sorption-desorption process to explain possible differences in equilibria. Koulermous (1989) demonstrated that desorption from contaminated sediments followed a Fickian diffusion-based model for single compounds. Means and McMillin (1993) demonstrated that a reversible diffusion-based model predicted the behavior of several polycyclic aromatic hydrocarbons (PAHs) measured simultaneously in a contaminated sediment bed.

The role of colloidal organic matter in the sorption of pesticides and PAHs was discovered by Means and co-workers (Means and Wijayaratne 1982; Wijayaratne and Means 1984a, 1984b) and has subsequently been confirmed in numerous other laboratories for PCBs (Brownawell and Farrington 1985; Hassett and Anderson 1979; Burgess et al. 1996) and chlorophenols (CPs) (Carter and Suffet 1982; Caron and Suffett 1989; Henry et al. 1989). Wijayaratne and Means (1984b) showed that the regression of aqueous solubilities for several PAHs versus colloidal organic matter K_{doc} values yielded results parallel to those obtained on sediment. That equation was

$$\log K_{doc} = -0.693 \log S + 4.851 \quad (r^2 = 0.985) \tag{11-4}$$

Similarly, the partition coefficient of HTO to dissolved organic matter (DOM), K_{dom} $(= C_{HTOb} / (C_{HTOf} \times C_{dom}))$, can be described (Henry et al. 1989) as

$$\log K_{dom} = 0.59 \log K_{ow} + 1.37 \tag{11-5}$$

with C_{HTOb} = bound HTO concentration, C_T, increased due to HTO binding to DOM, as follows:

$$C_T = C_{HTOf} + C_{HTOb} = C_{HTOf} (1 + C_{dom} \times K_{dom}) \tag{11-6}$$

HTO concentrations can thus be significantly higher when associated with dissolved or colloidal organic matter (e.g., Wershaw et al. 1969; Porrier et al. 1972; Carter and Suffett 1982; Kile and Chiou 1989; Rees 1991 and references therein). However, their bioavailability or toxicity is often changed in the presence of organic matter (McCarthy 1989; Alexander 1995; Campbell 1995).

The behavior of neutral hydrophobic molecules in association with colloidal organic materials appears to operate by similar mechanisms (e.g., hydrogen bonding and hydrophobic association). The observation in several laboratories that the sorptive capacity of aquatic organic matter may be highly variable suggests that a full understanding of the mechanisms has not yet been achieved. For example, Means and Wijayaratne (1982) observed that sorption K_{doc} values varied significantly for colloids collected in different parts of an estuary. They went on to show that sorption constants were influenced by both salinity and pH changes in the equilibration mixture for both slightly polar molecules (Means and Wijayaratne 1982) and neutral compounds (Wijayaratne and Means 1984a, 1984b). Kile and Chiou (1989) showed that sorption constants, as measured by solubility enhancement, varied significantly for several types of humic acid isolates. In the Mississippi River, significant proportions of atrazine and other herbicides were found in association with colloidal phases (Periera and Rostad 1990).

While the nature of the hydrophobic bond has been investigated with a variety of biomolecules, the nature of aquatic colloidal organic matter associations with contaminant molecules remains relatively unstudied. Hydrophobic interactions of biomolecules are understood to be about 10-fold stronger than van der Waals forces and operate in the range of 0–10 nm in diameter, with the magnitude of the interactive force decreasing exponentially with distance (Israelachvili and Pashley 1982). While hydrophobic binding constants (K_b) for substrate-enzyme binding sites in the range of 10^{+10} to 10^{+12} are relatively common, a number of investigators have argued that contaminant-colloid binding constants cannot exceed K_{oc} values derived from sediments.

Further improvements in predictions of partition coefficients of hydrophobic trace organics are based on the knowledge of a polarity index of organic matter, defined as a mass ratio of $(O + N) / C$, in addition to K_{ow} values of trace organic contaminants (e.g., Xing et al. 1994; Kile et al. 1995b). A relationship between K_{oc} and the polarity index (PI) for nonionic organic contaminants was established by Xing et al. (1994):

$$\log K_{oc} = 1.83 + 0.625 \log K_{ow} - 2.34 \text{ PI} \qquad (11\text{-}7)$$

For the trace contaminants they studied, Xing et al. (1994) found that values of K_{oc} decreased with increasing P.I. of organic matter, suggesting that values of K_{oc} may vary with the age and diagenetic status of the organic phase.

Sorption of Charged (Hydrophilic) Organic Molecules to NOM

The primary mechanisms proposed to explain the sorption of hydrophilic organic molecules to NOM and inorganic surfaces involve ligand exchange (Zierath et al. 1980; Means and Wijayaratne 1989; Gu et al. 1995), cation bridging (Baham and Sposito 1994), and entropy-driven physical interactions (Jardine et al. 1989; Means and Wijayaratne 1989). Early work by Hunter and Liss (1979) established that colloidally stable particles from a number of estuaries exhibited a negative charge at all salinities encountered naturally but that the magnitude of the negative charge did decrease with increasing salinity due to interactions with polyvalent solute ions such as Ca^{+2} and Mg^{+2}. Thus, it would be expected that positively charged solutes would be strongly attracted to such colloidal materials.

Bartha and Pramer (1967) reported that, while some of the herbicide-derived chloroanilines may be condensed in soils to form azobenzene derivatives by microbial oxidases and peroxidases, the bulk of the aniline moiety is tightly complexed to soil organic matter by physical and chemical sorption. As demonstrated with 3,4-dichloroaniline, the difficulty in extracting anilines by organic solvents or acid and alkaline hydrolyses (Chisaka and Kearney 1970; Bartha 1971) suggests that these compounds may be partially immobilized in the soil matrix. Decomposition studies of aniline derivatives in soils indicated that the dichloroanilines were more persistent than aniline or the monochloroanilines. It appears that anilines may be bound to humus or soil organic matter rather than to clay particles (Hsu and Bartha 1974).

Bond stabilities and model reactions with monomeric constituents of humic substances suggest covalent binding of the residue by at least two distinct mechanisms: in a hydrolyzable (probably aniline and anilinoquinone) and in a non-hydrolyzable (probably heterocyclic rings and ether bonds) manner (Hsu and Bartha 1974). Parris (1980) reported the results of binding experiments with aromatic amines and compounds that serve as models of humate functional groups (e.g., carbonyls and quinones). He postulated that primary ring substituted anilines bind covalently to soil organic matter (e.g., humates) via reactions with carbonyl and quinone moieties. The proposed mechanism of binding involves two phases. Initially, a reversible rapid equilibrium is established with the formation of an imine linkage with the humate carbonyls.

Wijayaratne (1982) studied the sorption of a variety of aromatic amines on estuarine colloidal organic matter. Basically, as the structural characteristics of the basic aromatic amine nucleus (aniline) were modified by the addition of one or more substituents (methyl groups or halide atoms), the behavior of the polar compound adhered increasingly to the behavior of a neutral hydrophobic molecule. Zierath et al. (1980) observed the same basic phenomenon on sediment NOM as pH was altered to yield different charged forms of the aromatic amine benzidine. Other possible mechanisms of interactions of charged organic molecules with NOM include ion exchange (anion and cation), proton exchange, hydrogen bonding, van der Waals interactions, and chemisorption (reversible or irreversible covalent bond formation). However, there is little documentation in the literature of these mechanisms (Baham and Sposito 1994).

Transport of NHOCs in the solution phase as a result of aqueous leaching from contaminated sediments is expected to be quite low. This is based on their low aqueous solubility (<100 ppb) and high octanol-water partition coefficient (log K_{ow} >3). These properties can be used in one of many available regression equations (Means et al. 1980) to predict sediment partition coefficients normalized to organic carbon (K_{oc}) values of >3.5. Many theories have been proposed to explain the "facilitated" transport of hydrophobic organics in aquatic systems, including association with colloids and/or microparticles (Means and Wijayaratne 1982, 1984; Sigleo and Means 1990) and combination with natural surfactants to form micelles. In these systems, a three-phase partitioning process is established that may influence both the flux rates and equilibria in the systems and the bioavailability of contaminants (Fig. 11-3). Organic colloids are likely the primary facilitators at contaminated sediment sites at which high dissolved organic matter concentrations are observed in interstitial waters. Two important questions to be addressed in extending the knowledge of solute transport and bioavailability in water-sediment systems to reflect the potential presence of colloids are: (1) what are the governing theories for the sorption processes? and (2) what are the kinetic and equilibrium characteristics of sorption and desorption?

CHEMODYNAMIC PROCESSES

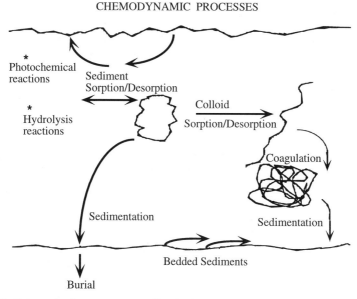

FIG. 11-3. *Schematic of processes controlling the fate and transport of trace metals and organics in dynamic hydrologic environments.*

Hydrophobic compounds that readily associate with sediment particles or colloids are nevertheless bioavailable to both benthic and pelagic organisms in aquatic systems. Benthic infaunal invertebrates rapidly accumulate residues of sediment-bound contaminants such as PCBs, chlorinated hydrocarbons (CHs), and PAHs in their tissues (Foster et al. 1987; McElroy and Means 1988; Means and McElroy 1997). Davis and Means (1989) observed that both the number and the type of benthic infaunal organisms profoundly influenced contaminant mobilization from bedded sediments in both the dissolved and particulate flux rates. These preliminary studies, however, raised many more questions concerning organism-sediment interactions as they relate to contaminant mobilization and bioavailability to other components in aquatic ecosystems. Among the issues raised were the degree to which infaunal communities influence bioavailability to other infaunal organisms, benthic organisms, epibenthic organisms, and pelagic organisms. Hawker and Connell (1986) reported that different bioaccumulation factors must be applied to different phyla or trophic levels of aquatic organisms in order to estimate bioaccumulation correctly from log P (P is the partition coefficient). They have further pointed out that the extrapolation of regressions of bioaccumulation factors (BAFs) versus log P to kinetic relationships between uptake and depuration rate constants leads to the experimentally unsupported conclusion that uptake rate constants must increase without an upper limit with increasing log P (Hawker and Connell 1985; Connell 1988). This conclusion breaks down experimentally due to a number of factors, including uptake efficiencies across membranes, steric limitations on membrane transport due to molecular size and shape, and mass flow characteristics of respiratory apparatus of aquatic organisms. The issue of food-chain transfers of contaminants between trophic levels is another important issue that has not been studied in controlled systems (Rubinstein et al. 1983). However, data from a limited set of cases suggest that food uptake is the major route of uptake of many NHOCs and several metals (Young et al. 1984). This is especially true when sediment is the food supply of the organism.

The bioavailability of contaminants from sediments is dependent upon numerous factors, including redox, pH, lipophilicity, organic carbon content, sediment particle size, water dynamics, organismic localization relative to sediments, contaminant localization in sediment, and sediment flux (Farrington 1991). Figure 11-3 shows an expanded view of some of the critical chemodynamic processes that determine the distribution of chemical species associated with contaminated sediments in solution, colloidal, suspended, and bedded sediment phases. Considering this model, it is clear that there is the potential for a variety of phase exchanges to occur during normal seasonal re-suspension events or during dredge material disposal operations, and these exchanges may involve both rapid and slow processes and therefore both equilibrium and non-equilibrium kinetics.

COLLOID-ASSOCIATED TRACE ELEMENTS AND TRACE ORGANICS

Trace Elements

Colloidal organic matter (COM) from Gulf coast estuaries such as Galveston Bay seems to have a relatively low titratable proton concentration. For example, Honeyman et al. (unpublished results) found that the proton concentration of COM_1 (>1 kDa) from Galveston Bay is only 1 meq g^{-1} organic matter, with pK_a values in the range of 4–5, 7, and 9 (Fig. 11-2). The total proton concentration of COM_1 is similar to that of polysaccharides, which agrees with the 50% polysaccharide composition of Benner et al. (1992) of Gulf coast COM.

Recent application of cross-flow ultrafiltration techniques has allowed for more detailed studies on phase speciation of trace elements (Buesseler et al. 1996; Wen et al. 1996). Some recent available results on colloidal trace metals in Gulf of Mexico estuaries are listed in Table 11-1 for data on colloidal phases and in Table 11-2 for data on isolated colloidal organic material. Data from estuaries of other part of the world are also given in Table 11-1 for comparison.

It is clear from recent studies in Gulf of Mexico estuaries that most trace metals measured are, at least to some extent, associated with NOM in a colloidal form (Tables 11-1 and 11-2). For example, up to 80–90% of dissolved Hg was associated with COM in six Texas estuaries (Stordal et al. 1996a) and in the

TABLE 11-1. A Comparison of Colloidal Trace Metal Concentration and Percentage in Riverine and Estuarine Waters

River or Estuary	Metal	[Me]	Prefilter Size (μm)	% in >10-kDa Colloids	% in >1-kDa Colloids	Reference
Ochlockonee Estuary	Hg	3.4–24 pM	0.4	3–40	37–88	Guentzel et al. (1996)
Ochlockonee Estuary	Fe	41–6980 nM	0.4	14–86	71–97	Powell et al. (1996)
	Mn	11–437 nM		0–30	1–30	
	Ni	2.5–5.5 nM		4–45	5–78	
	Cu	2–5.4 nM		1–15	15–66	
	Cd	36–54 pM		5–48	5–66	
Texas estuaries	Hg	0.12–13.6 pM	0.45	—	12–93	Stordal et al. (1996a)
Texas estuaries	Ag	37–59 pM	0.45	20–60	25–85	Wen et al. (1997b)
Galveston Bay	Cu	3.5–18 nM	0.45	1–23	48–66	Wen et al. (in review)
	Ni	4.6–32 nM		12–26	12–47	
	Pb	0.09–0.7 nM		2–57	52–88	
	Cd	0.03–0.17 nM		9–41	23–61	
	Co	0.37–0.6 nM		4–16	6–31	
	Zn	1.4–32 nM		14–73	82–97	
	Fe	4.7–1120 nM		38–98	49–99	
Medway River	Al	4070 nM	0.4	66	n.a.	Whitehouse et al. (1990)
	Fe	1700 nM		73	n.a.	
	Mn	890 nM		45	n.a.	
	Cu	11.8 nM		64	n.a.	
Rhone delta	Cd	0.09–0.3 nM	0.4	0–38	n.a.	Dai et al. (1995)
	Cu	3–23 nM		20–39	n.a.	
	Ni	5–21 nM		0–18	n.a.	
Venice Lagoon (Italy)	Cd	0.01–0.12 nM	0.4	1.2–63	n.a.	Martin et al. (1995)
	Cu	5.6–15.9 nM		21–59	n.a.	
	Fe	7.7–1125 nM		67–99	n.a.	
	Ni	6.6–20.9 nM		1.1–34	n.a.	
	Pb	0.05–1.08 nM		3.1–94	n.a.	
	Mn	11–44.2 Mn		24–76	n.a.	
Amazon	U	0.12–14 nM	0.45	15–92	n.a.	Swarzenski et al. (1995)
San Francisco Bay	Al	0.09–20 μM	0.2	3–99	n.a.	Sanudo-Wihelmy
	Ag	10–75 pM		0–93	n.a.	
	Cd	0.15–1.1 nM		0–9	n.a.	
	Cu	4.4–36 nM		1–18	n.a.	
	Fe	0.7–344 nM		4–88	n.a.	
	Mn	28–641 nM		2–20	n.a.	
	Ni	7–40 nM		0–2	n.a.	
	Sr	2.9–82 μM		0	n.a.	
	Zn	3.1–29 nM		1–3	n.a.	

TABLE 11-2. Comparison of Trace Metal Concentration in Estuarine and Coastal Colloidal Organic Materials

Location	Metal	Ultrafilter Size	$[Me]_{com}$ ($\mu g\ g^{-1}$)	Reference
Chesapeake Bay	Al	1 nm	$3.6–140 \times 10^{-3}$	Sigleo and Helz (1981)
	Fe		$3.4–71 \times 10^{-3}$	
	As		3–19	
	Ba		45–1290	
	Ce		5.8–140	
	Co		3.4–116	
	Cr		40–279	
	Mn		125–10190	
	V		4.3–296	
	Zn		143–608	
Galveston Bay	Al	10 kDa	$3–56 \times 10^3$	Benoit et al. (1994)
	Fe		$3–24 \times 10^{-3}$	
	Mn		80–145	
	Cu		13–33	
	Pb		29–69	
	Zn		270–650	
Vineyard Sound[a]	50 elements	1 kDa	n.a	Bertine and Vernon Clark (1996)
Galveston Bay	Cu	1 kDa	34.3	Guo et al. (in press)
	Pb		0.45	
	Zn		13.6	
	Cd		0.14	
	Co		1.90	
	Ni		21.2	
	Cr		15.0	
	Fe		255	
	Mn		8.14	
	V		1.93	
	Ba		19.8	

[a] Concentrations of metals in isolated colloids were measured but reported as pM, i.e., metals concentrations in seawater.

Ochlockonee estuary (Guentzel et al. 1996). Other trace metals that mostly complexed with COM include Ag, Cu, Zn, Cd, and Pb in Texas estuaries (Wen et al. in review, 1997b) and Ni, Cu, and Cd in the Ochlockonee estuary (Powell et al. 1996). Almost all dissolved Fe was also found in the high molecular weight (HMW) fraction (e.g., Powell et al. 1996; Wen et al. in review), but it is unclear whether Fe is complexed with COM or exists as Fe-oxyhydroxides. Even though one can strictly conclude from these ultrafiltration studies only that a certain trace metal is found in the colloidal fraction, the fact that concentrations of inorganic ions are usually low in this fraction allows one to assume that it is macromolecular organic matter that is the main carrier phase. For example, a strong linear relationship between COC and colloidal Hg was found in Galveston Bay (Stordal et al. 1996a). Significant relationships between colloidal metals (e.g., Ni, Cu, Zn, Co) and COM were also observed in Galveston Bay waters

(Wen et al. in review). Powell et al. (1996) reported a linear correlation between Ni and Cu with carbon in the medium molecular weight colloidal fraction in the Ochlockonee estuary in the Gulf of Mexico. Similar results were recently reported for other estuaries (Table 11-1 and references therein).

Examples of estuarine distributions of trace metals are given in Figs. 11-4 to 11-6. These figures show that a major fraction of all these metals (Cu, Pb,

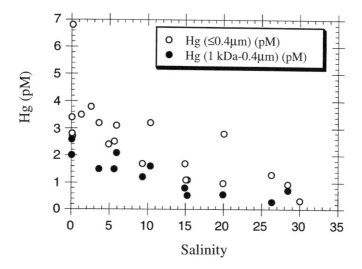

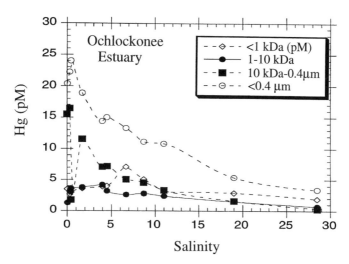

FIG. 11-4. *Examples of Hg phase speciation studies in Galveston Bay (modified from Stordal et al. 1996a) and Ochlockonee Bay (modified from Guentzel et al. 1996 with kind permission by Elsevier Science).*

July 1995

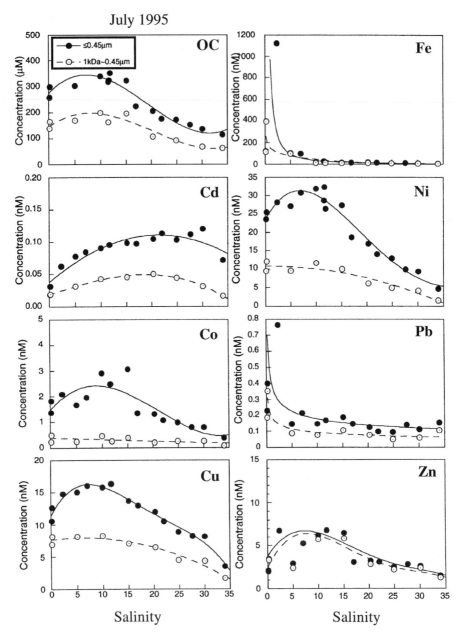

FIG. 11-5. Example of phase speciation study in Galveston Bay: Cu, Pb, Cd, Co, Fe, Ni, and Zn in terms of concentrations (from Wen et al. in review with kind permission by Elsevier Science).

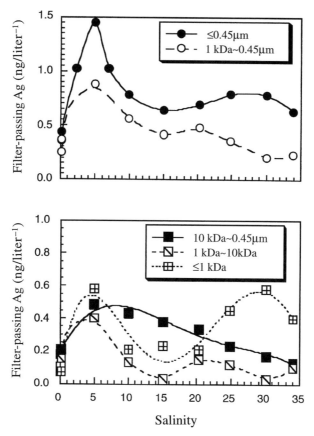

FIG. 11-6. *Example of phase speciation study in Galveston Bay: Ag (modified with permission from Wen et al. 1997b. Copyright 1997 American Chemical Society).*

Cd, Co, Fe, Hg, and Ag) is in the colloidal fraction and that these results likely have much broader significance. Contrary to earlier work, all recent studies involved strict sampling and processing protocols, which included con-tamination control with low blanks, good mass balance, and optimized condi-tions (e.g., Stordal et al. 1996a; Wen et al. 1996). The fact that even Ag, a metal that previously was found not to be organically complexed to the HMW fraction, that is, >10 kDa (Miller and Bruland 1995), indicates that functional groups other than the ordinary carboxylic and hydroxylic groups are involved. Wen et al. (1997b) invoked the presence of organic sulfhydryl groups as the major complexants for B metals such as Ag using a smaller pore size ultrafilter, that is, 1 kDa. Guentzel et al. (1996) also invoked thiol-type binding sites in the COC to explain the behavior of colloidal Hg.

Interestingly, there is a significant relationship between the colloidal metal fraction and the COC fraction (Fig. 11-7). The correlation between $[Me]_c/$

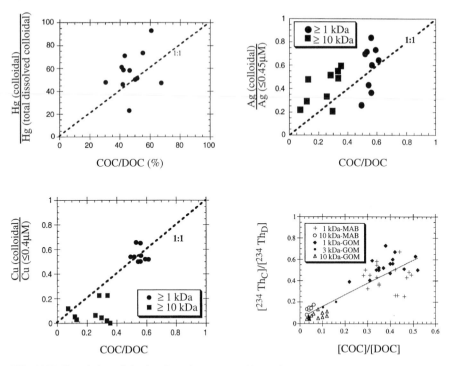

FIG. 11-7. *Correlation of the fraction of trace metal bound to macromolecular organic carbon to the ratio of COC to DOC for Hg (data from Stordal et al. 1996a), Ag (Wen et al. 1997b), Cu (Wen et al. in review), and Th (Guo et al. 1997a), indicating relatively homogeneous complexation of COC.*

[Me]$_d$ and COC/DOC suggests that the organic ligand groups appear to be relatively evenly distributed over the size (or molecular weight) spectrum, as is evident from the data shown for Hg, Ag, Cu, and Th (Stordal et al. 1996a; Guo et al. 1997a; Wen et al. in review, 1997b). Plots of the colloidal fraction of metals (Me$_c$/Me$_{c+d}$) agree, within the scatter of the measurements, with the colloidal fraction of DOC (COC/DOC). Even though there are significant correlations between Ag (or Pb) and Fe in colloids and suspended particulate matter, as shown in Fig. 11-8, this likely does not indicate that the carrier molecule is Fe oxides, but rather that Fe and the B metals such as Ag are complexed by the same functional groups, which are hypothesized to be sulfhydryl groups.

Studies on the phase speciation of metals have also allowed for the calculations of distribution coefficients between particles and the dissolved phase (K_d), particles and the truly dissolved phase (K_p), and colloids and the truly dissolved phase (K_c) in different estuarine environments (Table 11-3). While values of K_d, K_p, and K_c in estuaries and coastal waters of the Gulf of Mexico are very similar for the same element (ranging from 4.4 to 7.8), they are much

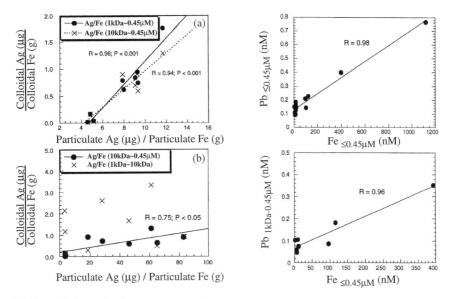

FIG. 11-8. *Relationship between Fe-specific concentrations of (a) Ag and (b) Pb in colloids and suspended particles (from Wen et al. in review, 1997b).*

more variable in other estuaries, such as the Venice Lagoon (Table 11-3), probably due to different particle or colloids composition or differences in sampling protocols. However, the similarity between values of K_d, K_p, and K_c (Table 11-3) suggests that the complexing strength for particles and colloids is similar for trace elements. This result is surprising, since specific surface areas or sites concentrations should increase with decreasing size of the particles or colloids. It thus appears that the specific surface site density of metal complexing ligands decreases with decreasing colloidal or particle size (Guo et al. 1997a). In addition, specific complexing sites for metals might also change as a function of the "freshness" of organic carbon in estuaries. To fully understand this, knowledge of the molecular and functional group composition of NOM is required.

The inverse relationship between a trace metal's K_d value and suspended matter concentration, which has been taken as an indicator for the presence of a colloidal fraction of the trace metal (e.g., Honeyman and Santschi 1988; Benoit et al. 1994; Sanudo-Wilhelmy et al. 1996), disappears when the surface reactive fraction is taken for the calculation of the K_d value and if one accounts for the presence of colloids. This is shown in Fig. 11-9 for Ag. This figure also shows the remarkable similarity of Ag concentrations in colloids to the concentrations in suspended matter. On average, Ag concentrations are only about a factor of 2 lower than those in suspended matter.

Measurements of trace metals on isolated estuarine colloidal material have also been reported recently (see Table 11-2). After complete desalting, isolated

TABLE 11-3. Comparison of Colloid-Water and Particle-Water Partition Coefficients of Metals in Estuarine and Coastal Environments

Location	Metal	log K_d (liter kg^{-1})	log K_p (liter kg^{-1})	log K_c (liter kg^{-1})	Reference
Mississippi	Pb	5.5 ± 0.1	n.a.	n.a.	Trefry et al. (1986)
River Delta	Cd	4.7 ± 0.2	n.a.	n.a.	
	Cu	4.2 ± 0.1	n.a.	n.a.	
	Fe	7.3 ± 0.2	n.a.	n.a.	
	Cr	5.4 ± 0.1	n.a.	n.a.	
	Ni	4.4 ± 0.2	n.a.	n.a.	
	Mn	5.9 ± 0.5	n.a.	n.a.	
Texas	[7]Be	5.5–6.1	n.a.	n.a.	Baskaran and
estuaries					Santschi (1993)
	[210]Pb	4.6–5.7	n.a.	n.a.	
	[234]Th	3.4–5.9	n.a.	n.a.	
Galveston Bay	Fe	4.5–7.3	n.a.	n.a.	Benoit et al. (1994)
	Al	4.1–7.2			
Galveston Bay	Hg	4.6–5.2	5.2–5.6	5.1–5.6	Stordal et al. (1996a)
Sabine-Neches	[7]Be	3.2–4.9	n.a.	n.a.	Baskaran et al. (1997)
	[210]Pb	3.4–4.6	n.a.	n.a.	
Gulf of Mexico	[234]Th	5.9 ± 0.4	6.3 ± 0.5	6.3 ± 0.2	Guo et al (1997a)
Galveston Bay	Cd	4.4 ± 0.3	4.7 ± 0.5	5.3 ± 0.2	Wen et al. (in review)
	Cu	4.1 ± 0.1	4.5 ± 0.2	5.5 ± 0.2	
	Fe	6.9 ± 0.5	7.8 ± 0.2	6.2 ± 0.5	
	Ni	4.2 ± 0.1	4.4 ± 0.1	5.2 ± 0.2	
	Pb	5.3 ± 0.3	5.8 ± 0.4	5.7 ± 0.3	
	Zn	5.3 ± 0.2	6.4 ± 0.4	6.4 ± 0.4	
Texas Rivers	Ag	5.0–5.5	5.0 ± 0.3	n.a.	Wen et al. (1997b)
Amazon Shelf	U	n.a.	2.5–5.2	n.a.	Swarzenski et al. (1995)
Venice Lagoon	Cd	n.a.	0.78–7.3	1.3–4.8	Martin et al. (1995)
	Cu	n.a.	1.5–36	1.3–8.8	
	Fe	n.a.	263–3525	12–2982	
	Ni	n.a.	0.35–1.5	0.06–2.8	
	Pb	n.a.	3.7–44	0.16–98	
	Mn	n.a.	2.8–126	0.9–16	
San Francisco	Cu	4.2–5.6	4.2–5.7	4.4–5.4	Sanudo-Wilhelmy et
Bay Estuary[a]					al (1996)
	Zn	4.8–6.2	4.8–6.2	4.3–4.9	
San Francisco	Cu	n.a.	2.8–3.6[b]	2.8–3.6	Phinney and Bruland
Bay					(1997)

[a] Values of K_c were estimated with measured COC concentrations.
[b] Values of log K_{om} (organic matter-water partitioning) for $Cu(DDC)_2$ and $Cu(Ox)_2$ species.

COM from the Mississippi River plume could contain up to 35–37% of organic carbon (Guo and Santschi 1997b) indicating that isolated colloidal materials are mostly organic in nature. Therefore, higher concentrations of trace metals in isolated estuarine colloidal materials indicate the complexation between COC and trace metals. Using metal:Al (Me:Al) ratios normalized to the Me:Al ratios in the upper continental crust (Wedepohl 1995), enrichment factor (EF) values of trace elements in estuarine colloidal materials are listed

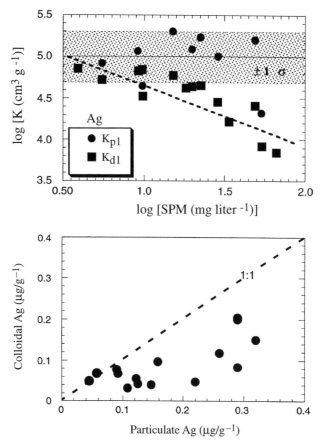

FIG. 11-9. *Particle concentration effect for Ag particle/solution partitioning data from Galveston Bay (from Wen et al. 1997b).*

in Table 11-4. While the EF values in Table 11-4 varied from element to element, higher EF values indicate that these trace elements are highly enriched in estuarine colloidal materials relative to the terrestrial inorganic particles. Higher Me : Al ratios in COM also imply that trace metals are bound to COM preferentially to Al in estuarine environments. Therefore, NOM may play a central role in regulating the fate and transport of many trace metals in estuarine waters.

Another interesting observation with respect to the sorption of metals in the colloidal phase may be derived by plotting trace element : Al ratios to examine the "enrichment" levels of those trace elements relative to expected geochemical abundances. Figure 11-10 shows the element : Al ratios for selected contaminant elements (Means, pers. comm.). While Cr is only slightly enriched in the colloidal phase, Cd, Hg, Pb, Sn, and Cu are all highly enriched. This suggests that there are elevated inputs of these elements in colloidal forms into the Gulf of Mexico from estuarine discharges.

TABLE 11-4. Enrichment Factors for Selected Elements in Estuarine and Coastal Colloidal Material

Location	Element	Me:Al (wt ratio)	EF[b]	Reference
Galveston Bay	Cu	0.62	3,357	Guo et al. (submitted)
	Pb	0.0082	373	
	Zn	0.25	373	
	Cd	0.0035	1,898	
	Co	0.035	233	
	Ni	0.38	1,582	
	Cr	0.27	597	
	Be	0.0019	47	
	Fe	4.63	12	
	Mn	0.15	22	
	V	0.035	51	
	Ba	0.36	4.2	
Chesapeake Bay[a]	As	0.00083	31 ± 10	Sigleo and Helz (1981)
	Ba	0.0202	27.1 ± 21	
	Co	0.00094	23.1 ± 21	
	Cr	0.028	62	
	Fe	0.944	2.4	
	Mn	0.129	19	
	V	0.0024	3.5	
	Sb	0.00037	92	
	Se	0.039	36,378	
	Zn	0.043	64	
Vineyard Sound (MA)	Ti	1.33	33	Bertine and Vernon Clark (1996)
	V	0.073	107	
	Cr	0.41	912	
	Mn	0.049	7.2	
	Fe	5.23	13	
	Ni	0.19	805	
	Co	0.049	328	
	Cu	1.46	7,901	
	Zn	0.21	310	
	As	0.14	5,674	
	Se	0.67	629,546	
	Mo	0.53	29,504	
	Ag	0.00022	312	
	Cd	0.0051	3,865	
	I	16	867,232	
	Ba	0.011	1.3	
	Pb	0.019	89	
San Francisco Bay	Cu	0.0071	38	Sanudo-Wilhelmy et al. (1996)
	Zn	0.0004	0.59	
	Mn	0.022	3.2	
	Fe	0.289	0.72	

[a]Data from summer stations only.
[b]Enrichment factor (EF) is the ratio of [Me/Al]$_{colloids}$ to [Me/Al]$_{reference}$.
Source: Data of element concentrations in the upper continental crust are from Wedepohl (1995).

RATIOS OF SELESCTED TRACE ELEMENTS WITH
ALUMINUM IN SURFACE WATER COLLOIDAL
PHASE SAMPLES FROM TRANSECT S1, CRUISE P92-1

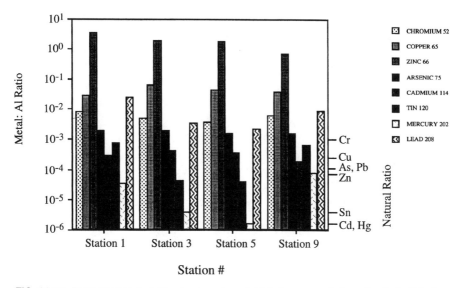

FIG. 11-10. *Ratio of trace metal to aluminum in colloidal phase sample from the Gulf of Mexico.*

Trace Organics

Means and co-workers showed that numerous trace organics, as well as certain trace elements, were partitioned into estuarine and marine colloidal phases and transported long distances in the Gulf coastal currents (Means and McMillin 1995; McMillin and Means 1996; Means and McMillin 1997). Among the compound classes measured that were found to be colloid-associated were PCBs, selected HMW PAHs, *s*-triazine herbicides, and certain chlorinated pesticides. The colloidal phase was observed to be the dominant transport phase for several *s*-triazine herbicides and selected PAHs (McMillin and Means 1996). For example, colloidal fluxes for some trace organics could be comparable with those of dissolved and suspended particulate phases in the coastal zone of the Gulf of Mexico. It is clear that the colloidal and dissolved phases contribute significantly to the transport of many neutral hydrophobic contaminants (Means and McMillin 1995; McMillin and Means 1996).

Near the mouth of the Mississippi River, the rapid mixing processes and hydrodynamics create some very interesting patterns of colloidally associated herbicides (Means and McMillin 1997). The herbicide data suggest that colloidal phase association of *s*-triazines increased through the estuarine mixing zone toward higher salinity. This might suggest different sorption properties of macromolecular organic carbon to HOC compounds in riverine, estuarine, and oceanic environments.

LABORATORY STUDIES WITH NOM

Trace Elements

In addition to the field studies discussed in previous sections, laboratory experiments have been used to test the validity of observed complexation parameters. Some available laboratory-derived K_d or K_c values are compared with those determined in the field (Fig. 11-11). The good agreement of laboratory-derived partition coefficients for a number of trace metals to colloids validates the extrapolation of results from laboratory experiments to field situations.

Macromolecules may not only be good complexants for trace metals, but some may also be surface active, as indicated by coagulation experiments carried out by Stordal et al. (1996b) and Wen et al. (1997c). Rate constants for particle uptake of colloidal Hg show a particle concentration dependency with a power of 0.3, as expected for a Brownian coagulation mechanism (Farley and Morel 1986; Honeyman and Santschi 1989). In addition, rate constants determined from radiotracer uptake experiments of colloidally bound trace metals correlate well with overall particle-water partition coefficients (K_d). Such correlations have been taken to suggest the presence of two distinctly different ligand groups, biopolymers and phytochelatins (Wen et al. 1997c). Biopolymers, as major components of the colloidal pool, would coagulate at a rate that is similar for all complexed trace metals, while phyto-

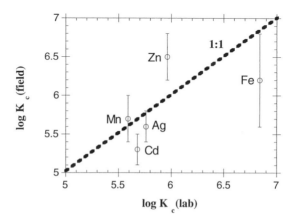

FIG. 11-11. Correlation between partition coefficients for selected trace metals determined experimentally in the laboratory and in the field.

chelatins would show kinetic association as well as coagulation constants that are different for each trace metal ion.

As a consequence of the presence of colloidal forms of trace metals in the filter-passing fraction, particle-water partition coefficients become a function of the suspended particulate matter concentration (e.g., Honeyman and Santschi 1988, 1989; Benoit et al. 1994; Sanudo-Wilhelmy et al. 1996). This effect can become the dominant factor in predicting the solid-solution partitioning of trace metals in Gulf coast estuaries (Benoit et al. 1994).

While trace metal toxicity and bioavailability to aquatic organisms can be generally well predicted by a free-ion model, exceptions are often related to the presence of COM (Campbell 1995). For example, trace metals (in radioactive forms) associated with COM isolated from Galveston Bay were shown to be bioavailable to penaeid shrimp (Carvalho et al. in review).

Interactions of Humic Substances with Trace Contaminants

Humic substances isolated from the Suwannee River, Georgia, have been thoroughly characterized for their structural characteristics and their interactions with various organic and inorganic contaminants (Averett et al. 1995). The Suwannee River, which discharges into the Gulf of Mexico, has its source in the Okefenokee Swamp (a wetland) in southeastern Georgia. The river water is characterized by low pH (4.0), large concentrations of fulvic acids, and a dark color attributed to the dissolved humic substances. The DOC concentration in this river water is about 35–50 mg liter^{-1}.

Kile et al. (1995a) have studied the solubility enhancements of various organic contaminants by humic substances isolated from the Suwannee River. They observed that the magnitude of solubility enhancement for various contaminants was controlled by the molecular size, polarity, and concentration of the DOM and by the inherent water solubility of the solute. The least soluble solutes (e.g., DDT) showed the greatest enhancement in solubility compared to relatively water-soluble compounds such as Lindane. In this study, the partition coefficients for given solutes between DOM and water (K_{dom}) were directly proportional to the solute's water solubility (or octanol-water partition coefficient, K_{ow}) and inversely related to the polarity (as indicated by the elemental analysis) of DOM. The interaction of DOM with organic contaminants (as manifested by the solubility enhancement) is accounted for by a partition-like interaction between the solute and the organic macromolecule.

Metal Binding Studies from Suwannee River

The carboxylic and phenolic functional groups present in the humic molecule, which primarily contribute to their acidic and metal binding character, have been quantified in Suwannee River humic substances by acid-base titration (Bowles et al. 1995). The fulvic acid is estimated to have a carboxyl content of 6.1 meq g^{-1} and a phenolic content of 1.2 meq g^{-1}. In a related study, using proton nuclear magnetic resonance, Noyes and Leenheer (1995) estimated the exchangeable-proton content in Suwannee River fulvic acid to be about 10.5 mmoles g^{-1}. McKnight and Wershaw (1995) studied the complexation of copper ion by Suwannee River fulvic acid. Two types of metal binding sites were identified in fulvic acid molecules. Strong, site-specific binding involved electron-transfer interactions and weak binding sites due to territorial binding of counter ions or electrostatic attraction. While the strong binding sites occur in small numbers, abundance of weak binding sites such as carboxylic groups makes possible common weak binding properties of metals by fulvic acid. However, the presence of trace, non-oxygen-containing functional groups such as groups that contain nitrogen or sulfur makes it possible for strong binding between these groups and several biologically significant metals such as Cd, Co, Cu, Fe, Hg, Pb, Mn, and Ni.

Metal Binding to Mississippi River Organic Matter

Due to the importance of the Mississippi River for draining more than 40% of the continental United States, a number of laboratory studies have been carried out with water samples from this river. For example, Sunda and Ferguson (1983), using bioassay techniques, showed that more than 99% of the Cu in the Mississippi River, as well as in coastal Florida water, is organically bound.

SUMMARY

Evidence is provided in this chapter to show that a large fraction of many trace elements and trace organics is associated with colloidal (or macromolecular) organic matter (COM). The ability of COM to bind both hydrophilic trace metals and hydrophobic trace organic compounds is made possible by its amphilicity. Theoretical and field evidence for organic association of selected trace elements and trace organics in Gulf coast estuaries is presented and compared with that for other estuaries in the world. It appears that trace substance association to organic matter is controlled by the abundance of suitable functional groups for trace metals and hydrophobic moieties for hydrophobic trace organics.

ACKNOWLEDGMENTS

We acknowledge the financial support by the Office of Naval Research (Grant N00014-93-1-0877), the National Science Foundation (OCE-9633125), a Texas Sea Grant, and the Texas Institute of Oceanography.

REFERENCES

Adamson, A. W. 1976. Physical chemistry of surfaces. Wiley.

Alexander, M. 1995. How toxic are toxic chemicals in soil? Environ. Sci. Technol. **29:** 2713–2717.

Averett, R. C., J. A. Leenheer, D. M. McKnight, and K. A. Thorn [eds.]. 1995. Humic substances in the Suwannee River, Georgia: Interactions, properties, and proposed structures. USGS Water-Supply Paper #2373.

Baham, J., and G. Sposito. 1994. Adsorption of dissolved organic carbon extracted from sewage sludge on montmorillonite and kaolinite in the presence of metal ions. J. Environ. Qual. **23:** 147–153.

Bailey, G. W., and J. L. White. 1970. Factors influencing the adsorption, desorption and movement of pesticides in soil. Res. Rev. **32:** 29–92.

Bartha, R. 1971. Fate of herbicide-derived chloroanilines in soil. J. Agric. Food Chem. **19:** 385–387.

———, and Pramer, 1967. Pesticide transformation to analine and azo-compounds. Soil. Sci. **156:** 1617–1618.

Baskaran, M., M. Ravichandran, and T. S. Bianchi. 1997. Cycling of ^{7}Be and ^{210}Pb in a high DOC, shallow, turbid estuary of southeast Texas. Estuar. Coast. Shelf Sci. **45:** 165–176.

———, and P. H. Santschi. 1993. The role of particles and colloids in the transport of radionuclides in coastal environments of Texas. Mar. Chem. **43:** 95–114.

Benner, R., J. D. Pakulski, M. McCarthy, J. I. Hedges, and P. G. Hatcher. 1992. Bulk chemical characteristics of dissolved organic matter in the ocean. Science **255:** 1561–1564.

Benoit, G. 1995. Evidence of the particle concentration effect for Pb and other metals in freshwaters based on ultra-clean technique analyses. Geochim. Cosmochim. Acta **59:** 2677–2687.

———, S. Oktay, A. Cantu, M. E. Hood, C. Coleman, O. Corapcioglu, and P. H. Santschi. 1994. Partitioning of Cu, Pb, Ag, Zn, Fe, Al, and Mn between filter-retained particles, colloids and solution in six Texas estuaries. Mar. Chem. **45:** 307–336.

Bertine, K. K., and R. Vernon Clark. 1996. Elemental composition of the colloidal phase isolated by cross-flow filtration from coastal seawater samples. Mar. Chem. **55:** 189–204.

Bowles, E. C., R. C. Antweiler, and P. MacCarthy. 1995. Acid-base titration and hydrolysis of fulvic acid from the Suwannee river, p. 115–127. *In:* R. C. Averett, J. A. Leenheer, D. M. McKnight, and K. A. Thorn [eds.], Humic substances in the Suwannee River, Georgia: Interactions, properties, and proposed Structures. USGS Water-Supply Paper #2373.

Brownawell, B. J., and J. W. Farrington. 1985. Partitioning of PCBs in marine sediments, p. 97–120. *In:* A. C. Sigleo and A. Hattori [eds.], Marine and Estuarine Geochemistry. Lewis.

Buesseler, K. O., J. Bauer, R. Chen, T. Eglinton, O. Gustafsson, W. Landing, K. Mopper, S. B. Moran, P. H. Santschi, R. VernonClark, and M. L. Wells. 1996. An intercomparison of cross-flow filtration techniques used for sampling marine colloids: Overview and organic carbon results. Mar. Chem. **55:** 1–31.

Buffle, J. 1990. Complexation reactions in aquatic systems. An analytical approach. Ellis Horwood.

——, R. S. Altmann, M. Filella, and A. Tessier. 1990. Complexation by natural heterogeneous compounds: Site occupation distribution functions, a normalized description of metal complexation. Geochim. Cosmochim. Acta **54:** 1535–1554.

Burgess, R. M., R. A. Mckinney, W. A. Brown, and J. G. Quinn. 1996. Isolation of marine sediment colloids and associated polychlorinated biphenyls: An evaluation of ultrafiltration and reverse-phase chromatography. Environ. Sci. Technol. **30:** 1923–1932.

Campbell, P. G. C. 1995. Interactions between trace metals in aquatic organisms: Critique of the free-ion activity model, p. 45–102. *In:* A. Tessier and D. R. Turner [eds.], Metal speciation and bioavailability in aquatic systems.

Caron, G., and I. H. Suffett. 1989. Binding of nonpolar pollutants to dissolved organic carbon: environmental fate modeling, p 117–130. *In:* I. H. Suffett and P. Maccarthy [eds.], Aquatic humic substances. American Chemical Society.

Carter, C. W., and I. H. Suffett. 1982. Binding of DDT to dissolved humic materials. Environ. Sci. Technol. **16:** 735–740.

Carvalho, R. A., M. C. Benfield, and P. H. Santschi. (In review). Comparative bioaccumulation studies of colloidally-complexed and free-ionic heavy metals in juvenile brown shrimp. Limnol. Oceanogr.

Chiou, G. T., D. E. Kile, T. I. Brinton, R. L. Malcolm, J. Leenheer, and P. MacCarthy. 1987. A comparison of water solubility enhancements of organic solutes by aquatic humic materials and commercial humic acids. Environ. Sci. Technol. **21:** 1231–1234.

Chisaka, H., and P. C. Kearney. 1970. Metabolism of propanil in soils. J. Soil. Sci. **18:** 854–858.

Connell, D. W. 1988. Bioaccumulation behavior of persistent organic chemicals with aquatic organisms. Rev. Environ. Contam. Toxicol. **101:** 118–154.

Curtis, G. P., M. Reinhard, and P. V. Roberts. 1986. Sorption of hydrophobic organic compounds by sediments, p. 191–216. *In:* I. H. Suffet and P. MacCarthy [eds.], Aquatic humic substances. American Chemical Society Symposium Series 219, American Chemical Society.

Dai, M.-H., J.-M. Martin, and G. Cauwet. 1995. The significant role of colloids in the transport and transformation of organic carbon and associated trace metals (Cd, Cu, and Ni) in the Rhone delta (France). Mar. Chem. **51:** 159–175.

Davis, W. R., and J. C. Means. 1989. Benthic-water contaminant exchange processes in bioturbated cohesive sediment, p. 216–224. *In:* Proceedings of the 21st European Marine Biology Symposium, Gdansk, Poland, Sept. 14–19, 1986. U.S. EPA, Ecol. Res. Ser.

DiToro, D. M., and L. M. Horzempa. 1982. Reversible and resistant components of PCB adsorption-desorption isotherms. Environ. Sci. Technol. **16:** 594–602.

Donat, J. D., and K. W. Bruland. 1995. Trace elements in the ocean, p. 247–281. *In:* B. Salbu and E. Steinnes [eds.], Trace elements in natural waters. CRC Press.

Dzombak, D. A., W. Fish, and F. M. Morel. 1986. Metal-humate interactions. 1. Discrete ligand and continuous distribution models. Environ. Sci. Technol. **20:** 669–675.

Farley, K. J., and F. M. M. Morel. 1986. Role of coagulation in the kinetics of sedimentation. Environ. Sci. Technol. **20:** 187–195.

Farrington, J. W. 1991. Biogeochemical processes governing exposure and uptake of organic pollutant compounds in aquatic organisms. Environ. Health Persp. **90:** 75–84.

Foster, G. D., S. M. Baksi, and J. C. Means. 1987. Bioaccumulation of a mixture of sediment-associated organic contaminants by the Baltic clam (*Macoma balthica*) and soft shell clam (*Mya arenaria*). J. Environ. Toxicol. Chem. **6:** 969–976.

Freeman, D. H., and L. S. Cheung. 1981. A gel permeation model for organic desorption from pond sediment. Science **214:** 790–792.

Grill, E., E. L. Winnacker, and M. H. Zenk. 1985. Phytochelatins: The principal heavy metal-complexing peptides of higher plants. Science **230:** 674–676.

———. 1987. Phytochelatins, a class of heavy metal-binding peptides from plants, are functionally analogous to metallothioneins. Proc. Natl. Acad. Sci. USA **84:** 439–445.

Gschwend, P. M., and S.-C. Wu. 1985. On the constancy of sediment-water partition coefficients of hydrophobic organic pollutants. Environ. Sci. Technol. **19:** 90–96.

Gu, B., J. Schmitt, Z. Chen, L. Liang, and J. F. McCarthy. 1995. Adsorption and desorption of different organic matter fractions on iron oxide. Geochim. Cosmochim. Acta **59:** 219–229.

Guentzel, J. L., R. T. Powell, W. M. Landing, and R. P. Mason. 1996. Mercury associated with colloidal material in an estuarine and an open ocean environment. Mar. Chem. **55:** 177–188.

Guo, L., and P. H. Santschi. 1997a. Composition and cycling of colloids in marine environments. Rev. Geophys. **35:** 17–40.

———. 1997b. Isotopic and elemental characterization of colloidal organic matter from the Chesapeake Bay and Galveston Bay. Mar. Chem. **59:** 1–15.

———, P. H. Santschi, and M. Baskaran. 1997a. Interactions of thorium isotopes with colloidal organic matter in oceanic environments. Colloids and Surfaces A **120:** 255–271.

———, P. H. Santschi, L. A. Cifuentes, S. Trumbore, and J. Southon. 1996. Cycling of high molecular weight dissolved organic matter in the Middle Atlantic Bight as revealed by carbon isotopic (^{13}C and ^{14}C) signatures. Limnol. Oceanogr. **41:** 1242–1252.

———, P. H. Santschi, and K. Warnken. (submitted). Trace metal composition of colloidal material in estuarine and marine environments. Limnol Oceanogr.

Hassett, J. P., and M. A. Anderson. 1979. Association of hydrophobic organic compounds with dissolved organic matter in aquatic systems. Environ. Sci. Technol. **13:** 1526–1529.

Hawker, D. W., and Connell, D. W. 1985. Relationships between partition coefficients, uptake rate constants, clearance rate constant, and time to equilibrium for bioaccumulation. Chemosphere **14:** 1205–1219.

———. 1986. Predicting the distribution of persistent organic chemicals in the environment. Chem. Australia **53:** 428–443.

Henry, L. L., I. H. Suffet, and S. L. Friant. 1989. Sorption of chlorinated hydrocarbons in the water column by dissolved and particulate organic material, p. 159–171. *In:* I. H. Suffet and P. MacCarthy [eds.], Aquatic humic substances. American Chemical Society Symposium Series 219, American Chemical Society.

Hering, J., and F. M. Morel. 1989. Slow coordination reactions in seawater. Geochim. Cosmochim. Acta **53:** 611–618.

Honeyman, B. D., and P. H. Santschi. 1988. Critical review: Metals in aquatic systems. Predicting their scavenging residence times from laboratory data remains a challenge. Environ. Sci. Technol. **22:** 862–871.

————. 1989. A Brownian-pumping model for trace metal scavenging: Evidence from Th isotopes. J. Mar. Res. **47:** 950–995.

Horvath, C., and W. Melander. 1978. Reverse-phase chromatography and the hydrophobic effect. Am. Lab. **10:** 17–36.

Hsu, T.-S., and R. Bartha. 1974. Interactions of pesticide-derived chloroaniline residues with soil organic matter. Soil Sci. **116:** 444–452.

Hunter, K. A., and P. S. Liss. 1979. The surface charge of suspended particles in estuarine and coastal marine waters. Nature **272:** 823–825.

Israelachvili, J., and R. Pashley. 1982. The hydrophobic interaction is long range, decaying exponentially with distance. Nature **300:** 341–342.

Jardine, P. M., N. L. Weber, and J. F. McCarthy. 1989. Mechanisms of dissolved organic carbon adsorption on soil. Soil Sci. Soc. Am. J. **53:** 1378–1385.

Karickhoff, S. W. 1984. Organic sorption in aquatic systems. J. Hydraulic Eng. **110:** 707–735.

————, D. S. Brown, and T. A. Scott. 1979. Sorption of hydrophobic pollutants on natural sediments. Water Res. **13:** 241–248.

Kenega, E. E., and C. A. Goring. 1980. Relationship between water solubility and concentration of chemicals in biota. p. 78–115. *In:* J. G. Eaton, P. R. Parrish, and A. C. Hendricks [eds.], Aquatic toxicology. ASTM STP 707, American Society for Testing Materials.

Kile, D. E., and C. T. Chiou. 1989. Water solubility enhancement of nonionic organic contaminants, p. 131–157. *In:* I. H. Suffet and P. MacCarthy [eds.], Aquatic humic substances. ACS Symposium Series 219, American Chemical Society.

————, C. T. Chiou, and T. I. Brinton. 1995a. Interactions of organic contaminants with fulvic and humic acids from the Suwannee river and with other humic substances in aqueous systems—with inferences pertaining to the structure of humic molecules, p. 21–32. *In:* R. C. Averett, J. A. Leenheer, D. M. McKnight, and K. A. Thorn [eds.], Humic substances in the Suwannee River, Georgia: Interactions, properties, and proposed structures. USGS Water-Supply Paper #2373.

————, C. T. Chiou, H. Zhou, H. Li., and O. Xu. 1995b. Partition of nonpolar organic pollutants from water to soil and sediment organic matters. Environ. Sci. Technol. **29:** 1401–1406.

Koulermous, A. 1989. An evaluation of the physico-chemical aspects of natural colloids with reference to their role in the transport and fate of organic pollutants in aquatic systems. M.S. thesis, Louisiana State University.

Leenheer, J. A., L. B. Barber, C. E. Rostad, and T. I. Noyes. 1995a. Data on natural organic substances in dissolved, colloidal and suspended-silt and -clay, and bed-sediment phases in the Mississippi River and some of its tributaries, 1991–1992. USGS, Water Resources Investigations Report 94-4191.

————, T. I. Noyes, and P. A. Brown. 1995b. Data on natural organic substances in dissolved, colloidal and suspended-silt and -clay, and bed-sediment phases in the

Mississippi River and some of its tributaries, 1987–1990. USGS, Water Resources Investigations Report 93-4204.

Mackay, D. 1979. Finding fugacity feasible. Environ. Sci. Technol. **13:** 1218–1223.

Mackay, D., and S. Patterson. 1981. Calculating fugacity. Environ. Sci. Technol. **15:** 1006–1014.

Mantoura, R. F. C., and E. M. S. Woodward. 1983. Conservative behavior of riverine dissolved organic carbon in the Severn Estuary: Chemical and geochemical implications. Geochim. Cosmochim. Acta **47:** 1293–1309.

———, A. Dickson, and J. P. Riley. 1978. The complexation of metals with humic materials in natural waters. Estuar. Coast. Shelf Sci. **6:** 387–408.

Martin, J.-M., M.-H. Dai, and G. Cauwet. 1995. Significance of colloids in the biogeochemical cycling of organic carbon and trace metals in a coastal environment—example of the Venice Lagoon (Italy). Limnol. Oceanogr. **40:** 119–131.

McCarthy, J. F. 1989. Bioavailability and toxicity of metals and hydrophobic organic contaminants, p. 263–280. *In:* I. H. Suffet and P. MacCarthy [eds.], Aquatic humic substances. American Chemical Society Symposium Series 219, American Chemical Society.

McElroy, A. E. and J. C. Means. 1988. Factors influencing the bioavailability of hexachlorobiphenyl to benthic organisms, p. 149–158, *In:* J. C. Means [ed.] Aquatic toxicology and hazard assessment, volume 10, ASTM STP 971.

McKnight, D. M., and R. L. Wershaw. 1995. Complexation of copper by fulvic acid from the Suwannee River—Effects of counter-ion concentration, p. 33–44. *In:* R. C. Averett, J. A. Leenheer, D. M. McKnight, and K. A. Thorn [eds.], Humic substances in the Suwannee River, Georgia: Interactions, properties, and proposed Structures. USGS Water-Supply Paper #2373.

McMillin, D. J., and Means, J. C. 1996. Spatial and temporal trends of pesticide residues in water and particulates in the Mississippi River plume and the northwestern Gulf of Mexico. J. Chromatog. A **754:** 169–185.

Means, J. C. 1995. Influence of salinity upon sediment-water partitioning of aromatic hydrocarbons. Mar. Chem. **51:** 3–16.

———, and A. E. McElroy. 1997. Bioaccumulation of tetra- and hexachlorobiphenyl by Yoldia Limatula and Nepthys Incisa from bedded sediments: Effects of sediment and animal related parameters. J. Environ. Toxicol. Chem. **16:** 1277–1294.

———, and D. J. McMillin. 1993. Fate and transport of particle-reactive normal, alkylated and heterocyclic aromatic hydrocarbons in a sediment-water-colloid system. Referred Monograph, USMMS OCS Study MMS-93-0018.

———. 1995. Pollutant transport, *In:* S. P. Murrey, [ed.], Mississippi River plume study. Referred Monograph, USMMS OCS Study MMS-95-0033, 101–120.

———. 1997. Fate and transport of selected organic and trace element pollutants in the Louisiana-Texas (LATEX) coastal shelf, *In:* S. P. Murrey [ed.], Mississippi River Plume study. Referred Monograph USMMS OCS Study MMS-97, 256–270.

———, and R. Wijayaratne. 1982. Role of natural colloids in transport of hydrophobic pollutants. Science **215:** 968–970.

———. 1984. Chemical composition of estuarine colloidal organic matter: Implications for sorptive processes. Bull. Mar. Sci. **35:** 449–461.

————. 1989. Sorption of benzidine, tolidine and azobenzene with natural estuarine colloids, p 209–222. *In:* I. H. Suffett and P. MacCarthy [eds.], Aquatic humic substances. American Chemical Society.

————, S. G. Wood, J. J. Hassett, and W. L. Banwart. 1980. Sorption properties of polynuclear aromatic hydrocarbons by sediments and soils. Environ. Sci. Technol. **14:** 1524–1528.

————. 1982. Sorption of amino- and carboxy-substituted polynuclear aromatic hydrocarbons on sediments and soils. Environ. Sci. Technol. **16:** 93–98.

Miller, L. A., and K. W. Bruland. 1995. Organic speciation of silver in marine waters. Environ. Sci. Technol. **29:** 2616–2121.

Noyes, T. I., and J. A. Leenheer. 1995. Proton nuclear-magnetic-resonance studies of fulvic acid from the Suwannee River, p. 129–139. *In:* R. C. Averett, J. A. Leenheer, D. M. McKnight and K. A. Thorn [eds.], Humic substances in the Suwannee River, Georgia: Interactions, properties, and proposed structures. USGS Water-Supply Paper #2373.

Parris, G. E. 1980. Environmental and metabolic transformations of primary aromatic amines and related compounds. Res. Rev. **76:** 1–30.

Perdue, E. M., and C. R. Lytle. 1983. Distribution model for binding of protons and metal ions by humic substances. Environ. Sci. Technol. **17:** 564–660.

Periera, W., and C. E. Rostad. 1990. Occurrence, distribution and transport of herbicides and their degradation products in the lower Mississippi River and its tributaries. Environ. Sci. Technol. **24:** 1400–1408.

Phinney, J. T., and K. W. Bruland. 1997. Effects of dithiocarbamate and 8-hydroxyquinoline additions on algal uptake of ambient copper and nickel in South San Francisco Bay water. Estuaries **20:** 66–76.

Porrier, M. A., B. R. Bordelon, and J. L. Laseter. 1972. Adsorption and concentration of carbon-14 DDT by coloring colloids in surface waters. Environ. Sci. Technol. **6:** 1033–1035.

Powell, R. T., W. M. Landing, and J. E. Bauer. 1996. Colloidal trace metals, organic carbon and nitrogen in a Southeastern U.S. estuary. Mar. Chem. **55:** 161–176.

Rees, T. F. 1991. Transport of contaminants by colloid-mediated processes, p. 165–184. *In:* F. O. Hutzinger [ed.], The handbook of environmental chemistry, reactions and processes, volume 2, part F. Springer Verlag.

Rubinstein, N. I., E. Lores, and N. R. Gregory. 1983. Accumulation of PCBs, mercury and cadmium by nereis virens, mercenaria mercenaria, and palaomontes pugio from contaminated sediments. Aquat. Toxicol. **3:** 249–260.

Sabljic, A. 1984. Prediction of the nature and strength of soil sorption of organic pollutants by molecular topology. J. Agric. Food Sci. **32:** 243–246.

Sanudo-Wihelmy, S., I. Rivera-Duarte, and A. R. Flagal. 1996. Distribution of colloidal trace metals in the San Francisco Bay estuary. Geochim. Cosmochim. Acta **60:** 4933–4944.

Santschi, P. H., L. Guo, M. Baskaran, S. Trumbore, J. Southon, T. S. Bianchi, B. D. Honeyman, and L. Cifuentes. 1995. Isotopic evidence for the contemporary origin of high-molecular weight organic matter in oceanic environments. Geochim. Cosmochim. Acta **59:** 625–631.

————, J. Lenhart, and B. D. Honeyman. 1997. Heterogeneous processes affecting trace contaminant distribution in estuaries: The role of natural organic matter. Mar. Chem. **58:** 99–125.

Sharp, J. H. 1973. Size classes of organic carbon in seawater. Limnol. Oceanogr. **18:** 441–444.

Sigleo, A. C., and G. R. Helz. 1981. Composition of estuarine colloidal material: Major and trace elements. Geochim. Cosmochim. Acta **45:** 2501–2509.

————, and J. C. Means. 1990. Organic and inorganic components in estuarine colloids: Implications for sorption and transport of pollutants. Rev. Environ. Contam. Toxicol. **112:** 123–147.

Stordal, M. C., G. A. Gill, L. S. Wen, and P. H. Santschi. 1996a. Mercury phase speciation in the surface waters of selected Texas estuaries: Importance of colloidal forms. Limnol. Oceanogr. **41:** 52–61.

————, P. H. Santschi, and G. A. Gill. 1996b. Colloidal pumping: Evidence for the coagulation process using natural colloids tagged with [203]Hg. Environ. Sci. Technol. **30:** 3335–3340.

Sunda, W. G., and R. L. Ferguson. 1983. Sensitivity of natural bacterial communities to additions of copper and to cupric ion activity: A bioassay of copper complexation in seawater, p. 871–891. *In:* C. S. Wong, E. Boyle, K. W. Bruland, J. D. Burton, and E. D. Goldberg [eds.], Trace metals in seawater. Plenum Press.

Swarzenski, P. W., B. A. McKee, and J. G. Booth. 1995. Uranium geochemistry on the Amazon shelf: Chemical phase partitioning and cycling across a salinity gradient. Geochim. Cosmochim. Acta. **59:** 7–18.

Tipping, E. 1993. Modeling ion binding by humic acids, p. 117–131. *In:* Th. F. Tadros and J. Gregory [eds.], Colloids in the aquatic environment. Elsevier.

————. 1994. WHAM—a chemical equilibrium model and computer code for waters, sediments, and soils incorporating a discrete site/electrostatic model of ion-binding by humic substances. Comp. Geosci. **20:** 973–1023.

————, and M. A. Hurley. 1992. A unifying model of cation binding by humic substances. Geochim. Cosmochim. Acta **56:** 3627–3641.

Trefry, J. H., T. A. Nelson, R. P. Trocine, S. Metz, and T. W. Vetter. 1986. Trace metal fluxes through the Mississippi River delta system. Cons. Int. Explor. Mer. **186:** 277–288.

Van den Berg, C. M. G., A. G. A. Merks, and E. K. Duursma. 1987. Organic complexation and its control of the dissolved concentration of copper and Zn in Scheld estuary. Estuar. Coast. Shelf Sci. **24:** 785–797.

Wedepohl, K. H. 1995. The composition of the continental crust. Geochim. Cosmochim. Acta **59:** 1217–1232.

Wen, L. S., P. H. Santschi, G. Gill, C. Paternostro, and R. Lehman. 1997b. Colloidal and particulate silver in river and estuarine waters of Texas. Environ. Sci. Technol. **31:** 723–731.

————, P. H. Santschi, C. Paternostro, and G. Gill. (In review). Estuarine trace metal distributions in Galveston Bay: Colloidal forms and dissolved phase speciation, Mar. Chem.

————, M. C. Stordal, G. A. Gill, and P. H. Santschi. 1996. An ultra-clean cross-flow ultrafiltration technique for the study of trace metal phase speciation in sea water. Mar. Chem. **55:** 129–152.

————, P. H. Santschi, and D. Tang. 1997c. Interactions between radioactively labeled colloids and natural particles: Evidence for colloidal pumping. Geochim. Cosmochim. Acta **61:** 2867–2878.

Wershaw, R. L., P. J. Burcar, and M. C. Goldberg. 1969. Interaction of pesticides with natural organic material. Environ. Sci. Technol. **3:** 271–273.

Whitehouse, B. G., P. A. Yeats, and P. M. Strain. 1990. Cross-flow filtration of colloids from aquatic environments. Limnol. Oceanogr. **35:** 1368–1375.

Wijararatne, R. D. 1982. Sorption of organic pollutants on natural estuarine colloids. Ph.D. Dissertation, University of Maryland. University Microfilms.

————, and J. C. Means. 1984a. Affinity of natural estuarine colloids for hydrophobic pollutants in aquatic environments. Environ. Sci. Technol. **18:** 121–123.

————. 1984b. Sorption of polycyclic aromatic hydrocarbons (PAHs) by natural colloids. Marine Environ. Res. **11:** 77–89.

Xing, B., W. B. McGill, and M. J. Dudas. 1994. Cross-correlation of polarity curves to predict partition coefficients of nonionic organic contaminants. Environ. Sci. Technol. **28:** 1929–1933.

Young, D. R., A. J. Mearns, and R. W. Gossett. 1984. Bioaccumulation and biomagnification of P,P'-DDE and PCB 1254 by a flatfish bioindicator from highly contaminated marine sediments of southern California, p. 159–169. *In:* R. A. Baker [ed.], Organic substances and sediments in water: Biological processes. American Chemical Society.

Zierath, D., J. J. Hassett, W. L. Banwart, S. G. Wood, and J. C. Means. 1980. Sorption of benzidine by sediments and soils. Soil Sci. **129:** 277–281.

Chapter 12

Particle-Reactive Radionuclides as Tracers of Biogeochemical Processes in Estuarine and Coastal Waters of the Gulf of Mexico

M. Baskaran

INTRODUCTION

The total number of nuclides is close to 1700, of which about 85% are radioactive and about 15% are stable (Faure 1986). The major sources of these radionuclides include the following: (1) primordial (the common ones are the ^{238}U-, ^{235}U-, and ^{232}Th- series radionuclides); (2) anthropogenic (e.g., ^{90}Sr, ^{137}Cs, ^{238}Pu, ^{239}Pu, ^{240}Pu, ^{241}Am); and (3) cosmogenic (such as ^7Be, ^{10}Be, ^{14}C, ^{26}Al, ^{32}P, and ^{33}P). These isotopes are broadly classified into two major categories, particle-reactive and non-particle-reactive nuclides. The transport

Biogeochemistry of Gulf of Mexico Estuaries, Edited by Thomas S. Bianchi, Jonathan R. Pennock, and Robert R. Twilley.
ISBN 0-471-16174-8 © 1999 John Wiley & Sons, Inc.

pathways of non-particle-reactive nuclides are relatively straightforward and are primarily mediated by water mass movements. The other group of radionuclides, particle-reactive ones, are adsorbed onto particles, and subsequently the fate of particles becomes the fate of particle-bound radionuclides. These radionuclides provide a set of powerful tracers useful for investigation of the biogeochemical processes that affect the dynamics of particles and particle-reactive pollutants in estuarine and coastal areas.

One of the more common techniques used in obtaining information on rates of geochemical processes is the application of radionuclides, either U-Th decay series radionuclides or other cosmogenic particle-reactive isotopes. The disequilibrium between parent and daughter nuclides or precise measurements of fluxes and standing crops of atmospherically delivered radionuclides have been used to determine removal rate constants. For example, the disequilibrium between ^{234}Th (half-life = 24.1 d) and ^{238}U (or ^{228}Th/^{228}Ra), and the determination of fluxes and standing crops of atmospherically delivered ^{210}Pb (half-life = 22.1 yr, derived from ^{222}Rn decay, ^{222}Rn half-life = 3.8 d) and cosmogenic ^{7}Be (half-life = 53 d), are powerful tools for estimating the rates of processes that affect the distribution of pollutants in estuarine waters (^{228}Th/^{228}Ra, ^{234}Th/^{238}U: Broecker et al. 1973; Aller and Cochran 1976; Li et al. 1979, 1981; Santschi et al. 1979, 1980; Minagawa and Tsunogai 1980; Aller et al. 1980; Kaufman et al. 1981; McKee et al. 1986; ^{210}Pb and ^{210}Pb/^{226}Ra: Rama et al. 1961; Bacon et al. 1976; Nittrouer et al. 1979; Santschi et al. 1979, 1980; Benninger and Krishnaswami 1981; Li et al. 1981; Baskaran et al. 1997; cosmogenic (^{7}Be): Aaboe et al. 1981; Tanaka and Tsunogai 1983; Olsen et al. 1986; Baskaran and Santschi 1993; Baskaran et al. 1997; anthropogenic (^{239,240}Pu, ^{137}Cs): Simpson et al. 1976; Olsen et al. 1978, 1980; Benninger and Krishnaswami 1981; review paper on various radionuclides: Turekian et al. 1977; Olsen et al. 1982; Cochran 1984; Honeyman and Santschi 1988; Moore 1992).

The particle-reactive radioisotopes are not only useful tools for understanding the fate and mobility of these nuclides, but also provide analogue information on other particle-reactive organic and inorganic pollutants. Inorganic pollutants include heavy metals, radionuclides, and inorganic substrates derived from the industrial wastes (Olsen et al. 1982). Important organic pollutants include organic material derived from petroleum and coal hydrocarbons, synthetic organics, municipal wastes, and biological organisms that can affect the cycling of nutrients and oxygen. Humic substances constitute a major portion of riverine-derived dissolved organic matter. These substances are more refractory and have a strong capacity for binding particle-reactive (class-A and transition) metals. Thus, the complexation of metals with humic substances in particular and dissolved organic matter in general could play a major role on the fate and transport of particle-reactive contaminants in estuarine regions (Schnitzer and Kahn 1972; Reuter and Perdue 1977; Means and Wijayaratne 1982).

Particle-reactive radionuclides can be utilized to obtain analogue information on the removal of pollutants from the water column. In coastal waters, when pollutants are discharged, they are dispersed on a relatively short time scale. Most pollutants in estuarine regions are derived from the discharge of rivers (from sewage, from industrial effluents, and/or through runoff), from oil-related activities along the coastal areas, and from the water mass transport from the shelf and slope areas of the coastal ocean. In coastal areas of the United States, discharges of pollutants through these sources have resulted in higher concentrations of heavy metals and organic pollutants in sediments (Valette-Silver 1993 and other papers in that issue of "Estuaries"). For example, Cr and Ni enter the coastal system of the United States from the discharge of effluents delivered from electroplating industries (Ravichandran et al. 1995b and references therein). Lead, derived from the usage of leaded gasoline, has entered marine and fresh-water systems since the 1950s (with maximum usage in the early 1970s and a decline since the late 1970s). Extensive use of zeolites, enriched in light rare earth elements (LREE), during the catalytic cracking process in oil refineries along coastal areas and rivers resulted in significant input of LREE to the coastal and estuarine areas (Olmez and Gordon 1985; Olmez et al. 1991; Ravichandran 1996). Oil drilling activity resulted in the discharge of Ba-bearing fluids in coastal waters of the United States, particularly in the Gulf of Mexico (Baskaran et al. submitted). Particle-reactive radionuclides such as ^{210}Pb and ^{239,240}Pu are commonly utilized to establish the sediment chronologies needed in the investigation of the historical changes in pollutant inputs in coastal areas.

Particle-reactive radionuclide-based chronologies are also useful to investigate the historical variations on the inputs of organic pollutants. Concentrations of most polycyclic aromatic hydrocarbons (PAHs) and polychlorinated biphenyls (PCBs) in coastal systems started to increase in the early 1940s and reached a maximum in the 1970s. The total PAH distribution in coastal sediments can also be used to track the transition of fuel from coal to petroleum, which is characterized by a large PAH concentration peak in the 1940–1950 period. In addition, PAH and PCB concentration variations in dated sediment cores reflect changes in anthropogenic activity and regulations (and/or bans) on industrial and vehicle exhaust emissions. The discharge of several organochlorine pesticides into estuarine and coastal waters in the 20th century has also changed due to anthropogenic activity, and these changes are reflected in downcore profiles of sediments. In addition, the removal rates of these organic pollutants can be obtained using short-lived ^{234}Th (Gustafsson et al. 1997).

The principal processes that affect the distribution of particle-reactive nuclides in an estuary area are the following: (1) removal of nuclides from the water column by ion exchange, precipitation, and/or hydrophobic interactions with particle surfaces; (2) complexation with organic substrates adsorbed onto particles; (3) flocculation of nuclides and nuclides bound onto colloidal organic matter; (4) nuclides sorbed onto iron and manganese oxide coatings that undergo co-precipitation; (5) removal by incorporation into organisms, fecal material, or mineral lattices; (6) desorption of nuclides from suspended sediments delivered by rivers; and (7) release of nuclides from bottom sediments by physical and/or biological mixing of sediments (Saxby 1969; Edzwald et al. 1974; Burton and Liss 1976; Sholkovitz 1976; Boyle et al. 1977; Li et al. 1977; Turekian 1977; Santschi et al. 1979, 1980; Duinkar 1980; Gearing et al. 1980; Baskaran and Naidu 1995). These processes which are important in controlling the distribution of nuclides in estuaries, are depicted in Fig. 12-1. Many of the earlier studies on the distribution of particle-reactive radionuclides did not specifically address any one of the aforementioned processes instead, they addressed combined effects of several processes that lead to the removal of nuclides. For example, variations in the dissolved concentrations of nuclides could be a result of several processes, such as desorption of nuclides from suspended particles, release of nuclides from bottom sediments via physical and/or biological mixing of sediments, removal by adsorption onto particles, and removal by coagulation. Thus, the relative role of each of these processes in the field conditions is difficult to evaluate. However, these groups of processes do provide some useful information on the fate of particles and particle-sorbed pollutants.

The particle reactivity of any nuclide can be characterized in terms of a distribution coefficient. Even though this parameter is only a measure of the equilibrium partitioning of a nuclide between dissolved and particulate phases, it provides a proxy for the particle affinity of any nuclide. The distribution

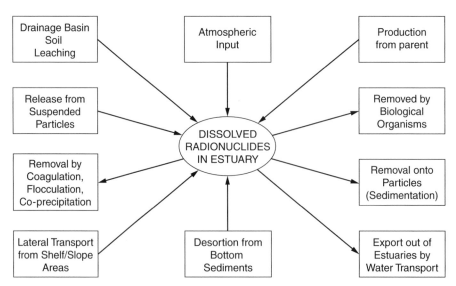

FIG. 12-1. *Various biogeochemical processes that control the distribution of particle-reactive nuclides in estuaries.*

coefficient for any pollutant can vary, depending on the chemical form of the pollutant. For selected redox-sensitive elements, the distribution coefficient can vary, depending on the oxidation state of that nuclide. For example, Pu can exist in the oxidized state (either V or VI), which has a lower particle affinity ($K_d = 0.6 - 2.5 \times 10^4$ cm^3 g^{-1}), or in the reduced form (either III or IV) with a strong affinity for particles, with K_d values ranging between 34 and 600×10^4 cm^3 g^{-1} (Sholkovitz 1983 and references therein). Similarly, different forms of chlorinated biphenyls have different distribution coefficients. For example, lower-chlorinated biphenyls (such as di- and tri-chlorobiphenyls) have lower distribution coefficients (10^4) than higher-chlorinated biphenyls (such as penta- and hexa-chlorobiphenyls) (10^5) (Clayton et al. 1977; Bopp et al. 1981; Olsen et al. 1982). It was predicted that any chemical species that has an experimentally determined distribution coefficient (K_d) on the order of 5–10×10^4 cm^3 g^{-1} (trace metals such as Th, Pb, Pu, Be, and particle-reactive hydrocarbons) has a 2- to 14-d residence time in the water column in shallow coastal areas with respect to transfer to the sediments (Santschi et al. 1984). For Narragansett Bay, it was estimated that 70–95% of the particle-reactive pollutants would be removed to the sediments if the residence time of these pollutants in the water column is about 30 days (Santschi et al. 1984).

The estuaries in the Gulf of Mexico, extending from the Rio Grande to the Florida Keys, are some of the most extensive estuarine systems in the world and are known for their productivity, fisheries, and various estuarine circulation and salinity regimes (Orlando et al. 1993). The Gulf-bound rivers drain nearly 40–50% of the contiguous United States and are major sources of fresh water to the Gulf of Mexico (Chapter 1 this volume). Most of these estuaries are shallow, between 1 and 5 m, except the Mississippi River. Some of these estuaries are organic-rich and hence would be ideal sites for the investigation of the fate of metals due to metal-organic complexation.

A list of isotopes that is useful for studying estuarine biogeochemical processes that control the fate and transport of particle-reactive radionuclides is presented in Table 12-1. In this chapter, I discuss the following subjects that have been investigated in estuarine waters of the Gulf of Mexico: (1) the fate of particle-reactive nuclides that are injected as a pulse in any dynamic estuarine system (such as during a heavy thunderstorm); (2) retention efficiency of metals in an organic-rich estuary for the past 100 years using radionuclides whose inputs are well defined; (3) the role of dissolved organic matter (DOM) in the removal of particle-reactive radionuclides in a shallow and organic-rich estuary; and (4) determination of the particle residence time in a shallow estuary. In addition, the role of particles in the removal of particle-reactive nuclides along the Gulf Coast estuaries will be assessed. The residence times and distribution coefficients in various estuaries along the Gulf Coast will also be compared.

TABLE 12-1. Radionuclides Used to Investigate Biogeochemical Processes in Estuaries and Adjacent Areas

Nuclide	Half-Life	Utility	Reference
^{7}Be	53 d	Residence times of particles and particle-sorbed pollutants, particle mixing, and short-term sedimentation rates	Aaboe et al. (1981); Baskaran and Santschi (1993); Baskaran et al. 1997.
^{137}Cs	30 yr	Watershed erosion, sedimentation, and mixing	Ritchie et al. (1974); Baskaran and Naidu (1995)
^{210}Pb	22.1 yr	Residence times of particles and particle-sorbed pollutants, particle mixing, and sedimentation rates	Rama et al. (1961); Baskaran and Santschi (1993); Baskaran et al. (1997).
^{224}Ra	3.66 d	Biogeochemical processes in salt marsh	Bollinger and Moore (1993)
^{234}Th	24.1 d	Rates of Removal of Th and particle-reactive pollutants from solution, short-term particle mixing, and sedimentation rates in estuaries	Aller et al. (1980); McKee et al. (1986)
^{228}Th	1.9 yr	Rates of removal of Th and particle-reactive pollutants from solution, short-term sediment accumulation	Kaufman et al. (1981); Minagawa and Tsunogai (1980)
^{238}U	4.5×10^9 yr	Removal of metals from low-salinity estuarine waters	Borole et al. (1982)
^{239,240}Pu	2.4×10^4 yr	Sedimentation and mixing	Benninger et al. (1979); Ravichandran et al. (1995a)

FATE OF PARTICLE-REACTIVE RADIONUCLIDES INJECTED AS A PULSE IN A DYNAMIC COASTAL SYSTEM

Particle-reactive radionuclides delivered to the water column, either through direct atmospheric fallout, riverine discharge, or in situ production from parent nuclides, become associated with fine particles in coastal waters; thus, the rate, pattern, and extent of removal from the coastal waters are highly variable. Theoretically, one could add a very large amount of particle-reactive tracer to an estuary and follow the pathway of this tracer in any dynamic system. This method is very difficult to use in large natural estuarine systems, as enormous amounts of spikes (essentially "point source" pollution for the experiment) would be needed. However, during heavy rain events, very high concentrations of atmospherically delivered radionuclides are injected into estuarine and/or coastal systems. Thus, when these systems are "tagged" with a set of particle-reactive tracers, the time-series distribution of these tracers can be utilized to obtain information on the fate and transport pathways of these radionuclides. Such studies also provide valuable information on the importance of sediment re-suspension in the transport of particle-reactive radionuclides and, by analogy, of other pollutants from coastal areas. Relatively few studies have been carried out utilizing these "opportunistic" tracers in estuarine systems (Olsen et al. 1989a; Baskaran and Santschi 1993). In Galveston Bay, particulate and dissolved concentrations of ^{7}Be and ^{210}Pb were measured for about 30 h after a pulsed rain input in estuarine water, off Galveston, Texas (Baskaran and Santschi 1993).

Concentrations of ^{7}Be and ^{210}Pb in a time-series study from Galveston Bay after a rain event are plotted in Fig. 12-2. In less than 1 h after the pulsed rain had stopped, most of the ^{7}Be (74–86% of the total), as well as ^{210}Pb (80–89%), were associated with particles. These results showed that particle and radionuclide concentrations decreased by a factor of 3 to 5 over the 28-h duration of the experiment. The exponential decrease in activity with time indicated a removal residence time of at most 1–1.5 d for both nuclides. The suspended particle (<0.4 μm) concentration decreased with a removal residence time of 1.5 d. The standing crop of ^{7}Be in the Galveston Bay water column after the rain was found to be only 17% of the amount injected by the rain, suggesting faster removal from the water column. The steady-state residence time for the sample collected 0.5 h after the rain had stopped is approximately 4 h for ^{7}Be. Similar studies can be carried out in any other coastal waters in the Gulf of Mexico or elsewhere.

The apparent distribution coefficient, K_d, has been used as a measure of the partitioning of a nuclide between filter-passing and filter-retained particulate phases and can be calculated using Eq. (1)

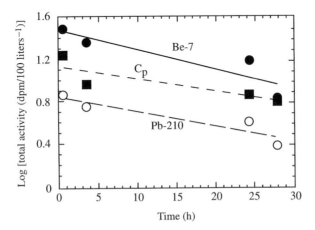

FIG. 12-2. *Log of total (dissolved + particulate) ^{7}Be, ^{210}Pb$_{xs}$, and suspended particle concentrations are plotted against time after a heavy pulsed rain input into Galveston coastal water (data taken from Baskaran and Santschi 1993 and replotted).*

$$K_d = \frac{C_p}{C_w} \qquad (12\text{-}1)$$

where C_p is the concentration of the nuclide per g of suspended material and C_w is the concentration of the dissolved ($<0.4\ \mu$m) nuclide per cm^3 of water. In a dynamic estuarine system, there could be sorption and release of radionuclides onto suspended particles; hence, the temporal variations of K_d will provide information on the kinetics of particle dynamics. A plot of K_d with the suspended particle concentration clearly showed that there was a general decrease in the K_d value of both ^{7}Be and ^{210}Pb with particle concentration after their pulse input into the water, coinciding with a decrease in particle concentration with time (Fig. 12-3). The range of distribution coefficients for Be and Pb following a thunderstorm can be compared to those reported in the literature for other estuaries and coastal waters (see Table 12-3). Of all the estuaries for which the K_d data for Pb and Be are available in the Gulf, the Sabine-Neches estuary appears to have generally lower K_d. The concentration of dissolved organic carbon in this estuary is one of the highest among all the Gulf coast estuaries (up to 21 mg liter^{-1}; Bianchi et al. 1997). Thus, it appears that lower distribution coefficients for these nuclides are related to the concentration and composition of DOM (Baskaran et al. 1997).

As mentioned earlier, suspended particles play a major role in the biogeochemical cycling of particle-reactive pollutants in estuaries. In coastal waters off Galveston, suspended particulate matter contained $>70\%$ of the total ^{210}Pb and ^{7}Be in the water column. A fraction of these suspended particles and associated nuclides could eventually be transported and dispersed to the shelf

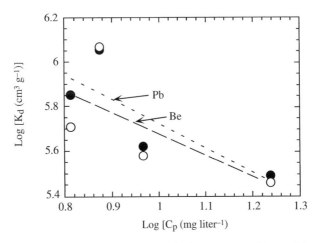

FIG. 12-3. *Log of the distribution coefficient, K_d (cm^3 g^{-1}), versus log of the particle concentration, C_p (mg liter^{-1}), for both ^{7}Be and ^{210}Pb after a pulse injection of ^{7}Be and ^{210}Pb isotope into Galveston coastal water (data taken from Baskaran and Santschi 1993 and replotted).*

and slope areas (Baskaran et al. 1997). From an analysis of the size distribution of suspended particles after a storm event, particle turnover or residence time could be calculated. A major portion of the injected ^{7}Be and ^{210}Pb was removed onto suspended particles in less than 1 h after the pulse input stopped injecting ^{7}Be and ^{210}Pb into the water (Baskaran and Santschi 1993). The particle-residence time was calculated from the mean water depth and the average Stokes settling velocity for median particle size. The residence time of the nuclide is given by the ratio of the residence time of the particles to that of the fraction on particles. Since major portions of ^{210}Pb and ^{7}Be were associated with particles, and since the residence times of the nuclides and particles are comparable, it was observed that ^{210}Pb and ^{7}Be were re-suspended several times before they disappeared from the water column (Baskaran and Santschi 1993). The same is likely true for many other particle-reactive organic as well as inorganic pollutants in dynamic coastal systems.

As a result of the many dynamic processes in some of the coastal systems, the pollutants introduced are not quantitatively removed and are retained in situ. One could potentially use some of the atmospherically delivered radionuclides (such as ^{239,240}Pu, ^{137}Cs, and ^{90}Sr) to obtain information on how much is retained in situ. The atmospheric fallout of radionuclides derived from weapons testing was introduced into the environment around 1952, with a maximum fallout in 1963. If a constant fraction of these weapons-test-derived radionuclides that are discharged (from direct atmospheric fallout plus riverine discharge) to the surface of estuarine and/or coastal waters is removed, then one would still expect the peak fallout record to be preserved. Plutonium is an excellent analogue for studies of heavy metal transport through the environment because it is particle-reactive, is effectively retained in aquatic and marine sediments, and has a well-defined input function (Santschi et al. 1980; Ravichandran et al. 1995b and references therein). The vertical distribution of ^{239,240}Pu concentrations in sediment cores from the Sabine-Neches Estuary indicated that peak fallout of Pu in 1963 was retained in sediments (Fig. 12-4). The presence of a sharp peak of Pu concentration but lower total concentrations (total concentration per unit area, inventory) than expected from atmospheric fallout could be due to partial removal of nuclides discharged to the estuary from direct atmospheric fallout and riverine input.

RETENTION EFFICIENCY OF HEAVY METALS IN GULF ESTUARIES

As was mentioned earlier, most of the estuaries in the Gulf coast are shallow, usually between 1 and 5 m. Some of the estuaries (e.g., the Sabine-Neches on the Texas-Louisiana border) are organic-rich, and the hydraulic residence times are short (see Table 12-3). In such estuaries, most of the particle-reactive nuclides could form complexes with dissolved organic ligands, resulting in longer residence times. As a consequence, the discharged particle-reactive

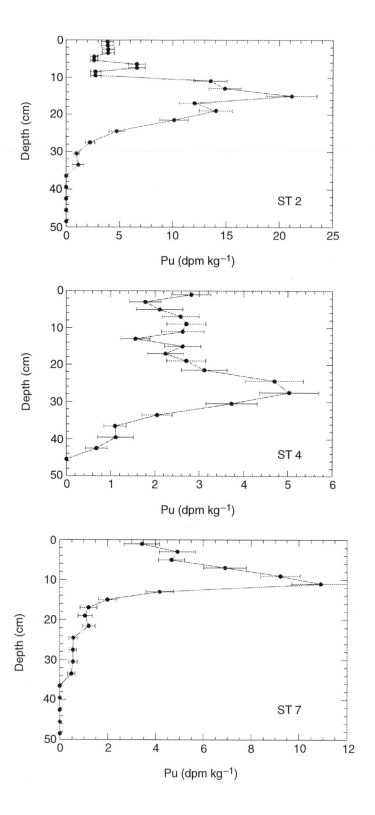

inorganic and organic species may not be quantitatively removed within the estuary and can be exported to the adjoining coastal areas.

As discussed before, heavy metals are delivered to estuaries from a variety of sources. These include (1) discharge through rivers, from sewage and industrial effluents, and/or through runoff; (2) atmospheric delivery; (3) oil-related and industrial activities along coastal areas; and (4) water mass transport from the shelf and slope areas of the marine environment to the estuary. In restricted and shallow estuaries (such as the Sabine-Neches Estuary), the input derived from the shelf and slope areas is likely negligible. To obtain information on the geochemical fate and transport of various metals, one can obtain proxy information using input functions that are well defined. ^{239,240}Pu and ^{210}Pb are excellent analogues for studies of heavy metal transport through the estuarine environment (Santschi et al. 1980; Ravichandran et al. 1995b and references therein). From a knowledge of the inputs and measurements of what is retained in the sediments, one can determine the retention efficiency of these nuclides within the estuary. In areas where there is only partial retention of nuclides, various processes that control the fate of the particle-reactive nuclides need to be investigated. One such likely process is the complexation of metals with certain types of DOM.

In organic-rich estuaries, DOM will likely play a major role in the removal of particle-reactive nuclides. Guentzel et al. (1996) reported that in waters with high levels of dissolved organic carbon (DOC) (5 to 15 mg DOC liter^{-1}) of the Ochlockonee estuary in northern Florida, 80–90% of particle-reactive nuclides, Fe, and Hg, are complexed by surface sites of colloids (size: 1 nm to 1 μm; Buffle et al. 1992). In San Francisco Bay, the fraction of dissolved Ni and Cu existing in organic complexes has been observed to range between 30 and 70% and between 80 and 92%, respectively (Donat et al. 1994). In most estuaries along the Gulf coast, DOC concentrations varied between 3 and 8 mg liter^{-1}, except in Lavaca Bay (~10 mg liter^{-1}; data compiled in Chapter 9 this volume) and the Sabine-Neches estuary (21 mg liter^{-1}; Bianchi et al. 1997). When particle-reactive nuclides undergo complexation with DOM, their fate is linked to the fate of the DOM. Recently, Baskaran et al. 1997 utilized two particle-reactive radionuclides, ^{7}Be and ^{210}Pb, in the Sabine-Neches estuary to investigate their particle reactivity in organic-rich waters. They suggested that these nuclides formed complexes with DOM and that this complexation could be responsible for the relatively long residence times of these nuclides in the water column.

One can determine the average retention efficiency of Pb and Pu (and possibly other heavy metals) in the sediments over a period of several decades to a century using excess ^{210}Pb and ^{239,240}Pu. This requires precise quantification

---

FIG. 12-4. *Vertical distribution of ^{239,240}Pu in sediment cores from select stations in the Sabine-Neches Estuary plotted against depth. (Data taken from Ravichandran et al. 1995b and replotted.)*

of the inputs of these nuclides to the estuary. The two major inputs to estuaries are the direct atmospheric fallout on surface waters of the estuary and watershed erosion of these nuclides with subsequent discharge through rivers. Although the major sources of ^{210}Pb and ^{239,240}Pu (present atmospheric deposition of Pu is negligible) in estuarine areas are generally assumed to be derived primarily from atmospheric deposition, significant contributions to the estuarine sediments could result from soil erosion following locations within a drainage basin: (1) where the residence times of ^{210}Pb and ^{239,240}Pu are relatively short and the ratio of drainage basin/estuary area is large and (2) where the residence times of riverine ^{210}Pb and ^{239,240}Pu in the water column are short compared to their hydraulic residence time in the estuary (Benninger 1978; Benninger et al. 1979; Wan et al. 1987; Appleby and Oldfield 1992; Baskaran and Santschi 1993). Under the commonly used assumption that these nuclides (in the case of ^{210}Pb, the excess ^{210}Pb derived from the atmospheric fallout retained on the particle surfaces) are removed from the drainage basin at a constant rate (Smith et al. 1987), the peak fallout of ^{239,240}Pu around 1963 should be clearly observed in the soil profile, as well as in the estuarine sediment cores, provided that the nuclides are removed completely or partially (but a constant fraction of the annual input) from the water column. The well-defined soil inventories of these nuclides in the drainage area and the precise measurements of their concentrations in river water aid in establishing rates of erosion of the drainage basin, as well as the residence times of these nuclides in the drainage basin. Assuming that 0.023% (i.e., half-removal time = 3000 yr) of the fallout Pu and atmospherically derived ^{210}Pb in the drainage basin are eroded each year (Smith et al. 1987), the inventories of ^{239,240}Pu and ^{210}Pb derived from the drainage basin (I_d) can be estimated from the following equations:

$$I_d \, (^{239,240}\text{Pu}) = \left(\frac{A_d}{A_{es}}\right) \Sigma \, I^t_f * f_e \tag{12-2}$$

$$I_d \, (^{210}\text{Pb}) = \left(\frac{A_d}{A_{es}}\right) * I_f * \frac{f_e}{\lambda} \tag{12-3}$$

where A_d is the area of the drainage basin, A_{es} is the area of the estuary, I^t_f λ is the inventory of ^{239,240}Pu in soil from 1954 ($t = 0$) to the present, I_f is the fallout inventory of ^{210}Pb (for Houston, Texas, 33 dpm cm^{-2} for ^{210}Pb; Baskaran et al. 1993), f_e is the fraction of the fallout that is eroded each year (0.023% assumed), and λ is the decay constant of ^{210}Pb (0.03114 yr^{-1}). It must be pointed out that the f_e values could vary as a function of climate, relief, amount of precipitation, and so on. We do not have the measured value for this site, and hence we use the value published in literature (Smith et al. 1987). The expected fallout inventory (i.e., direct atmospheric deposition) for ^{239,240}Pu can be calculated based on ^{90}Sr fallout measurements made in Houston, Texas, by the Environmental Measurements Laboratory (1977; reports EML #415, 457, 533 and HASL-329; 1977–1991), assuming Pu concentration = ^{90}Sr concentration * 0.02 (Joseph et al. 1979). The depositional fluxes of ^{210}Pb were measured for 3 yr (Baskaran et al. 1993), and based on the average value for this period, the sediment/soil inventory of ^{210}Pb can be determined. Using the above equations, the maximum input derived from the drainage basin was calculated for the Sabine-Neches estuary: 0.29 dpm cm^{-2} and 50 dpm cm^{-2} for ^{239,240}Pu and ^{210}Pb, respectively (Baskaran et al. 1997). Due to the nature of the fallout pattern, most of the drainage basin contribution of Pu would have been added to the sediments after 1965. One can determine the values of f_e (as well as the residence time in the drainage basin) for Pb and Pu if the concentrations of dissolved and particulate ^{210}Pb and ^{239,240}Pu in the river water samples are measured over a period of 1 yr.

The concentration profiles of ^{239,240}Pu and ^{210}Pb in a sediment core are useful for dating (sediment accumulation as well as mixing) that core. In addition, radionuclides in sediments provide information on the source(s) and extent of removal of these nuclides from the water column. Any significant deviation from the expected inventory values could provide information on the removal mechanism(s) and/or source(s) of these nuclides. We used such comparisons of the expected inventories to the measured values (Table 12-2) and obtained the retention efficiency of these nuclides for the Sabine-Neches estuary in southeast Texas. It is evident that a major portion of what has been coming into the estuary is not removed and is retained in the sediment record. Since Pb and Pu can be used as analogues for other heavy metals, it is likely that most other heavy metals discharged into this estuarine system are not retained within the estuary. This has been confirmed by other trace metal and rare earth element studies on the sedimentary cores from the Sabine-Neches estuary (Ravichandran et al. 1995b; Ravichandran 1996). Similar investigations can be extended to any other shallow, turbid, and organic-rich estuaries in the Gulf or elsewhere.

RESIDENCE TIME OF PARTICLE-REACTIVE RADIONUCLIDES AND SUSPENDED PARTICLES

Most of the inorganic as well as organic compounds [such as hydrophobic organic compounds, (HOCs), PCBs, chlorinated pesticides (e.g., DDT, mirec), and others generated as by-products of industrial production (e.g., hexachlorobenzene, HCB)] are carried by fine particles (Wong et al. 1995); thus, the fate of fine particles is tied to the fate of these pollutants. A recent study by Gustafsson et al. (1997) indicated that the vertical fluxes of PAHs thorough coastal and shelf waters can be obtained using ^{234}Th, a particle tracer. Thus, quantifying the residence times of dissolved pollutants, as well as those of fine particles that serve as carrier phases for these pollutants, will be important

TABLE 12-2. Measured and Theoretical Inventories of Excess ^{210}Pb and ^{239,240}Pu and Their Retention Efficiencies in Sediment Cores from Sabine-Neches Estuary

Station	^{210}Pb Inventory (dpm cm^{-2})	^{239,240}Pu Inventory (dpm cm^{-2})	Theoretically Expected ^{210}Pb Inventory[a] (dpm cm^{-2})	Theoretically Expected ^{239,240}Pu Inventory[b] (dpm cm^{-2})	^{210}Pb Retention Efficiency[c] (%)	^{239,240}Pu Retention Efficiency[c] (%)
2	8.18	0.24	83	0.49	9.9	49.0
4	12.8	0.14	83	0.49	15.4	28.6
6	12.3	0.11	83	0.49	14.8	22.4
7	28.6	0.09	83	0.49	34.5	18.4

[a] Total theoretical ^{210}Pb inventory = atmospheric fallout inventory (33 dpm cm^{-2}) + calculated erosional input from the watershed (50 dpm cm^{-2}).

[b] Total theoretical ^{239,240}Pu inventory = atmospheric fallout inventory (0.20 dpm cm^{-2}) + calculated erosional input from the watershed (0.29 dpm cm^{-2}).

[c] The differences in the retention efficiencies between Pb and Pu are not statistically significant. If the differences are real, this could be due to (1) recent changes in the removal behavior of these nuclides due to changes in input (^{239,240}Pu is a non-steady-state tracer, while ^{210}Pb is a steady-state tracer) and (2) differences in the chemical properties of Pu and Pb (Baskaran et al., 1997).

Note: The reverine inputs of ^{210}Pb and ^{239,240}Pu were estimated (Baskaran et al. 1997).

to a better understanding of the biogeochemical processes in estuaries that affect particle-reactive species.

As discussed before, the hydraulic residence times of the estuaries that discharge into the Gulf of Mexico vary considerably (Chapter 2 this volume). In areas where DOM concentrations are high (such as the Sabine-Neches and Ochlockonee estuaries), the residence times of particle-reactive pollutants in the water could be considerably longer, due to complexation of pollutants (such as heavy metals) with the DOM. In such a case, it is likely that shorter hydraulic residence times will result in the net export of particle-reactive nuclides and pollutants to coastal waters from the estuary.

Several short-lived particle-reactive radionuclides have been utilized to determine the residence times of these nuclides and to obtain analogous information for other organic and inorganic pollutants (Table 12-1). The well-defined production rates (from parent) and/or input functions (atmospherically delivered radionuclides) will enable us to calculate the residence times of these nuclides. For example, ^{234}Th is produced from dissolved ^{238}U at a known constant rate. The atmospheric fallout of ^{210}Pb (or ^{7}Be) for any given latitude is fairly well known (Turekian et al. 1977). This information on residence time is useful in investigating how quickly a pollutant will be removed from the water column and possibly enter the aquatic food chain.

From a knowledge of the inputs of particle-reactive nuclides, such as Th, Pb, or Be, one can calculate the dissolved (τ_s) and particulate (τ_p) residence times of these nuclides. Assuming a simple box model for ^{234}Th, one can calculate residence times using the following equations (Baskaran and Santschi 1993; Baskaran et al. 1996; Baskaran in press and references cited therein):

$$\tau_s = \frac{[^{234}Th_d]}{[(^{238}U - {}^{234}Th_d)\,\lambda_{Th}]} \tag{12-4}$$

and

$$\tau_p = \frac{[^{234}Th_p]}{\{[(^{238}U - {}^{234}Th_d) - {}^{234}Th_p]\,\lambda_{Th}\}} \tag{12-5}$$

where λ_{Th} is the ^{234}Th decay constant (0.029 d^{-1}), $[^{238}U_d]$ and $[^{234}Th_d]$ are the ^{238}U and ^{234}Th dissolved activity concentrations (dpm liter^{-1}), respectively, and ^{234}Th$_p$ is the particulate ^{234}Th concentration. Similarly, if in situ production of ^{210}Pb from ^{222}Rn is negligible compared to atmospheric fallout, then, the residence times of ^{210}Pb (or ^{7}Be) can be calculated as follows (assuming a steady state):

$$\tau_{Bo/Pb} = \ln 2 \times A_{Pb/Be} \times \frac{h}{I_{Pb/Be}} \tag{12-6}$$

where $A_{Pb/Be}$ is the total activity of ^{210}Pb or ^{7}Be (dpm m^{-3}), $I_{Pb/Be}$ is the atmospheric input rate of ^{210}Pb or ^{7}Be (dpm m^{-2} d^{-1}), and h is the mean depth (m) of the well-mixed coastal system. These residence times can be compared to values obtained in other coastal waters (Table 12-3, Broecker et al. 1973; Santschi et al. 1979; Baskaran and Santschi 1993; Baskaran et al. 1997). Equation 6 does not consider inputs from rivers or oceans (by lateral transport, some amount of particle-reactive nuclides reach the coastal areas) or export of nuclides from the estuary to the ocean by water mass transport. A more detailed calculation including some of these input and output terms is presented in Baskaran et al. 1997 for the Sabine-Neches estuary in southeastern Texas.

Residence times of ^{7}Be and ^{210}Pb calculated for various estuaries along the Gulf coast, along with a few other estuaries on the east coast of the United States, are given in Table 12-3. The ^{7}Be residence time along the Gulf coast (0.6–6.4 d) is the lowest compared to the east coast estuaries (2–33 d). The distinctl y lower residence times in Gulf coast estuaries are likely due to their shallow water compared to the east coast estuaries reported in Table 12-3. There are no data on the residence time of ^{210}Pb for the east coast estuaries; hence, the effect of water depth on the residence time cannot be inferred. However, it is likely that residence time depends on the depth of the water column, similar to the case of ^{7}Be. Of all the estuaries in the Gulf of Mexico for which the residence times are presented in Table 12-3, the Sabine-Neches estuary has the longest residence time. This estuary has one of the highest DOC concentrations, up to 21 mg liter^{-1}; hence, it is suggested that DOM forms a complex with these nuclides, which results in longer residence time (Baskaran et al. 1997, Bianchi et al. 1997). As discussed earlier, the role of particles in the biogeochemical cycling of particle-reactive metals is very significant; hence, the determination of particle residence time in an estuary is particularly useful.

PARTICLE RESIDENCE TIMES

In an estuarine system where particle residence times are relatively long and are comparable to hydraulic residence times, a significant portion of the suspended matter can be exported out of the system. In some of the Gulf coast estuaries, the hydraulic residence times are short (Table 12-3). As discussed before, most of the organic and inorganic pollutants are particle-reactive; hence, particles play a crucial role in determining the fate of these pollutants. In order to determine the particle residence time in any shallow estuary, a model can be employed (Baskaran et al. 1997). From a mass balance of particulate activity of a nuclide, one can determine the sediment re-suspension rate. The box model for particulate ^{210}Pb$_{xs}$ is given in Fig. 12-5. The inputs for particulate ^{210}Pb$_{xs}$ (which refers to excess ^{210}Pb in the case of sediments and particles) are the following: (1) production from parent in the

TABLE 12-3. Total Residence Times and Distribution Coefficients of ^{7}Be and ^{210}Pb in Different Estuarine Systems

Name of Estuary	Water Depth (m)	DOC (mg liter^{-1})[a]	Suspended Particle Conc. (mg liter^{-1})[b]	Hydraulic Residence Time (d)[c]	^{7}Be Residence Time (d)[d]	^{210}Pb Residence Time (d)[d]	^{7}Be-Kd (10^4 cm^3 g^{-1})	^{210}Pb-Kd (10^4 cm^3 g^{-1})
[1]Copano Bay, Texas	1.1	ND	30.4	ND	NM	1.9	NM	0.71
[1]San Antonio Bay, Texas	1.4	4.0–5.8	43.5	39	NM	0.2	NM	2.9
[1]Aransas Bay, Texas	2.4	ND	16.4	360	NM	3.7	NM	2.0
[1]Baffin Bay, Texas	2.4	ND	206	ND	NM	9.1	NM	0.11
[1]Corpus Christi Bay, Texas	3.2	6.7–7.6	17.0	356	NM	5.7	NM	2.2
[1]Laguna Madre, Texas	1.4	ND	21.2	ND	NM	2.8	NM	8.2
[1]Cedar Pass, Texas	0.5	ND	7.6	ND	NM	1.8	NM	NM
[1]Galveston Bay, Texas	2.0	5.0–5.8	11.0	41	NM	0.87	29–117	31–113
[2]Upper Chesapeake Bay, Maryland	8	ND	ND	—	5–33	NM	2.3–20	NM
[2]James River, Virginia	3.5	ND	ND	—	2–4	NM	0.76–16	NM
[2]Hudson River, New York	6.0	ND	ND	—	8–17	NM	1.3–7.5	NM
[2]Raritan Bay, New York	6.0	ND	ND	—	7–17	NM	3.9–16	NM
[3]Sabine-Neches Estuary	1.8	4.8–21.0	34.7	9	0.6–6.4	3.5–27	0.15–8.7	0.26–3.7

[a] The DOC concentrations were taken from the compiled data in Chapter 9 this volume (Table 9-1); ND = no data available.

[b] The suspended particle concentration is the average value of two to three seasons wherever data are available (Baskaran and Santshi 1993).

[c] The hydraulic residence time is taken from Chapter 2 of this volume.

[d] The Texas estuaries were sampled during winter and summer months, and the residence time is the average during these two seasons; the Chesapeake and Raritan Bays and the James and Hudson Rivers were sampled during summer, and the Sabine-Neches estuary was mainly sampled during spring and fall. NM = not measured.

1: Baskaran and Santschi (1993); 2: Olsen et al. (1986); 3: Baskaran et al. 1997.

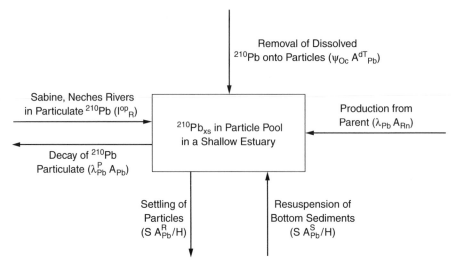

FIG. 12-5. *Box model showing the inputs and removal of particulate ^{210}Pb for any shallow estuary.*

particulate phase, ^{222}Rn ($\lambda_{Pb} A_{Rn}$); (2) riverine particle-sorbed (I^{op}_R) input; (3) removal of dissolved ^{210}Pb [total dissolved ^{210}Pb (A^{dT}_{Pb}) = riverine dissolved input + direct atmospheric input + production from dissolved ^{222}Rn] onto suspended particles ($\psi^o_c A^{dT}_{Pb}$); and (4) re-suspension of bottom sediments [$S A^s_{Pb} / H$, where S is the sediment re-suspension rate (g cm^{-2} yr^{-1}), A^s_{Pb} is the ^{210}Pb concentration in surficial sediments that is re-suspended, and H is the mean depth of the estuary (cm)]. Losses of particulate ^{210}Pb are due to (1) decay of ^{210}Pb in the particle reservoir ($\lambda_{Pb} A^P_{Pb}$, where A^P_{Pb} is the total particulate ^{210}Pb concentration in the particulate pool); (2) settling of particles containing ^{210}Pb ($S A^R_{Pb} / H$), where A^R_{Pb} is the ^{210}Pb concentration in suspended particles); and (3) desorption of $^{210}Pb_{xs}$ from particles. The desorption of ^{210}Pb is assumed to be negligible. The mass balance equation for particulate ^{210}Pb can be written as

$$\lambda_{Pb} A_{Rn} + I^{op}_R + \psi^o_c A^{dT}_{Pb} + \frac{S A^s_{Pb}}{H} = \lambda_{Pb} A^P_{Pb} + \frac{S A^R_{Pb}}{H} \quad (12\text{-}7)$$

Rearranging (Eq. 7),

$$S = \frac{H[\lambda_{Pb} A^P_{Pb} - \psi^o_c A^{dT}_{Pb} - I^{op}_R]}{[A^s_{Pb} - A^R_{Pb}]} \quad (12\text{-}8)$$

The term $\lambda_{Pb} A_{Rn}$ can be assumed to be negligible (Baskaran et al. 1997). In Eqs. 7 and 8, ψ^o_c is the composite rate constant corresponding to the removal by particles, as well as removal of these nuclides by the outflow to the ocean.

In Eqs. 7 and 8, the loss of particulate ^{210}Pb by the outflow of water, as well as input of ^{210}Pb from the ocean into the estuary, can also be assumed negligible. If there is a small export of ^{210}Pb from the estuary into the ocean, then the S value calculated will be an underestimate. In such case, an additional term for the export of nuclides to the ocean can be included in the mass balance equation. The particle residence time was determined from the following relation: the particle residence time (τ_p) = (suspended particle concentration (g cm^{-3}) × height of the water column (cm))/re-suspension rate (g cm^{-2} d^{-1}). The following values are used in the calculation of the particle residence time: λ_{Pb} = 0.03136 yr^{-1}; A^P_{Pb} is the particulate ^{210}Pb concentration (in dpm cm^{-3}) and A^{dT}_{Pb} is the concentration of dissolved ^{210}Pb (in dpm cm^{-3}); ψ^0_c = 0.693 / dissolved residence time; A^s_{Pb} is the excess ^{210}Pb (^{210}Pb$_{xs}$) concentration in surficial sediments, and A^R_{Pb} is the ^{210}Pb$_{xs}$ concentration in suspended particles (= particulate ^{210}Pb$_{xs}$ (dpm cm^{-3}) / suspended particle concentration (g cm^{-3})), dpm g^{-1}. By measuring these parameters, one can calculate the residence time of particles for any estuary. It must be pointed out that the particle-residence time is only an estimate. The particle residence time calculated for the Sabine-Neches estuary varied between 1 and 2 d. This range is distinctly higher than the values obtained for Galveston coastal waters following the pulse input of ^{7}Be and ^{210}Pb (Baskaran and Santschi 1993). We do not have the particle residence time data for any other estuary. The residence time of particles could vary considerably, depending on the size, nature, and composition of the particles. For the Gulf coast estuaries, longer particle residence times could facilitate the removal of particles and associated nuclides by the flushing of water into the Gulf of Mexico.

SUMMARY

The estuarines systems in the Gulf of Mexico are some of the most extensive in the world, with diverse circulation and salinity regimes. Radionuclides continue to provide insight into many estuarine biogeochemical processes, and in this chapter some of the recent applications of particle-reactive radionuclides in the Gulf of Mexico are highlighted. Using mass balance of radionuclides in suspended particulate matter, it is possible to determine the residence times of suspended particles in an estuary. In organic-rich estuaries, the concentration and composition of DOM appear to play a major role in the removal of particle-reactive nuclides. The distribution coefficient for particle-reactive nuclides in organic-rich estuaries appears to be the lowest, probably due to the complexation of nuclides with DOM. In estuaries where the hydraulic residence time is comparable to the residence times of particulate matter, a significant portion of the particles and particle-bound nuclides can be exported out of the estuary. Sediment inventories of ^{210}Pb and ^{239,240}Pu can be utilized to determine the retention efficiency of these nuclides and, by implication, those of other heavy metals.

FUTURE RESEARCH

Radionuclides have been utilized in almost all branches of geoscience as tracers and chronometers. Examples include sediment accumulation and mixing, scavenging reactions, determination of the retention of heavy metals in an estuary, and water mass mixing. For a better understanding of their utility in estuarine biogeochemical processes, the source functions need to be precisely determined. In most estuaries, the relative amounts of inputs from erosion of the watershed are not known. These amounts depend on the nature of drainage basins such as the nature of the substrate, porosity, climate, and relief. The amount of precipitation also plays a significant role. Another source function that is not well characterized is the lateral input from the ocean. It has been documented that suspended particles from the continental margins reach the coastal areas and pick up particle-reactive radionuclides during their transit; hence, these particles could be another potential source of radionuclides (Olsen et al. 1989b). The "macroscopic" inference, in terms of either determination of rate constants or the distribution of species, is a result of many processes, and further research should focus on quantifying each of these processes. These problems warrant extensive laboratory studies combined with field observations. This sort of investigation may also require further improvement and/or development of our analytical capabilities. Only a quantitative understanding of these individual processes will lead us to develop a better predictive capability.

ACKNOWLEDGMENTS

I thank Dr. Thomas Bianchi for a review of and editorial assistance with this manuscript and Dr. Bill Landing for a thorough review of this manuscript.

REFERENCES

Aaboe, E., E. P. Dion, and K. K. Turekian. 1981. ^{7}Be in Sargasso Sea and Long Island Sound. J. Geophys. Res. **86:** 3255–3257.

Aller, R. C., L. K. Benninger, and J. K. Cochran. 1980. Tracking particle-associated processes in nearshore environments by use of ^{234}Th/^{238}U disequilibrium. Earth Planet. Sci. Lett. **47:** 161–175.

———, and J. K. Cochran. 1976. ^{234}Th/^{238}U disequilibrium in near-shore sediment: Particle reworking and diagenetic time scales. Earth Planet. Sci. Lett. **29:** 37–50.

Appleby, P. G., and F. Oldfield. 1992. Applications of lead-210 to sedimentation studies, p. 731–778. *In:* M. Ivanovich and R. S. Harmon [eds.], Uranium-series disequilibrium. Oxford Science.

Bacon, M. P., D. W. Spencer, and P. G. Brewer. 1976. $^{210}Pb/^{226}Ra$ and $^{210}Po/^{210}Pb$ disequilibria in seawater and suspended particulate matter. Earth Planet. Sci. Lett. **32:** 277–296

Baskaran, M. In press. Scavenging of thorium isotopes in the Arctic regions: Implications for the fate of particle-reactive pollutants. Mar. Poll. Bull.

————, C. H. Coleman, and P. H. Santschi. 1993. Atmospheric depositional fluxes of ^{7}Be and ^{210}Pb at Galveston and College Station, Texas. J. Geophys. Res. **98:** 20,555–20,576.

————, and A. S. Naidu. 1995. ^{210}Pb-derived chronology, and the fluxes of ^{210}Pb and ^{137}Cs isotopes into continental shelf sediments, East Chukchi Sea, Alaskan Arctic. Geochim. Cosmochim. Acta **59:** 4435–4448.

————, R. J. Presley, S. Asbill, and P. H. Santschi. (Submitted). Reconstruction of historical contamination of trace metals in Mississippi River delta, Tampa Bay, and Galveston Bay sediments. Environ. Sci. Technol.

————, M. Ravichandran, and T. S. Bianchi. 1997. Cycling of ^{7}Be and ^{210}Pb in a high DOC, shallow, turbid estuary of southeast Texas. Estuar. Coast. Shelf Sci. **45:** 165–176.

————, and P. H. Santschi. 1993. The role of particles and colloids in the transport of radionuclides in coastal environments of Texas. Mar. Chem. **43:** 95–114.

————, P. H. Santschi, L. Guo, T. S. Bianchi, and C. Lambert. 1996. $^{234}Th:^{238}U$ disequilibria in the Gulf of Mexico: The importance of organic matter and particle concentration. Cont. Shelf. Res. **16:** 353–380.

Benninger, L. K., 1978. ^{210}Pb balance in Long Island Sound. Geochim. Cosmochim. Acta **42:** 1165–1174.

————, R. C. Aller, J. K. Cochran, and K. K. Turekian. 1979. Effects of biological sediment mixing on the ^{210}Pb chronology and trace metal distribution in a Long Island Sound sediment core. Earth Planet. Sci. Lett. **43:** 241–259.

————, and S. Krishnaswami. 1981. Sedimentary processes in the inner New York Bight: Evidence from excess ^{210}Pb and $^{239,240}Pu$. Earth Planet. Sci. Lett. **53:** 158–174.

Bianchi, T. S., M. Baskaran, J. DeLord, and M. Ravichandran. 1997. Carbon cycling in a shallow turbid estuary of southeast Texas: The use of plant pigment biomarkers and water quality parameters. Estuaries **20:** 404–415.

Bollinger, M. S., and W. S. Moore, 1993. Evaluation of salt marsh hydrology using radium as a tracer. Geochim. Cosmochim. Acta **57:** 2203–2212.

Bopp, R. F., H. J. Simpson, C. R. Olsen, and N. Kostyk. 1981. Polychorinated biphenyls in sediments of the tidal Hudson River, New York. Environ. Sci. Technol. **15:** 210–216.

Borole, D. V., S. Krishnaswami, and B. L. K. Somayajulu. 1982. Uranium isotopes in rivers, estuaries and adjacent coastal sediments of western India: Their weathering, transport and oceanic budget. Geochim. Cosmochim. Acta **46:** 125–137.

Boyle, E. A., J. M. Edmond, and E. R. Sholkovitz. 1977. The mechanism of iron removal in estuaries. Geochim. Cosmochim. Acta **41:** 1313–1324.

Broecker, W. S., A. Kaufman, and R. M. Trier. 1973. The residence time of thorium in surface seawater and its implications regarding the fate of reactive pollutants. Earth Planet. Sci. Lett. **20:** 35–44.

Buffle, J., D. Perret, and M. Newmann. 1992. The use of filtration and ultrafiltration for size fractionation of aquatic particles, colloids, and macromolecules, p. 171–229. *In:* J. Buffle and Van Leeuwen [eds.], Environmental analytical chemistry series, volume I, chap. 5. Lewis.

Burton, J. D., and N. S. Liss. 1976. Estuarine chemistry. Academic Press.

Clayton, J. R., S. P. Paviou, and N. F. Breitner. 1977. Polychlorinated biphenyls in coastal marine zooplankton: Bioaccumulation by equilibrium partitioning. Environ. Sci. Technol. **11:** 676–682.

Cochran, J. K. 1984. The fates of uranium and thorium decay series nuclides in the estuarine environment, p. 179–220. *In:* V. C. Kennedy [ed.], The estuary as a filter. Academic Press.

Donat, J. R., K. A. Lao, and K. W. Bruland. 1994. Speciation of dissolved copper and nickel in South Francisco Bay: A multi-method approach. Anal. Chim. Acta **284:** 547–571.

Duinkar, J. C. 1980. Suspended matter in estuaries: Adsorption and desorption processes, p. 121–151. *In:* E. Olausson and I. Cato [eds.], Chemistry and biochemistry of estuaries. Wiley and Sons, New York.

Edzwald, J. K., J. B. Upchurch, and C. R. O'Melia. 1974. Coagulation in estuaries. Environ. Sci. Technol. **8:** 58–63.

EML (Environmental Measurements Laboratory) Reports. 1977. Final tabulation of monthly ^{90}Sr fallout data: 1954–1976. HASSL-329. Also EML Reports 415 (1983), 457 (1987), and 533 (1991).

Faure, G. 1986. Principles of isotope geology (2nd ed.). Wiley.

Gearing, P. J., J. N. Gearing, R. J. Pruell, T. S. Wade, and J. G. Quinn. 1980. Partitioning of No. 2 fuel oil in controlled estuarine ecosystems. Sediments and suspended particulate matter. Environ. Sci. Technol. **14:** 1129–1136.

Guentzel, J. L., R. T. Powell, W. M. Landing, and R. P. Mason. 1996. Mercury associated with colloidal material in an estuarine and an open ocean. Mar. Chem. **55:** 177–188.

Gustafsson, O., P. M. Gschwend, and K. O. Buesseler. 1997. Using [234]Th disequilibria to estimate the vertical removal rate of polycyclic aromatic hydrocarbons from the surface ocean. Mar. Chem. **57:** 11–23.

Health and Safety Laboratory. 1977. Final tabulation of monthly Sr-90 fallout data: 1954–1976. HASL-329. Energy Research and Development Administration.

Honeyman, B. D., and P. H. Santschi. 1988. Metals in aquatic systems. Environ. Sci. Technol. **22:** 862–871.

Joseph, A. B., P. F. Gustafson, I. R. Russell, F. A. Schuert, H. L. Volchok, and A. Tamplin. 1979. Sources of radioactivity and their characteristics, p. 6–41. *In:* Radioactivity in the marine environment. National Academy of Sciences.

Kaufman, A., Y.-H. Li, and K. K. Turekian. 1981. The removal rates of [234]Th and [228]Th from waters of the New York Bight. Earth Planet. Sci. Lett. **54:** 385–392.

Li, Y.-H., H. W. Feely, and P. H. Santschi. 1979. [228]Th-[228]Ra radioactive disequilibrium in the New York Bight and its implications for coastal pollution. Earth Planet. Sci. Lett. **42:** 13–26.

———, G. Mathieu, P. Biscaye, and H. J. Simpson. 1977. The flux of Ra-226 from estuarine and continental shelf sediments. Earth Planet. Sci. Lett. **37:** 237–241.

———, P. H. Santschi, A. Kaufman, L. K. Benninger, and H. W. Feely. 1981. Natural radionuclides in waters of the New York Bight. Earth Planet Sci. Lett. **55:** 217–228.

McKee, B. A., D. J. DeMaster, and C. A. Nittrouer. 1986. Temporal variability in the partitioning of thorium between dissolved and particulate phases on the Amazon shelf: Implications for the scavenging of particle-reactive species. Cont. Shelf Res. **6:** 87–106.

Means, J. C., and R. Wijayaratne. 1982. Role of natural colloids in the transport of hydrophobic pollutants. Science **215:** 968–970.

Minagawa, M., and S. Tsunogai. 1980. Removal of ^{234}Th from a coastal sea: Funka Bay, Japan. Earth Planet. Sci. Lett. **47:** 51–64.

Moore, W. S. 1992. Radionuclides of the uranium and thorium decay series in the estuarine environment, p. 396–422. *In:* M. Ivanovich and R. S. Harmon [eds.], Uranium series disequilibrium—applications to earth, marine and environmental sciences (2nd ed.). Clarendon Press.

Nittrouer, C. A., R. W. Sternberg, R. C. Carpenter, and J. T. Bennett. 1979. The use of Pb-210 geochronology as a sedimentological tool: Application to the Washington continental shelf. Mar. Geol. **31:** 297–316.

Olmez, I., and G. E. Gordon. 1985. Rare earths: Atmospheric signatures for oil fired power plants and refineries. Science **229:** 966–968.

———, E. R. Sholkovitz, D. Hermann, and R. P. Eganhouse. 1991. Rare earth elements in sediments off southern California: A new anthropogenic indicator. Environ. Sci. Technol. **25:** 310–316.

Olsen, C. R., P. E. Biscaye, H. J. Simpson, R. M. Trier, N. Kostyk, R. F. Bopp, Y.-H. Li, and H. W. Feely. 1980. Reactor-released radionuclides and fine-grained sediments transport and accumulation patterns in Barnegat Bay, New Jersey, and adjacent shelf waters. Estuar. Coast. Shelf Sci. **10:** 119–142.

———, N. H. Cutshall, and I. L. Larsen. 1982. Pollutant-particle associations and dynamics in coastal marine environments: A review. Mar. Chem. **11:** 501–533.

———, I. L. Larsen, P. D. Lowry, N. H. Cutshall, and M. M. Nichols. 1986. Geochemistry and deposition of ^{7}Be in river-estuarine and coastal waters. J. Geophys. Res. **91:** 896–908.

———, I. L. Larsen, P. D. Lowry, R. I. McLean, and S. L. Demoter. 1989a. Radionuclide distributions and sorption behavior in the Susquehanna-Chesapeake Bay system. Maryland Power Plant and Environmental Review Division, Report PPER-R-1. Department of Natural Resources.

———, H. J. Simpson, R. F. Bopp, S. C. Williams, T.-H. Peng, and B. L. Deck. 1978. Geochemical analysis of the sediments and sedimentation in the Hudson Estuary. J. Sed. Petrol. **48:** 401–418.

———, M. Thein, I. L. Larsen, P. D. Lowry, P. J. Mulholland, N. H. Cutshall, J. T. Byrd, and H. L. Windom, 1989b. Plutonium, lead-210, and carbon isotopes in the Savannah Estuary: Riverborne versus marine sources. Environ. Sci. Technol. **23:** 1475–1481.

Orlando, S. P., Jr., L. P. Rozas, G. H. Ward, and C. J. Klein. 1993. Salinity characteristics of Gulf of Mexico estuaries. National Oceanic and Atmospheric Administration, Office of Ocean Resources Conservation and Assessment.

Rama, H., M. Koide, and E. D. Goldberg. 1961. Lead-210 in natural waters. Science **134:** 98–99.

Ravichandran, M. 1996. Distribution of rare earth elements in sediment cores of Sabine-Neches estuary. Mar. Poll. Bull. **32:** 719–726.

————, M. Baskaran, P. H. Santschi, and T. S. Bianchi. 1995a. Geochronology of sediments in the Sabine-Neches estuary, Texas, USA. Chem. Geol. **125:** 291–306.

————. 1995b. History of trace metal pollution in Sabine-Neches estuary, Beaumont, TX. Environ. Sci. Technol. **29:** 1495–1503.

Reuter, J. H., and E. M. Perdue. 1977. Importance of heavy metal-organic matter interactions in natural waters. Geochim. Cosmochim. Acta **41:** 325–334.

Ritchie, J. C., J. A. Spraberry, and J. R. McHenry. 1974. Estimating soil erosion from the redistribution of fallout Cs-137. Soil Sci. Soc. Am. Prof. **38:** 137–139.

Santschi, P. H., D. Adler, M. Amdurer, Y.-H. Li, and J. J. Bell. 1980. Thorium isotopes as analogues for "particle-reactive" pollutants in coastal marine environments. Earth Planet. Sci. Lett. **47:** 327–335.

————, Y.-H. Li, and J. J. Bell. 1979. Natural radionuclides in Narragansett Bay. Earth Planet. Sci. Lett. **47:** 201–213.

————, S. Nixon, M. Pilson, and C. Hunt. 1984. Accumulation rate of sediments, trace metals and total hydrocarbons in Narragansett Bay, Rhode Island. Estuar. Coast. Shelf Sci. **19:** 427–449.

Saxby, J. D. 1969. Metal-organic chemistry of the geochemical cycle. Rev. Pure Appl. Chem. **19:** 131–150.

Schnitzer, M., and U. Kahn. 1972. Characterization of humic substances by chemical methods, chapter 6, p. 203–251. *In:* M. Schnitzer and U. Kahn [eds.], Humic substances in the environment. Marcel Dekker.

Sholkovitz, E. R. 1976. Flocculation of dissolved organic and inorganic matter during the mixing of river water and seawater. Geochim. Cosmochim. Acta **40:** 831–845.

————. 1983. The geochemistry of plutonium in fresh and marine water environments. Earth Sci. Rev. **64:** 95–161.

Simpson, H. J., C. R. Olsen, R. M. Trier, and S. C. Williams. 1976. Man-made radionuclides and sedimentation in the Hudson River estuary. Science **194:** 179–183.

Smith, J. N., K. M. Ellis, and D. M. Nelson. 1987. Time-dependent modeling of fallout radionuclide transport in a drainage basin: Significance of "slow" erosional and fast hydrological components. Chem. Geol. **63:** 157–180.

Tanaka, N., and S. Tsunogai. 1983. Behavior of ^{7}Be in Funka Bay, Japan, with reference to those of insoluble nuclides, ^{234}Th, ^{210}Po and ^{210}Pb. J. Geochem. **17:** 9–17.

Turekian, K. K. 1977. The fate of metals in the ocean. Geochim. Cosmochim. Acta **41:** 1139–1144.

————, Y. Nozaki, and L. K. Benninger. 1977. Geochemistry of atmospheric radon and radon products. Ann. Rev. Earth Planet. Sci. **5:** 227–255.

Valette-Silver, N. J. 1993. The use of sediment cores to reconstruct historical trends in contamination of estuarine and coastal sediments. Estuaries **16:** 577–588.

Wan, G. J., P. H. Santschi, M. Sturm, K. Farrenkothen, A. Lueck, E. Werth, and C. Schuler. 1987. Natural and fallout (^{137}Cs, ^{239,240}Pu, ^{90}Sr) radionuclides as geochemical tracers of sedimentation in Greifensee, Switzerland. Chem. Geol. **63:** 181–196.

Wong, C. S., G. Sanders, D. R. Engstrom, D. T. Long, D. L. Swackhamer, and S. J. Eisenreich. 1995. Accumulation, inventory, and diagenesis of chlorinated hydrocarbons in Lake Ontario sediments. Environ. Sci. Technol. **29:** 2662–2672.

Section V

Summary

Chapter *13*

Biogeochemistry of Gulf of Mexico Estuaries: Implications for Management

Thomas S. Bianchi, Jonathan R. Pennock, and Robert R. Twilley

INTRODUCTION

The field of biogeochemistry involves the study of how biological, chemical, and geological processes interact to determine the fate and effects of materials that influence the metabolism of ecosystems. An understanding of the role that biogeochemical and physical processes play in regulating the chemistry and biology of estuaries is fundamental to evaluating complex management issues such as those found in the Gulf of Mexico. As we have described in this book, biogeochemistry links the processes that control the fate of sediments, nutrients, organic matter, and trace metals in estuarine ecosystems. Therefore, this discipline requires an integrated perspective on estuarine dynamics associated with the introduction, transport, and either accumulation or export of materials that largely control the primary productivity. The metabolism of in situ primary production, and indirectly the utilization of allochthonous organic

Biogeochemistry of Gulf of Mexico Estuaries, Edited by Thomas S. Bianchi, Jonathan R. Pennock, and Robert R. Twilley.
ISBN 0-471-16174-8 © 1999 John Wiley & Sons, Inc.

matter, are also linked to patterns of secondary productivity and fishery yields in estuaries in the Gulf of Mexico. As humans alter the way regional watersheds and local landscapes of estuaries produce and process natural and synthetic chemicals, principles of biogeochemistry will continue to influence how we manage these unique coastal ecosystems.

Estuaries in the Gulf of Mexico basin possess physical characteristics that are as diverse as those of any coastal region in the world, and offer a unique opportunity to compare the role that physical factors play in regulating estuarine biogeochemical processes. Recent research on biogeochemical processes in Gulf systems also provides new insights into the similarities and differences between warm-temperate and sub-tropical estuaries—such as those found in the Gulf—and cool-temperate estuaries on which much of our current understanding of estuarine biogeochemistry is based. Deegan et al. (1986) argue that the Gulf of Mexico is an excellent region in which to compare the characteristics and processes among estuaries for the following reasons: (1) there is a large number of estuaries (64); (2) the climate ranges from tropical to temperate and from humid to arid; (3) the area encompasses a wide range of riverine influences, from systems with almost no riverine input to the Mississippi River; and (4) the size of estuarine areas (in terms of both water and inter-tidal area) varies from very small to the largest in North America (Table 13-1). Deegan et al. (1986) compared the influence of physical factors among Gulf estuaries, such as river discharge and size of estuary, to vegetation distribution and fishery harvest. The chapters of this book have instead compared the biogeochemical properties of selected estuaries in the Gulf of Mexico. Both analyses are predicated on a similar assumption: that the physical and geomorphological template of an estuary—modified by the evolution of coastal landscapes—constrains the patterns of biological and chemical processes that we observe today.

Linkages between the biogeochemistry and biotic resources of an estuary should provide managers with insights to recognize the unique properties that are responsible for sustaining water quality and economic conditions in individual estuaries. For example, Deegan et al. (1986) found that the fishery harvest and area of an estuary are strongly related to fresh-water input and physiography, and that fishery harvest per unit of open water in the southern Gulf is highly correlated with river discharge. The biogeochemistry of these estuaries can provide insights into the mechanisms by which fresh-water delivery of nutrients, sediment, and organic matter supports higher trophic levels. This should provide alternatives to managing the sensitive issues of fresh-water delivery and water quality in many regions of the Gulf of Mexico.

MANAGEMENT ISSUES

Estuaries of the Gulf of Mexico can be loosely divided into four regions based on the geological, climatological, and physical characteristics of the areas: (1) the Eastern Gulf—from Florida Bay to the Suwannee River, (2) the Northern Gulf—from Apalachicola Bay to the Atchafalaya/Vermilion Bays, (3) the Western Gulf—from Calcasieu Lake to the Lower Laguna Madre, and (4) the Southern Gulf—from Laguna de Tamiahua to the Rio Lagartos lagoon in Mexico (Chapters 1 and 2 this volume). Within each of these regions, the dominant biogeochemical processes that are observed directly influence critical management issues.

In the Eastern Gulf, urban development and conversion of wetlands to residential and recreational land use has changed the quantity and quality of waters discharged into estuaries along the southwest coast of Florida. Increased agriculture and urban demand for water has also had major impacts on the delivery of fresh water to areas such as the Everglades and estuaries of the western Florida shelf. In addition, this region has some of the largest phosphate mining operations in the world. This industry has historically contributed to severe eutrophication in Tampa Bay but has undergone significant cleanup over the past decade.

The transition from the Eastern Gulf to the Northern Gulf from the Florida panhandle region to Mississippi includes rivers that drain forested coastal plain watersheds. In recent years, land-use changes in the forested watersheds of Florida, Alabama, Georgia, and Mississippi associated with timber processing have resulted in increased erosion and the introduction of chemical pollutants to estuarine environments in the central Gulf of Mexico. The coastal areas of this region are also becoming increasingly populated with commercial activity, which requires maintenance dredging of ship channels and residential housing.

There have been major environmental changes in the watershed of the Mississippi River that have led to significant patterns of sedimentation and nutrient cycling in estuaries in the Northern Gulf. Over the last century, there has been a quadrupling of river nitrate concentrations and a decrease in suspended load in the last several decades (Walsh et al. 1981; Turner and Rabalais 1991). Discharge from the Mississippi River is approximately 577 km^3 yr^{-1} and enters the Gulf of Mexico via two principal routes: either directly to the shelf or through the shallow bay/wetland complex. The manner in which this river water reaches the Gulf through these two routes has changed significantly over the past 100 yr; historically, a large fraction of the river entered shallow inner shelf waters after passing through a diverse array of bays and an extensive wetland flood plain, such as the region of Fourleague Bay (Madden et al. 1988; Day et al. 1994). Presently—as a result of a levee system that confines most discharge to the main river channel—more than 65% of the river water enters the Gulf as discharge in plumes directly onto the middle and outer shelf, where there is strong interaction with only the upper water column. The increase in nitrate concentration and the decreased sediment load have thus occurred simultaneously with changes in the way in which materials from the river are processed in estuarine ecosystems along the Louisiana coast.

TABLE 13-1. Dimensions, Climatic, Geologic, and River Discharge Characteristics of Gulf of Mexico Estuaries

Map	Estuary Name	Submerged Vegetated Area (ha)	Emergent Vegetated Area (ha)	Saltflat Area (ha)	Open Water Area (ha)	Mean Annual River Discharge (CMS)	Width Along Coast (km)
1	Florida Bay	105104	86473	0	245518	283.1	46.0
2,3	Ten Thousand Islands	1955	72095	0	42000	9.5	117.0
5	Charlotte Harbor	21558	26181	0	112463	86.0	43.0
6	Sarasota Bay	3080	1647	0	14067	2.3	14.0
7	Tampa Bay	11985	8517	0	123855	43.8	49.5
8	Suwanne Sound	3277	13114	0	16084	311.2	14.0
9	Apalachee Bay	12806	28337	0	28440	86.2	33.2
10	Apalachicola Bay	3796	8623	0	68814	763.6	23.3
11	St. Andrew Bay	5245	21017	0	47651	15.1	39.7
12	Choctawahatches	1251	1140	0	34937	204.6	50.0
13	Pensacola bay	3202	4216	0	61300	268.0	49.3
14	Perdido Bay	0	433	0	6989	26.5	13.0
15	Mobile Bay	2024	8693	0	115255	1664.0	28.0
16	Mississippi Sound	12000	27087	0	175821	715.0	78.0
19	Deltaic Plain	100	771193	0	1505814	22897.7	201.0
23	Calcasieu	0	102073	0	93815	157.8	52.9
24	Sabine	0	17199	0	22605	474.0	26.4
25	Trinity-San Jacinto	7327	93684	179	143210	73.1	39.8
27	Matagorda Bay	2850	48582	4532	118057	68.8	12.9
28	Guadalupe estuary	6619	10121	5723	56161	60.8	59.6
29	Aransas Bay	8552	18218	—	46279	5.7	36.3
30	Nueces estuary	5161	18218	—	44451	24.5	36.6
31	Laguna Madre-Upper	77327	101214	66400	150060	1.3	109.3
32	Laguna Madre-Lower	—	8461	—	200978	25.0	206.0
33	Laguna Tamiahua	—	908	—	63430	—	60.0
34	Laguna Tuxpan	—	2393	0	5852	—	7.5
35	Barra de Tecoluta	—	851	—	1000	45.0	2.5
36	Laguna Alvardo	—	56550	0	1428	—	53.0
37	Grijalva-Usamacinta	—	26634	0	10224	1900.0	10.0
38	Laguna Terminos	10000	130000	0	160000	200.0	108.0
39	Punta Sanita	—	119	1073	0	0.0	20.2
39	La Ensenada	—	119	1073	0	0.0	12.2
39	MOA	—	255	2300	0	0.0	20.2
39	Santa Juana	—	102	920	0	0.0	12.2
39	Huaymil 1	—	62	562	0	0.0	6.2
39	Isla Piedra	—	107	971	0	0.0	7.5
39	Huaymil 2	—	272	2453	396	0.0	15.0
39	El Nemate	—	141	1277	0	0.0	6.2
39	Estero Yaltun	—	346	3118	1135	0.0	18.7
39	Estero Celestun	—	550	4957	0	0.0	32.5
39	Parque Celestun	—	3379	30411	0	0.0	35.0
39	Estero Yukalpeten	—	2460	22148	2089	0.0	37.5
39	Laguna Rosada	—	2420	21781	1901	0.0	21.2
39	San Crisanto	—	1033	9311	0	0	21.2
39	Bocas Dzilan	—	19291	192	2755	0	31.2
39	Estero Rio Lagarto	—	35062	3895	26066	0	52.5

*PE = Potential Evapotranspiration, P = Precipitation, AE = Actual Evapotranspiration, D = Average Daily Water Deficit S = Average Daily Water Surplus.   Dashes indicate no data. (Modified from Deegan et al. 1986).

Climatic Water budget* (mm)					Distance to Contour (kn)			Mean Tide Range (m)	Freeze Free Period (days)	Mean Depth (m)
PE	P	AE	D	S	50 m Upland	10 Fathom	100 Fathom			
1295	1309	1257	38	52	600.0	86.4	259.2	0.84	365	1.3
1253	1368	1234	19	134	450.00	54.0	270.0	1.17	365	1.4
1253	1368	1234	19	134	150.00	43.2	226.8	0.55	365	2.3
1127	1359	1112	15	247	67.50	32.4	205.2	0.67	330	1.7
1186	1278	1171	15	107	30.00	43.2	205.2	0.77	330	3.3
1078	1455	1076	2	379	112.50	64.8	226.8	1.00	270	1.6
1025	1465	1022	3	433	82.50	64.8	237.6	1.00	270	1.3
1080	1432	1071	9	361	142.50	43.2	172.8	0.72	300	2.9
1070	1487	1062	8	425	37.50	10.8	97.2	0.47	270	3.6
1065	1519	1057	6	462	11.25	21.6	97.2	0.18	270	4.1
1063	1519	1057	6	462	7.50	21.6	64.8	0.44	270	5.9
1063	1519	1057	6	462	22.50	21.6	86.4	0.15	300	2.6
1040	1614	1039	1	575	52.50	32.4	108.0	0.36	270	2.5
1079	1578	1077	2	471	52.50	43.2	151.2	0.52	300	3.0
1098	1533	1095	2	437	425.00	43.2	108.0	0.36	300	2.0
1073	1448	1054	19	394	217.50	64.8	216.0	0.61	300	1.5
1086	1339	1061	25	278	210.00	64.8	216.0	0.67	300	1.4
1125	1137	1040	85	97	75.00	43.2	172.8	0.30	300	2.3
1173	888	888	285	0	75.00	21.6	108.0	0.18	300	1.1
1173	888	888	285	0	52.50	21.6	108.0	0.18	300	1.1
1173	888	888	285	0	45.00	32.4	108.0	0.42	300	1.1
1118	677	677	441	0	71.25	21.6	108.0	0.39	300	1.2
1243	686	686	557	0	105.00	10.8	86.4	0.42	330	1.0
1283	761	761	522	0	25.00	10.8	75.6	0.42	330	—
1327	1239	1209	118	31	2.50	10.8	32.4	0.42	365	—
1334	1311	1249	85	62	11.25	10.8	43.2	0.52	365	—
1358	1489	1230	178	258	6.25	10.8	43.2	0.52	365	—
1392	2322	1269	123	1053	65.00	10.8	32.4	0.39	369	—
1401	2879	1356	45	1523	70.00	21.6	64.8	0.50	365	—
1586	1738	1471	115	267	52.50	43.2	129.6	0.50	365	3.5
1476	1019	1019	457	0	12.50	64.8	171.8	0.50	365	—
1476	1019	1019	457	0	12.50	64.8	172.8	0.50	365	—
1476	1019	1019	457	0	15.00	64.8	172.8	0.50	365	—
1476	1019	1019	457	0	18.75	64.8	172.8	0.50	365	—
1476	1019	1019	457	0	40.00	54.0	205.2	0.50	365	—
1476	1019	1019	457	0	42.50	54.0	194.4	0.50	365	—
1476	1019	1019	457	0	28.00	54.0	205.2	0.50	365	—
1476	1019	1019	457	0	33.75	43.2	205.2	0.50	365	—
1476	1019	1019	457	0	33.75	32.4	216.0	0.50	365	—
1444	466	466	978	0	35.00	43.2	205.2	0.50	365	—
1444	466	466	978	0	35.00	43.2	205.2	0.55	365	—
1444	466	466	978	0	36.25	32.4	183.6	0.55	365	—
1444	466	466	978	0	36.25	32.4	194.4	0.55	365	—
1444	466	466	978	0	36.25	32.4	194.4	0.55	365	—
1502	898	898	604	0	42.50	21.6	194.4	0.55	365	—
1502	898	898	604	0	50.00	32.4	237.6	0.55	365	—

To the west of the Mississippi River, fresh-water discharge rates decrease significantly (Chapter 1 this volume); as a result, management efforts in the Western Gulf are primarily focused on water resources. In south Texas, river inputs are fully regulated, and loss of fresh-water input has significantly altered estuarine food webs and fisheries production. In addition, the initiation and persistence of a "brown tide" over the past decade has had a strong negative impact on seagrass ecosystems of the southern Texas coast. In addition, pollutant discharges associated with concentrated industry in the Trinity-San Jacinto Bay region continue to have an important impact on fisheries harvest and human use of the coast.

The Southern Gulf has estuaries that display characteristics nearly as great as those in the other three regions, although significantly less research has been carried out there. In the Campeche region of Mexico, the three main management issues are fisheries, petroleum production, and coastal plain agriculture, similar to those described for the central Gulf of Mexico (Yanez-Arancibia and Day 1982, 1988). The region of Terminos Lagoon represents a transition in geomorphology from carbonate escarpments in the Yucatan Peninsula to the east and the second largest river system of the Gulf of Mexico in the Usumancita-Grijalva delta to the west. Offshore is located the Bay of Campeche, an area of broad continental shelf that is the second largest fishery in the Gulf of Mexico. The economy of this region has gradually shifted from the 1950s from being largely dependent on shrimp and offshore fisheries to industrial activity with the discovery of huge petroleum deposits in the Bay of Campeche. More recently, there has been increased development of agriculture in the watersheds above the fall line of rivers in the Tabasco region, as well as in regions adjacent to Terminos Lagoon. In the drier coast of the Yucatan estuaries (including Celestun) east of Terminos Lagoon, fresh-water delivery by ground water is a key management issues of coastal resources in that region. Tourism in the Yucatan Peninsula, along with commercial and residential development, will impact ground water quantity and quality. This shoreline is a microcosm of how tropical estuaries with distinct geomorphology and geophysical processes will respond to human alterations of land use.

PATTERNS OF BIOGEOCHEMISTRY

Physical Characteristics of Gulf Estuaries

Estuaries of the Gulf of Mexico are located along a 4000-km shoreline, with unique gradients in geomorphology and geophysical processes that strongly influence patterns of biogeochemistry. The geomorphology and circulation patterns of Gulf estuaries are distinctly different from those of coastal plain estuaries—from which much of the traditional estuarine literature is based (Pritchard 1955, 1967). The most common geomorphology found in Gulf estuaries is a bar-built system or a combination of bar-built and coastal plain systems. Many of these systems are very shallow (<3 m), with a broad, flat topography. Tides in the Gulf are relatively small (ca. 30 cm) and occur as diurnal or mixed tides, resulting in inputs of tidal mixing energy that are much lower than those found in most regions of the world. As a result, variability in physical energy caused by local wind forcing, and the interaction of local and far-field wind-driven sea-level slopes, play a more important role in biogeochemical cycling in Gulf estuaries than in other, more tidally dominated systems.

The fresh-water residence time of an estuary is an important regulator of both biological and chemical reaction rates. Residence times of water in Gulf of Mexico estuaries are controlled by precipitation and evaporation rates and by the rate of influx of saline water due to tides and fresh-water inflow (Chapter 1 this volume). The Tabasco plain—located along the Isthmus of Tehuantepec in Mexico—and the northern Gulf—from the Mississippi delta to the Florida panhandle—are the two regions on the Gulf coast with the highest precipitation, with 500 and 130 cm yr^{-1} of rainfall, respectively. In contrast, rainfall rates can be as low as 50 to 60 cm yr^{-1} along the southern Texas and Yucatan coasts. Despite the high variability in rainfall rates along the Gulf coast, the range in evaporation rates across these different regions is smaller, ranging from 120 to 160 cm yr^{-1}. As a result of these factors and of the large differences in the area of individual drainage basins, fresh-water inflow rates to the Gulf are highly variable.

In general, astronomical tides are relatively small but are observed across the majority of Gulf estuaries. In many places where astronomical tidal effects are essentially nonexistent (e.g., Laguna Madre), meteorologically induced tides dominate residence time (Chapter 1 this volume). However, in most Gulf estuaries, fresh-water runoff is the dominant regulator of estuarine residence time. For example, the Mississippi and Atchafalaya Rivers' discharge of 15,092 m^3 s^{-1} accounts for approximately 55% of the total inflow to the Gulf, followed by the Rio Grijalva-Usumacinta system in Mexico. The Mississippi River drainage basin is the largest and represents 61% of the total Gulf drainage area. Conversely, the drainage areas of the Florida estuaries, particularly from Tampa Bay and south, are among the smallest, and fresh-water discharge is generally <6 m^3 s^{-1}. Groundwater inputs are also locally important in the delivery of fresh water to Gulf estuaries, particularly in the carbonate systems along the Florida and Yucatan Peninsulas (Herrera-Silveira 1996; Chapter 5 this volume).

Another factor affecting estuarine mixing and circulation along the Gulf coast is channelization. In some cases, estuaries are directly connected via the Gulf Intracoastal Waterway (GIWW). These channels tend to be more vertically stratified than the majority of Gulf estuaries, which are shallow and typically well mixed. Average salinities in Gulf coast estuaries are also quite extreme due to the extreme differences in precipitation and fresh-water inflow. For example, the coastlines of Texas and Florida are typically >20 g $liter^{-1}$ compared to <9 g $liter^{-1}$ along the northern Gulf coast, which is strongly influenced by river-dominated estuaries. Using many of the aforementioned parameters in their residence time models, Solis and Powell (Chapter 1 this volume) estimated that residence times of Gulf estuaries varied from <5 d to >300 d. These extreme differences in residence time are likely to be particularly important when comparing biogeochemical cycles across different estuarine systems.

Estuarine sediments are derived from fluvial, shoreline erosion, marine, eolian, and biological sources (Chapter 1 this volume). While detailed analysis of sedimentary processes is not available for most Gulf estuaries, McKee and Baskaran (Chapter 3 this volume) describe sedimentary processes that control the distribution of terrigenous materials in Gulf estuaries, with an emphasis on the Mississippi/Atchafalaya River system. Emphasis was placed on this region because of the dominance of fresh-water inflow from this region. Overall, it has been estimated that approximately 90% of sediments discharged by rivers in the United States enter estuaries that are bordering the Gulf of Mexico (Milliman and Meade 1983), mostly in the Northern Gulf. Sediment discharge measurements for the Mississippi and Atchafalaya Rivers have been recorded since 1950 and 1973, respectively. The suspended sediment load of rivers that empty into the Gulf of Mexico has been dramatically altered over the past 200 yr due to numerous anthropogenic perturbations (i.e., agriculture, land management, dams, levees). In fact, concentrations of suspended sediments in the Mississippi River decreased from >900 mg liter^{-1} in the early 1950s to <200 mg liter^{-1} in the 1990s. There are also seasonal changes in the delivery of sediments to estuaries and the continental shelf in the Gulf. In the Mississippi River, there is a significant increase in the accumulation of sediments in the channel at downriver stations compared to upriver stations. However, during high discharge stages, these sediments are re-distributed to other regions of the estuary or exported to the shelf. While there is an observed short-term episodic pulsing in sediment accumulation for much of the time, there is no net sediment accumulation on the decadal scale or longer. For example, in the shallow, wind-dominated estuaries of the northern Texas coast (Sabine-Neches and Trinity-San Jacinto Bay), it is common to observe no excess ^{210}Pb in sediment cores, suggesting that no net sedimentation has occurred in the last 100 yr. While this may be true, conclusive evidence is lacking without detailed temporal measurements that can account for short-term episodic pulsing. Thus, it is necessary to develop a technology that accurately measures these short-term episodic events in order to understand fully the sedimentary dynamics of these stochastic systems.

Nutrient Dynamics

The loading of suspended particulate matter (SPM) and dissolved nutrients to estuarine systems has become an important topic because of the worldwide problems associated with the eutrophication of estuaries. The overall loading of dissolved and particulate nutrients to Gulf of Mexico estuaries is highly variable and largely a function of the fresh-water input, the nutrient concentration, and the drainage area:volume ratio of the estuary (Chapter 4 this volume). For most estuaries of the Gulf, non-point source inputs of N and P from agricultural, timber, and range lands dominate total nutrient loading estimates. However, in particular systems (e.g., Tampa Bay: Chapter 5 this volume), point-source nutrient loading has resulted in excess nutrient concentrations and degradation in water quality due to changes in the composition and abundance of phytoplankton.

Gulf estuaries are characterized by a high degree of variance in both nutrient and phytoplankton dynamics (Chapters 5 and 6 this volume). River-dominated estuaries of the Northern Gulf such as Mobile Bay and Apalachicola Bay display strong seasonal cycles in nutrient concentration. However, maximum dissolved inorganic nitrogen (DIN) and PO_4 concentrations are generally 5- to 10-fold lower than in similar systems along the U.S. east coast and in western Europe, where significant point-source inputs are found. In estuaries that lack dominant river forcing, alternate sources of nutrients are important. For example, nutrient dynamics in Florida Bay are closely tied to accumulation and re-mineralization of seagrass-derived organic matter, while lagoons of the northern Yucatan display nutrient dynamics that are alternately tied to ground water inputs, benthic nutrient re-mineralization, and input from waterfowl rookeries. Low rates of benthic nutrient regeneration of PO_4 and NH_4 to the water column are linked to distribution of submersed macrophytes in the Eastern and Western Gulf, compared to more elevated rates in the river-dominated estuaries in the Northern Gulf (Chapter 6 this volume). There are also distinct patterns among these regions in rates of benthic nitrification coupled to denitrification, which occurs in the Eastern and Western Gulf estuaries, but direct denitrification of NO_3 from river discharge is more prevalent in estuaries of the Northern Gulf. However, as observed for nutrient concentrations, benthic nutrient regeneration in estuaries of the Gulf of Mexico displays generally lower rates for NH_4 and PO_4 (particularly for PO_4) than for estuaries along the U.S. east coast and in Europe (Chapter 6 this volume).

Much of the Gulf coast also has extensive wetland systems where chemical transformations and storage of nutrients can occur (Chapter 7 this volume). As a result, dominant physical factors such as wind forcing can change wetland to water interactions, which can substantially alter the particulate and dissolved inputs to the Gulf estuaries. Despite significant losses of wetlands in recent years, 14,500 km^2 of estuarine wetlands remain along the Gulf of Mexico. South Florida, Louisiana, and southern Mexico contain the dominant portion of these wetlands in the Gulf of Mexico. Wetland-water column interactions can be important in determining SPM and dissolved nutrient inputs to Gulf estuaries (Chapter 4 this volume). Most of the data that exist on wetland-water exchange in Gulf of Mexico estuaries are from mangrove forests, although Childers et al. (Chapter 7 this volume) found no significant differences in wetland-water column exchange of nutrients between mangroves and marshes for Gulf estuaries. Wetlands proximal to open water release more dissolved organic carbon (DOC) and take up more particulate organic carbon (POC) than wetlands adjacent to other wetland systems.

Organic Matter Cycling

A diverse array of allochthonous and autochthonous particulate organic matter (POM) sources (terrestrial, submersed macrophytes, plankton) enter into riverine and estuarine ecosystems. The composition and reactivity of POM, along with the residence time of this material, influences the potential for net heterotrophy across a range of Gulf of Mexico estuaries (Chapter 8 this volume). The issue of organic matter reactivity is critical to our understanding of autotrophic-heterotrophic balance in estuaries. Net heterotrophy in estuaries is thought to occur over long (annual or inter-annual) time scales and is generally supported by reactive allochthonous organic matter inputs. Using the simple ratio of respiration (R) and production (P), Smith and Hollibaugh (1993), among others, suggested that estuaries are generally net heterotrophic. Cifuentes et al. (Chapter 8 this volume) use a simple model to determine if organic matter remains for a sufficient period of time to allow for net heterotrophy in Gulf of Mexico estuaries. In contrast to predictions from previous literature, many of the Gulf estuaries, particularly in south Texas, are net autotrophic and not heterotrophic. This model is based only on primary production and bacterial degradation that are likely to account for some of these differences; however, many estuarine food webs are dominated by microbial processes. Conversely, the Lake Pontchartrain and Sabine-Neches estuaries were found to be net heterotrophic. Despite the extreme differences in residence times, organic matter processing produced net heterotrophy. However, because of the potential de-coupling between autotrophic and heterotrophic production in time and/or space, these estuaries may not be net heterotrophic when integrated over their entire area. These findings suggest that it is critical to have an appropriate spatial and temporal sampling regime when examining the interaction of residence time and reactivity of organic matter in estuaries.

DOC represents one of the largest global pools of organic carbon in aquatic ecosystems. Information on the distribution and cycling of DOC, including the colloidal fractions using various biomarkers and tracers, indicates that there is a wide range of non-conservative and conservative mixing patterns of DOC in Gulf of Mexico estuaries (Chapter 9 this volume). In many of the Texas estuaries, DOC was removed (via coagulation, flocculation, etc.) during the mixing of fresh water and seawater. Exceptions to these patterns along the Texas coast included Trinity-San Jacinto Estuary, where non-conservative behavior dominates, largely due to the flux of DOC from sediments. Conservative DOC mixing behavior has also been observed in Mobile Bay and the Mississippi River plume, where high fresh-water inputs at the head of the estuary have established a distinct gradient. In general, the annual flux of DOC to Gulf of Mexico estuaries decreased from the highest fresh-water inflow (i.e., Mississippi River plume) to the lowest (i.e., Colorado River estuary). DOC concentrations in the Gulf estuaries discussed by Guo et al. (Chapter 9 this volume) ranged from 60 to 1500 μM, with the highest concentrations in the Sabine-Neches in Texas and in two Florida estuaries (Ochlocknee and Rookery Bay). Seasonal changes in DOC concentrations can vary substantially, depending on the residence time of the system.

A significant fraction (35–66%) of the total DOC pool in Gulf estuaries is represented by colloidal organic material (COM) (>1 kDa). As discussed in Chapters 10–12, COM has also been shown to be important in determining the fate and transport of trace contaminants in Gulf estuaries. Large differences in the colloidal fractions across different Gulf estuaries are likely the result of differences in riverine inputs, sediment fluxes, and primary productivity—as well as differences in the ultrafiltration procedures used. In general, the carbon sources of COM in Gulf of Mexico estuaries varied from younger, terrestrially derived sources in the upper estuary to older, more labile source in the lower estuary. However, internal cycling, such as selective removal of the labile components in COM and DOC, can alter bulk isotopic and elemental signatures. Further work is needed that allows for compound-specific analyses of isolated components of COM to determine accurately the ages and sources of these materials.

Trace Element/Organic Cycling

Due to the strong chemical gradients (i.e., ionic strength, pH, redox) that exist in estuarine systems, the interactions of trace elements tend to be more dynamic than those in other aquatic environments. These unique interactions have allowed trace element cycling to be used as an important parameter in understanding the time scales of sediment input, biological uptake, and flocculation processes in estuaries. Wen et al. (Chapter 10 this volume) provide the first review of trace element cycling in Gulf of Mexico estuaries. Making inter-estuarine comparisons between trace metal cycling in Gulf estuaries is very difficult because of the absence of complementary work on biogeochemical and hydrodynamic processes. The Trinity-San Jacinto and Mississippi River estuaries exemplify differences in the chemical behavior of trace elements (Cu, Ni, Cd) that can occur in distinctly different yet well-characterized systems (Chapter 10 this volume). The distributions of these dissolved trace metals generally behaved conservatively and non-conservatively in the Mississippi and Trinity-San Jacinto Estuaries, respectively. The non-conservative behavior observed in the Trinity Estuary was largely due to desorption from re-suspended particles or benthic fluxes, particularly in the mid-estuarine region. Re-suspension is common in many of the shallow Gulf estuaries, where these events are principally due to wind-driven forces and, to a lesser extent, bioturbation. The cycling behavior of trace elements has also been shown to be linked to nutrient cycles that exist as organic complexes in aquatic systems. Significant positive correlations were found between Cu and Ni and nutrients (phosphate and silicate) in the Mississippi and Trinity-San Jacinto River Estuaries. However, because complex co-variables can cause these correlations (COM, suspended sediments, etc.), further work is needed that specifically examines the relative importance of each parameter. For example, because COM has a greater surface area than SPM, it may be more important than SPM in controlling trace element cycling. The binding of trace elements to COM is discussed in further detail in Chapter 11 of this volume. The flux of trace elements to Gulf of Mexico estuaries via rivers is dependent on reliable dissolved trace element data; the Mississippi River estuary probably has the largest available dataset. A positive net internal flux for dissolved Co, Zn, Cd, Ni, and Cu resulted in significant export of these elements from the estuary; trace elements complexed in COM more than likely enhanced these fluxes. Sediment fluxes of trace elements to the Trinity-San Jacinto Estuary were important and were strongly affected by advective mixing from bioturbation.

Trace metals and organics have been shown to have significant interactions with natural organic matter (NOM), which includes both POM and COM. Santschi et al. (Chapter 11 this volume) describe the fundamentals of how NOM complexes with trace substances, and how this affects the cycling of trace organics and metals in Gulf estuaries. The amphilic nature of macromolecular NOM, which has both hydrophobic and hydrophilic parts, strongly affects its ability to complex with charged and uncharged molecules and ions. Functional group concentrations and proton reactive sites in NOM are the key binding sites for trace metals, while the hydrophobic and hydrophilic portions of NOM typically sorb to trace organics through van der Waals and electrostatic interactions, respectively. Most of the trace metals in Gulf of Mexico estuaries are associated with the colloidal fraction (>1 kDa) of NOM. Percentages of a few select trace metals found in COM from the Galveston Bay (Texas) and Ochlocknee (Florida) estuaries are as follows: (Cu) 15–66%, (Cd) 5–66%, (Pb) 52–88%, and (Fe) 49–97%. The functional groups most responsible for the complexation of these trace metals in COM are carboxylic and hydroxylic groups, and some recent work includes sulfhydryl groups. The distribution coefficients of trace metals between different phases such as particles and the dissolved phase (K_d), particles and the truly dissolved phase (K_p), and colloids and the truly dissolved phase (K_c) range from 4.4 to 7.8 in Gulf estuaries. However, more work is needed to further understand what is controlling the partitioning between the different phases on the temporal and spatial changes in the functionality of NOM. While only a limited number of studies have examined trace organics in Gulf of Mexico estuaries, or for that matter in any estuary, it has been shown that these organics are strongly sorbed to COM. The compounds found to be associated with COM in the lower Mississippi River were polychlorinated biphenyls, polycyclic aromatic hydrocarbons, triazine herbicides, and chlorinated pesticides. The percentage of colloidal-bound herbicides actually increased toward the Gulf with increasing salinity, which may suggest a change in the functionality of riverine versus marine COM. Gulf of Mexico estuaries clearly provide an ideal location for examining the importance of COM complexation with trace metals and organics because of the wide range of DOC concentrations found across this region.

Particle-reactive radionuclides have been employed to investigate numerous biogeochemical processes in estuarine and coastal waters. Many of the earlier studies focused on the determination of residence time of particle-reactive radionuclides such as ^{210}Pb, ^{7}Be, and ^{234}Th isotopes. Similar to work described in Chapters 10 and 11, recent work with radionuclides have also been extended to include effects of DOM on the fate and transport of particle-reactive nuclides. Baskaran (Chapter 12 this volume) discusses the historical applications of radionuclides in estuaries, as well as some of the key factors that control the biogeochemical cycling of particle-reactive nuclides in selected Gulf estuaries. Particle-reactive nuclides provide a useful tool for understanding the fate and transport of inorganic and organic pollutants. Particle reactivity of nuclides is typically assessed using distribution coefficients (K_d). Most particle-reactive nuclides that are delivered to the estuarine water column via atmospheric fallout, riverine discharge, and in situ production from parent radionuclides have been shown to become associated with fine-grained particles. Sediment inventories of ^{210}Pb and ^{239,240}Pu can be utilized to determine the retention efficiency of these nuclides, which can then be used for other heavy metals. The retention efficiency of Pb and Pu in the Sabine-Neches (Texas) Estuary was observed to be considerably lower than expected, assuming that most of the nuclide would adsorb onto fine particles and be removed to the sediments. It was determined that most of these nuclides are complexed in COM rather than suspended particulates, which results in the rapid export of nuclides out of the estuary. The Sabine-Neches has some of the highest observed concentrations of DOC (400 to 1350 μM) (Chapter 9 this volume) and shortest hydraulic residence times (7–10 d) (Chapter 2 this volume). The residence times of ^{7}Be in Gulf of Mexico estuaries (0.6–6.4 d) is significantly lower than those in selected coastal plain estuaries along the east coast (2–33 d) because of the shallow features of Gulf estuaries.

SUMMARY

The susceptibility of estuaries to trends of eutrophication and declining patterns of sustainable fisheries is of major concern to increasingly populated areas of the coastal zone. Eutrophication leads to increased hypoxia and harmful algal blooms, as well as increased bioaccumulation of toxic inorganic and organic compounds in higher trophic levels of estuarine food webs. The susceptibility of estuaries in the Gulf of Mexico to eutrophication and the status of this problems were evaluated based on projected nutrient loadings and water residence times by a joint effort of NOAA and EPA (NOAA/EPA 1989). The report includes a caveat that projected evaluations of nutrient susceptibility are limited by other factors that influence fate of nutrients, such as physical controls of distribution and processes that control concentration, many of which were described in the previous chapters. Susceptibility to eutrophication should now be assessed, using our understanding of estuarine biogeochemistry to provide insights into these assessments of eutrophication in estuaries of the Gulf of Mexico (as elsewhere).

The results of the relative susceptibility classification for Gulf of Mexico estuaries suggest a regional pattern similar to that described for the biogeochemistry of these estuaries. Those estuaries that generally show low susceptibility (classification of VII to IX) are in the central region of Gulf of Mexico, including the Mississippi River, Mobile Bay, and Atchafalaya/Vermilion Bay. More susceptible eutrophication classes occur in both east and west directions along the Gulf shoreline. Estuaries most susceptible to eutrophication (I to III) include Tampa Bay, Ten Thousand Island and Perdido Bays in Florida, and Corpus Christi Bay, Aransas Bay, and San Antonio Bay in Texas. These indices are based purely on the physical properties of these estuaries, such as fresh-water loading and residence time (Chapter 2 this volume) and nutrient discharge (Chapter 4 this volume). When compared to actual evidence of eutrophication, such as the occurrence of hypoxia, there are some discrepancies with this analysis of susceptibility. For example, two of the estuaries assigned a low susceptibility index, the Mississippi River and Mobile Bay, both have documented occurrences of hypoxia over the last 50 years (May 1973; Pennock et al. 1994). The loading of nutrients combined with the occurrence of stratification may account for the difference in predicted and observed patterns of eutrophication (Pennock et al. 1994). In addition, both Mobile Bay and the Mississippi River have the highest rates of benthic phosphorus flux among the estuaries surveyed in the Gulf of Mexico, indicating the significance of nutrient loading and recycling to patterns of eutrophication (Chapter 6 this volume). Thus, the biogeochemistry of estuaries is important in understanding the susceptibility of estuaries to eutrophication.

In addition to the physical processes of estuaries that control susceptibility to eutrophication, there are unique chemical properties that have to been considered in this type of evaluation. Concentrations of COM in the estuaries of the Gulf of Mexico can also reduce the retention of nutrients and trace elements, and thus reduce residence times based purely on physical calculations. The association of COM and DOC concentrations with several river systems, such as the Sabine and Rookery Bay, can be linked to the influence of floodplain and inter-tidal wetlands, respectively. Thus, the coupling of vegetated regions of estuarine watersheds in the Gulf, which include some of the largest wetland regions in the United States, can modify the fate of nutrients and trace elements in Gulf of Mexico estuaries. As indicated by tracers including natural isotope abundance, biomarkers, and radionuclides, there are complex patterns of biogeochemical processes that control the susceptibility of Gulf estuaries to eutrophication and pollutants. Although the low residence time of water can account for much of the variation among Gulf estuaries, the chemical nature of the organics in these ecosystems can also explain both the distribution of trace elements and the metabolism of estuaries in the Gulf region.

REFERENCES

Day, J. W., Jr., C. J. Madden, R. R. Twilley, R. F. Shaw, B. A. McKee, M. J. Dagg, D. L. Childers, R. C. Raynie, and L. J. Rouse. 1994. The influence of Atchafalaya River discharge on Fourleague Bay, Louisiana (USA), p. 151–160. *In:* K. R. Dyer and R. J. Orth [eds.], Changes in fluxes in estuaries. Olsen and Olsen.

Deegan, L. A., J. W. Day, Jr., J. G. Gosselink, A. Yan;ti;ez-Arancibia, G. S. Chavez, and P. Sanchez-Gil. 1986. Relationships among physical characteristics, vegetation distribution and fisheries yield in Gulf of Mexico estuaries, p. 83–100. *In:* D. A. Wolfe [ed.], Estuarine variability. Academic Press.

Herrera-Silveira, J. A. 1996. Salinity and nutrients in a tropical coastal lagoon with groundwater discharges to the Gulf of Mexico. Hydrobiologia **321:** 165–176.

Madden, C., J. Day, and J. Randall. 1988. Coupling of freshwater and marine systems in the Mississippi deltaic plain. Limnol. Oceanogr. **33:** 982–1004.

May, E. B. 1973. Extensive oxygen depletion in Mobile Bay, Alabama. Limnol. Oceanogr. **18:** 353–366.

Milliman, J. D., and R. H. Meade. 1983. World-wide delivery of river sediment to the oceans. J. Geol. **91:** 1–21.

NOAA/EPA. 1989. Strategic assessment of near coastal waters: Susceptibility and status of Gulf of Mexico estuaries to nutrient discharges. Strategic Assessment Branch, NOS/NCA.

Pennock, J. R., J. H. Sharp, and W. W. Schroeder. 1994. What controls the expression of estuarine eutrophication? Case studies of nutrient enrichment in the Delaware Bay and Mobile Bay estuaries, USA, p. 139–146. *In:* K. R. Dyer and R. J. Orth [eds.], Changes in fluxes in estuaries. International Symposium Series. Olsen & Olsen.

Pritchard, D. V. 1955. Estuarine circulation comparisons. Proc. Amer. Soc. Civil Eng. **81:** 1–11.

Pritchard, D. W. 1967. What is an estuary: Physical viewpoint, p. 3–5. *In:* G. H. Lauff [ed.], Estuaries, volume 83. AAAS.

Smith, S. V., and J. T. Hollibaugh. 1993. Coastal metabolism and the oceanic organic carbon balance. Rev. Geophys. **31:** 75–89.

Turner, R. E., and N. N. Rabalais. 1991. Changes in Mississippi River water quality in this century. BioScience **41:** 140–147.

Walsh, J. J., G. T. Rowe, R. L. Iverson, and C. P. McRoy. 1981. Biological export of shelf carbon is a sink of the global CO_2 cycle. Nature **291:** 196–201.

Yañez-Arancibia, A., and J. W. Day, Jr. 1982. Ecological characterization of Terminos Lagoon, a tropical lagoon-estuarine system in the Southern Gulf of Mexico. Oceanolog. Acta 431–440.

―――. 1988. Ecology of coastal ecosystems in the southern Gulf of Mexico: The Terminos Lagoon region. Universidad Nacional Autonoma de Mexico, Ciudad Universitaria.

Index